DISCARDED
FROM
UNH LIBRARY

P9-EMP-251
3 4600 01033 3117

Braving the Elements

Harry B. Gray
CALIFORNIA INSTITUTE OF TECHNOLOGY

John D. Simon
UNIVERSITY OF CALIFORNIA, SAN DIEGO

William C. Trogler
UNIVERSITY OF CALIFORNIA, SAN DIEGO

University Science Books
Sausalito, California

University of New Hampshire
Library

Chem
QD
37
.G64
1995

University Science Books
55 D Gate Five Road
Sausalito, CA 94965
Fax (415) 332-5393

Production Manager: *Superscript Editorial Production Services*
Manuscript Editor: *Constance Day*
Designer: *Robert Ishi*
Compositor: *LM Graphics*
Printer and Binder: *Maple-Vail Book Manufacturing Group*

This book is printed on acid-free paper.

Copyright © 1995 by University Science Books

Reproduction or translation of any part of this work beyond that
permitted by Section 107 or 108 of the 1976 United States Copyright Act
without the permission of the copyright holder is unlawful. Requests for
permission or further information should be addressed to the Permissions
Department, University Science Books.

Library of Congress Cataloging-in-Publication Data

Gray, Harry B.
    Braving the elements / Harry B. Gray, John D. Simon, and William
C. Trogler.
        p.   cm.
    Includes bibliographical references and index.
    ISBN 0-935702-34-2
    1. Chemistry—Popular works.   I. Simon, John D. (John Douglas).
1957–   . II. Trogler, William C., 1952–   .   III. Title.
QD37.G64   1995
540—dc20   94–45685
                                                            CIP

Printed in the United States of America
10 9 8 7 6 5 4 3 2

Braving the Elements

Contents

Everyone—doctors, lawyers, media people, even physicists!—should know some chemistry. What used to be a "shake and bake" science has been revolutionized by computers, lasers, and structure determination machines such as X-ray diffractometers, nuclear magnetic resonance spectrometers, and scanning tunneling microscopes.

The computer has had great impact on the work of inorganic and organic chemists. In the old days (the sixties!), our style would have been to examine some references in books and journals and begin mixing things, hoping for publishable results. Now, we head not only to the bookshelf and laboratory bench, but to a computer workstation. At a workstation, three-dimensional simulations of all sorts of molecules can be displayed. We can command the computer to insert new atoms or to remove existing atoms. We can change whole groups of atoms, if that is what the problem requires. And, after changing the entire molecular architecture, we can ask if the new structure will be stable or not. If the calculations predict reasonable stability, the chances are good that we can go back into the lab and actually create the substance we have modeled. In organic synthetic problems we can even ask the computer to review known reaction paths and suggest the best way to make a compound from readily available starting materials.

The new synthesis technology is of increasing importance to pharmaceutical companies, who have coined the phrase "rational drug design." Many new drugs are being conceived on a workstation video screen. After the planning stage, structure determination machines allow us to characterize new compounds rapidly. Solutions and solids that contain nuclei with

magnetic moments can be studied by nuclear magnetic resonance spectroscopy; and the arrangements of atoms in crystalline materials can be precisely defined by X-ray diffraction
experiments. Structure determinations that took months of work
in the early 60s can now be completed in a day or two.

The ability to design and synthesize whole new classes of
molecules is producing dramatic breakthroughs in materials science. Many of these breakthroughs are discussed in our book.
One that is not, but that is coming, is molecular memory. Today,
all high-speed computers are based on solid-state devices (silicon chips). But there could be a much more powerful memory,
one in which information is stored at the atomic level in individual molecules. Indeed, Richard Feynman predicted that molecular memory will allow us to increase the speed of today's
supercomputers a million-fold or even more!

The importance of fundamental research in chemistry will
be highlighted in this decade by responses to many environmental issues. Chemists will be called on time and time again to
replace dangerous chemicals with safer ones. Research that may
lead to the replacement of hydrofluoric acid by less toxic acids
(solid superacids?) and CFCs by molecules that are less threat-
ening to stratospheric ozone is being done because of very real
concerns about the environment. And more and more energy
research will be directed toward the production of fuels that can
reduce the risk of global warming.

We will see renewed research initiatives in artificial photosynthesis. Here the goal is to use chemistry to convert sunlight
and abundant raw materials into fuels that we can use to heat
homes and run cars and airplanes. Those of us who work in this
field hope that by the end of the century some of this research
will have reached an advanced stage of development and will
have begun to have an impact.

Nowhere are the fruits of research in chemistry more evident
than in medicine. Anesthetics have revolutionized surgical procedures; anti-inflammatory drugs have made motion possible in
arthritics; antiulcer drugs have produced dramatic cure rates;
and antibiotics have wiped out many terrible bacterial infections. Steroid hormones have been used to build tissue in burn
patients; and hormone therapy has been employed to control

biological function (birth control pills) and correct hormone deficiencies (human growth hormone treatments for dwarfism). Antihypertensive drugs are being used to reduce high blood pressure and the attendant risks of heart disease and stroke associated with old age; chemotherapy has been employed successfully to arrest some forms of cancer; and drugs that combat osteoporosis are just beginning to be released.

We have written *Braving the Elements* for nonchemists. We think it will be useful in courses for nonmajors and for others interested in learning about modern chemistry and how it relates to the environment, energy, health, and other areas of human concern. Even physicists may want to see how chemistry is doing these days—after all, it has been over sixty years since a prominent physicist announced that (with the invention of quantum mechanics) "all of chemistry is solved."

Chemistry is alive and well. Read our book and see.

Acknowledgments

We thank Peter Armbruster, Rudy Baum, Arnold Beckman, Ralph Barnhard, Geoff Blake, Raymond Chang, Jonah Colman, Peijun Cong, Bob Dean, Melanie Dean, Carol Dempster, Ed Deutsch, Mark De Rosch, Robert Dunn, Robert Doolen, Jim Espenson, Peter Gantzel, Daniel Gerrity, Mike Gray, Diane Hawley, Bruce Hudson, Ron Henderson, John Hogg, James T. Hynes, Richard Kanner, Susan Kegley, Dave Kliger, Rachel Kliger, Andy Kummel, Jack Kyte, Clark Landis, Nate Lewis, John Lighty, Joe MacNeil, James McGarrah, Bo Malmström, Marie Messmer, Audrey Miller, Stanley Miller, Fred Pecoroni, Charles Petit, Andrew Prokopovitsh, Fae Ryan, Jack Roberts, Mike Sailor, Paul Saltman, Gail Silver, Thomas Simon, Ken Suslick, Diane Szaflarski, Mark Thiemens, Peggy Thompson, Margaret Tolbert, Dave Tyler, Dotty Vasquez, Judy Verbanets, Rob Whitnell, Sunney Xie, and Terry Young for many helpful comments. Eugene Volker and his class of elementary education majors at Shepherd College, the participants in the Program for Teacher Enhancement in Science at UCSD (1992), and the students of Chem 11 at UCSD are acknowledged for class testing various versions of the book.

Special thanks go to our editor, Jane Ellis, and to Ann Knight and the production crew at Superscript for their expert help. The patience, suggestions, and support of our publisher, Bruce Armbruster, are much appreciated.

CHAPTER 1

The Periodic Table

The Origins of Chemistry

Early Greek philosophers were the first to introduce the idea that matter consists of simple building blocks. Leucippus and Democritus, who lived around 400 B.C, wondered whether the water in the sea was continuous or whether it might be like sand, which appears continuous from a distance but on closer inspection is seen to consist of small grains. Democritus proposed that all matter is made up of small indivisible particles, and he called these particles atomos (Greek for "atoms"). The atoms of Democritus had no weight, and no tendency to combine, until random motions brought them together to form the world.

> *Nothing exists except atoms and empty space; everything else is opinion.*
>
> Democritus of Abdera (ca. 460–370 B.C.)

Unlike Democritus, Empedocles (ca. 450 B.C.) viewed matter as a continuous substance. Empty space did not exist for Empedocles. He believed that all matter could be made by mixing four *elementa* (Latin for "elements")—earth, air, fire, and water. Aristotle (384–323 B.C.) embraced and extended the ideas of Empedocles. In his treatise *On the Heavens*, Aristotle discussed the four material elements and their interrelations. Aristotle paid particular attention to the question of elemental change, whereby one element can be altered to form another. The idea that elements can be transmuted into one another gave birth to a new science, *Chëmeia*.

The Greek philosopher Epicurus (341–270 B.C.) described his atomic view of matter in *Physics*, a 37-volume set. Unlike Democritus, Epicurus believed that atoms had weight. The atoms of Epicurus varied in size, but none was large enough to be visible. The followers of Epicurus were content to take his postulates for granted. No particular efforts were made to explain, test, or even understand his theories. (This complacency was to change just before the birth of Christ.) In the philosophical work *De rerum natura (On the Nature of Things)*, a 7500-line poem, the Roman philosopher Lucretius (ca. 96–55 B.C.) expounded an atomic view of matter. Epicureanism was to exert great influence on future philosophers. Central to the Epicurean explanation of nature was the claim that atoms and the laws governing them determine all things. This left no place for the role of gods in worldly events. For centuries, the Christian Church regarded this theory as inconsistent with its religious beliefs.

Experimental methods for studying matter developed as well. The Greeks and Egyptians experimented with the transformation of inexpensive metals, such as lead and mercury, into gold. After conquering Egypt in A.D. 640, the Arabs adopted this science and termed it *Alchëmeia*. Similar developments occurred in Indian and Chinese cultures, as they too wrestled with explanations for changes in matter. Europeans did not discover Latin translations of the Arabic writings on alchemy until about A.D. 1150, but this science took root in Europe and flourished during the Middle Ages. In some instances, science became linked with astrology. Alchemy served a useful purpose, because it spurred the discovery of new substances (alcohols, acids, metal salts) and the development of such techniques as sublimation, crystallization, and distillation. Sadly, the alchemists never achieved their aim, and the pursuit became tainted with fraud during its later stages.

Interest in the atomic view of matter was rekindled by the French philosopher and scientist Pierre Gassendi (1592–1655), who revived Epicureanism and reconciled atomic concepts with Christian beliefs. In 1661 the British scientist Robert Boyle (1627–1691) attacked Aristotle's theory of the four elements in his book *The Sceptical Chymist: or Chymico-Physical Doubts & Paradoxes, touching the Spagyrist's Principles commonly call'd Hypostatical, as they are wont to be Propos'd and Defended by*

Matter exists in the form of atoms and combinations of atoms. Void and matter are mutually exclusive; matter is solid and eternal. No void exists in the atom; hence it is both indestructible and indivisible. It does consist of least parts, however; but these have never existed separately.

Lucretius
(ca. 96–55 B.C.)
"On Nature"

the Generality of Alchymists. During the Middle Ages, European alchemists had postulated that the four elements of Aristotle appeared in matter as the fundamental principles salt, sulfur, and mercury. Boyle argued that there was no experimental evidence for the four Greek elements. They could not be extracted from matter or combined to form all other substances. He also discredited the three fundamental principles of alchemists. Creating a practical definition, Boyle said that an element was a substance that could not be resolved into simpler substances. He insisted that experimental facts must support claims for elements. Boyle is often credited with introducing rigorous experimental methods in chemistry. But although Boyle laid the groundwork for defining elements, he did not provide a table of substances he regarded as elements. The next great advance would come from across the English Channel.

The French scientist Antoine Lavoisier (1743–1794) published the first extensive list of elemental substances based on Boyle's definition. Lavoisier was an extremely talented experimentalist. He gained international recognition when he proved incorrect the prevailing belief that water is converted into earth by repeated distillation. He was also the first to recognize the element oxygen as the gaseous component of air that supports combustion. In 1789 Lavoisier published his views on the elements, chemical reactions, and scientific inquiry in his book *Traité élémentaire de Chemie.* Antoine Lavoisier is often spoken of as "the father of modern chemistry."

Many concepts used in chemistry today result from work begun in the nineteenth century. In 1803 John Dalton (1766–1844), a British chemist–physicist, proposed a set of laws to explain the relationship between an element and an atom. He maintained that each element consists of many small, identical, and indestructible particles that he called atoms. To explain the differences among the elements, Dalton suggested that different elements are composed of atoms of different weight. Each element is characterized by the weight of its atom. Dalton further reasoned that atoms of different elements may combine in specific numerical ratios to form chemical compounds. The great variety of compounds in nature arises from the diverse ways in which the atoms of different elements can be combined. As insightful as it was, however, Dalton's theory provided no information on *how* atoms are organized in chemical compounds.

In 1811 the Italian physicist Amedeo Avogadro proposed that gaseous chemical compounds are built from a fundamental unit, which he called a molecule. A molecule is a particle that contains two or more atoms. Avogadro hypothesized that equal volumes of gases at the same temperature and pressure contain equal numbers of molecules. Molecular theory gained wide acceptance in 1858 after an Italian chemist, Stanislao Cannizzaro, showed how it unified all of organic and inorganic chemistry. Consider water: Dalton knew that water vapor is a compound composed of the elements hydrogen (H) and oxygen (O) in a specific weight ratio, 1(H):8(O). But he guessed incorrectly that hydrogen atoms had $\frac{1}{8}$ the mass of oxygen atoms and that water contained one hydrogen atom for each oxygen atom. Avogadro's method for determining numbers of atoms made it possible to deduce atomic weights, and hydrogen atoms were shown to have $\frac{1}{16}$ the mass of oxygen atoms. The weight ratio for water revealed that it contained H and O atoms in a ratio of 2:1. Cannizzaro concluded correctly that water in its liquid, solid, and vapor forms is composed of large numbers of identical molecules. Each molecule is made of three atoms: two hydrogen atoms and one oxygen atom (H_2O). Of course a molecule is much too small to see, and conclusive proof for the molecular nature of chemical compounds awaited the development of new experimental techniques.

Working in Cambridge, England, in the early 1900s, the father-and-son team of William Henry and William Lawrence Bragg discovered how to locate atoms in solids by using the technique of X-ray diffraction. They passed an X-ray beam through a small crystal of table salt and analyzed the way the beam scattered. By means of a mathematical analysis of the scattering pattern, they were able to make a three-dimensional model of how the atoms pack in the crystal. For their work, the Braggs were awarded the Nobel Prize for Physics in 1915. Their technique was initially applied to salts and metallic solids, which contain an extended network of connected atoms. A few years later, in 1923, the first X-ray structure of a solid built of discrete molecules was reported by two chemists at Caltech,

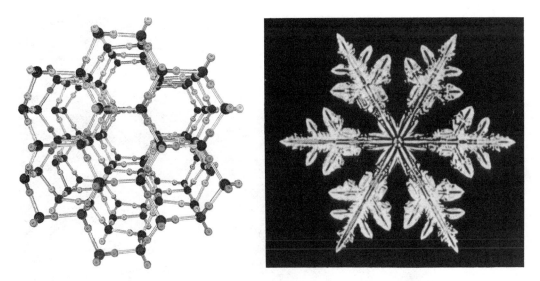

Figure 1.1.   *The packing of water molecules in an ice crystal shown next to a snowflake. Note the similarity between the hexagonal crystal pattern and the shape of the snowflake. (Left, computer model copyright © 1976 by W. G. Davies and J. W. Moore. Right, photomicrograph by W. A. Bentley in Dover reprint of* Snow Crystals *by W. A. Bentley and W. J. Humphreys, 1931.)*

Roscoe G. Dickinson and Albert L. Raymond (the structure was that of the organic compound hexamethylenetetramine). Today, with the aid of high-speed computers, the X-ray method for determining crystal structure is applied routinely. Figure 1.1 shows how water molecules pack in crystalline ice. The hexagonal, star-shaped arrangement of water molecules in the crystal accounts for the characteristic shape of snowflakes.

In 1981 Gerd Binnig and Heinrich Rohrer, at the IBM research laboratories in Zurich, Switzerland, developed the scanning tunneling microscope (STM). This technique, for which they shared the 1986 Nobel Prize for Physics, can be used to image atoms. An STM picture of a graphite surface is shown in Figure 1.2. The carbon atoms of graphite appear as a series of regularly aligned, round bumps. Chemists use the STM and many other modern instruments to determine the atomic and molecular structure of matter. There is no longer any doubt about the existence of the atoms of Democritus and the molecules of Cannizzaro.

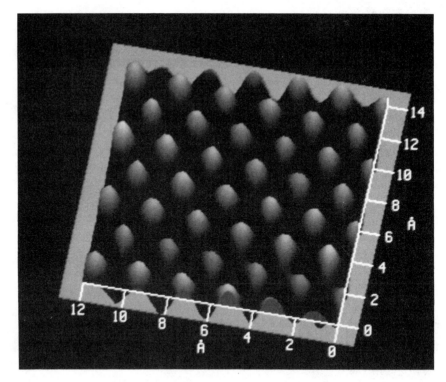

Figure 1.2.
A graphite surface as seen with a scanning tunneling microscope (STM). Each carbon atom appears as a rounded bump. (Courtesy of Chun Yan and A. C. Kummel.)

Atoms

An understanding of chemistry requires a knowledge of the inner structure of atoms. Consider the atoms that make up gold, one of the few elements that occurs in nature as a pure metal. Gold is soft, and it can be drawn into extremely thin wires. Its lasting beauty has captured the imagination of everyone from peasants to royalty for ages.

Figure 1.3 depicts what we would see if we examined a piece of gold with a powerful microscope. Solid gold consists of a regular array of spherical gold atoms packed together closely. Common light microscopes, offering a maximum magnification of 1000, are unable to resolve this atomic structure. A magnification of 100 million is necessary to reveal the stacking pattern of the individual gold atoms that make up the metal. With an additional 10-fold magnification, the atom appears to consist of a thin haze of negatively charged particles surrounding a massive positively-charged core.

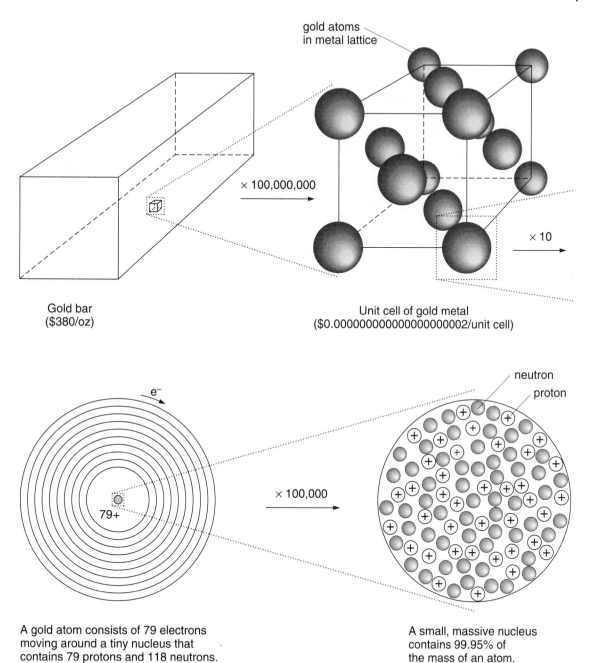

gold atoms
in metal lattice

× 100,000,000

× 10

Gold bar
($380/oz)

Unit cell of gold metal
($0.00000000000000000002/unit cell)

neutron
proton

e⁻

79+

× 100,000

A gold atom consists of 79 electrons
moving around a tiny nucleus that
contains 79 protons and 118 neutrons.

A small, massive nucleus
contains 99.95% of
the mass of an atom.

Figure 1.3.   *The successive magnifications of a gold bar that are
required to show the atoms, the structure of a gold atom, and finally
the contents of the gold nucleus.*

Another magnification by 100,000 is needed to reveal the detailed structure of the nucleus itself. Knowledge about subatomic species is indirect, because the very act of observing these particles moves them. In 1911 Ernest Rutherford, Hans Geiger, and Ernest Marsden reported experiments suggesting that an atom consists of a positively charged, massive nucleus circled by negatively charged particles of low mass. The negatively charged particles, called electrons, move around the nucleus at speeds approaching that of light, which is 186,000 miles per second.

Rutherford's experiments suggested that the nucleus, atom, and electron compare in relative size to a ladybug (the positively charged nucleus) in the center of a large football stadium (the atom) with a few gnats (the negatively charged electrons) flying so rapidly that they create an imperceptible blur throughout the stadium. Most of the atom is empty space within which the electrons travel. The attraction between the negatively charged electrons and the positively charged nucleus prevents electrons from escaping the atom.

The negative charge of the electron is defined as −1. Close examination of the positively charged nucleus shows that it consists of two different types of particles: neutrons and protons. A neutron has no charge. The positive charge of the nucleus is created by the protons. Each proton has a charge of +1. This charge is exactly equal to, and opposite, that of the electron. Particles with the same charge repel one another, but nuclei are stable because there are strong attractive forces between protons and neutrons. The neutrons act as glue, holding the nucleus together. Unlike an atom, the dense nucleus contains little free space between its particles.

Mass and Moles of Atoms and Molecules

The sum of the masses of the protons, neutrons, and electrons is the total mass of an atom. Because the mass of an electron is very small (about 0.05% that of a proton or neutron), the mass of an atom approximately equals the mass of its nucleus. The mass of a single proton is defined as 1 amu (atomic mass unit). The mass of the neutron is slightly greater than 1 amu. To a good approximation, the mass of an atom in amu is equal to the total mass of the protons and neutrons in the nucleus.

Because atoms are so small, a measurable mass corresponds to a tremendous number of atoms. Chemists use a unit called the mole to measure quantities of matter large enough to weigh on a scale. A mole equals 602,000,000,000,000,000,000,000 (6.02 × 10^{23}) particles, and the number of particles in a mole is called Avogadro's number. It is a staggeringly large number: a billion times larger than the U.S. national debt in cents! One mole of M&Ms spread over the surface of the continental United States would be about 52 miles thick! A mole of helium gas consists of 6.02 × 10^{23} helium atoms, and a mole of hydrogen gas contains 6.02 × 10^{23} hydrogen molecules. The mass of a mole of molecules in grams approximately equals the sum of the atomic mass units for the nuclei in the constituent atoms. The mass of a carbon atom is 12 amu, and a mole of carbon weighs 12.001 grams. A mole of helium atoms weighs only 4.003 grams. Chemists use metric units: grams to measure mass and liters to measure volume. A pound is about 450 grams, and a liter is a little more than a quart.

Hydrogen and Helium

The simplest atom is hydrogen; it contains one proton and one electron. Hydrogen atoms are the building blocks for the element hydrogen. The next more complex element, helium, is made up of helium atoms, each consisting of two protons in the nucleus. The helium nucleus also contains two neutrons. Two electrons move rapidly around the helium nucleus to balance the charge of the two protons. Overall, the atom has no net charge.

A flask of helium gas contains many individual helium atoms, which move rapidly throughout the container and bounce off its walls. Helium gas is completely unreactive; it cannot combine with other substances. The element hydrogen also exists as a gas, but it differs considerably from helium. The reason why the elements hydrogen and helium behave differently arises from the difference in their atoms. Hydrogen atoms are reactive as individual atoms; they spontaneously bind together to form hydrogen molecules, H_2. A flask of hydrogen gas contains many such molecules, which rapidly move throughout the container and bounce off its walls. Unlike helium, hydrogen gas undergoes many chemical reactions and explodes when ignited in air. The reactivity of hydrogen molecules arises because the bond between the two atoms breaks and the free atoms can then bind to other atoms. A helium atom cannot make bonds and therefore is unreactive. The difference in the chemistry of hydrogen and helium can have dramatic consequences.

The Hindenburg

Hundreds of men and women and a few children craned their necks skyward as the leviathan cruised toward them at 70 mph. Nearly 250 ground crew scurried about the U.S. Naval air station at Lakehurst, New Jersey. Most had been waiting several hours for the thundershowers to abate so that the airship could dock. The drone of four 16-cylinder diesel engines stilled the chatter on the ground. Mahogany–walnut laminate propellers 20 feet in diameter thrust the hulk toward the crowd. Many of the 37 passengers had relatives or American hosts waiting below. Some waved excitedly from the windows of rectangular cabins built into the massive airship's belly. The image of an object almost three football fields long pressed into the consciousness of the earthbound gawkers.

Aloft, a crew of 54 prepared hurriedly for descent. Teams of two mechanics monitored each engine. Others readied trail lines, spider cables, and hand ropes, which would be dropped shortly. Cabin attendants briefed the passengers on departure procedures. Docking almost seemed routine to them now. Most of the crew had worked the world's first regular transatlantic air route for exactly a year. Ads in *The New York Times* regularly touted their service: "Europe by Air in 2½ days—$400 incl. berth, meals, tips." In several minutes the tired crew would leave their floating workplace for some food and sleep. Sailing above the Atlantic Ocean at a 700-foot altitude loses its novelty after several voyages.

In the forward control gondola, Captain Max Pruss examined the activity on the ground to verify the radio message that gave him the go-ahead to dock. No other man had the ability to maneuver the Promethean dirigible as if it were a small toy. A repaired tail fin, evidence of a previous skipper's mistake, certified the claim. Captain Pruss vented some hydrogen gas through valves on top of the ship and ordered the engines reversed to brake the craft. By dropping water ballast, he trimmed the floating sausage's nose up a few degrees in preparation for docking. Within a few minutes, the ship drifted a mere 75 feet above the ground. Telephone orders went out to the crew at 7:21 P.M. to drop the 300-foot-long forward trail lines and slow the engines. On the ground, teams of men raced to retrieve these ropes and attach them to two mooring cars on rails that encircled the massive steel mooring mast that served to anchor the nose. Unexpectedly, the craft's momentum caused it to glide several hundred yards past the mast. One mooring line snagged taut and threw the airship off balance. The engines were revved up and

The Hindenburg Explosion. (The Bettman Archive.)

Captain Pruss yelled to the crew to pay out the lines.

Whether it was sabotage, structural failure, a faulty vent valve, or simply a rip in a gelatin-painted cotton gas cell, hydrogen gas leaked out near the vessel's stern. This produced little effect, until the hydrogen ignited—perhaps just from static electricity. Hydrogen molecules inside the 16 gas cells within the ship were poised to combine with atmospheric oxygen and produce water vapor. All they needed was a spark of energy to start the reaction. Thereafter, the reaction energy itself helped propagate the chemical change. (Ironically, the ship's flotation cells were designed to use helium, an unreactive gas. However, under the Helium Control Act the world's sole supplier—the United States—limited trade to favored nations.)

When the trail lines dropped, Herb Morrison recorded a radio broadcast of the year's first transatlantic flight by a passenger airship. "Passengers are looking out of the windows waving; the ship is standing still now. The vast motors are just holding it, just enough to keep it from…. It's broken into flames! It's flashing! Flashing!" Within seconds a white ball of flame mushroomed into the sky.

Oxygen molecules were drawn in from the surrounding air by the updraft of hot gases racing skyward. The craft's outer linen skin burned away to reveal duraluminum alloy girders and bracing wire. Thunder punctuated the sky when enough air mixed with pockets of hydrogen to create explosive mixtures. Hydrogen gas containing as little as 6% air can explode. With the disintegration of the dirigible's tail section, the loss of aft lift caused the nose to shoot 500 feet into the air. The sixteen gas bags inside the shell ruptured sequentially and released seven million cubic feet of hydrogen to fuel the fire storm. A wall of flame erupted from stern to stem in 34 seconds.

Frantic ground crew and onlookers raced to escape the hell above. One didn't make it. Terrified passengers leaped from the windows, with flames at their backs. The twisted aluminum skeleton, with intact passenger pods beneath, collapsed to earth in a smoking ruin. Of the 97 on board, 22 crew members and 13 passengers died. Many more were severely injured on that overcast May evening in 1937 when the pride of Nazi Germany, the largest aircraft ever flown, fell victim to the chemical energy stored in a hydrogen molecule.

The Periodic Table

Organizing the elements into tabular form posed a difficult problem for chemists. Early tables simply listed the elements in order of increasing atomic weight. This arrangement, however, provided no insight into the chemistry of the elements. Other schemes were proposed, but no single arrangement was widely accepted by the chemical world until 1869, when the Russian chemist Dmitri Ivanovitch Mendeleev organized the known elements in columns of similar reactivity. Mendeleev's table was based on the idea of a periodic trend, whereby elements exhibit a repeating pattern of chemical reactivity with increasing atomic weight. Mendeleev's original table had 17 columns and listed about 60 elements. The layout of today's periodic table, shown in Figure 1.4, is based on the ideas of Mendeleev, though in Mendeleev's time, no one knew why the elements exhibited periodic behavior.

The periodic table collects elements in irregular rows that read from left to right in order of increasing atomic number. An element's atomic number tells how many protons there are in the nucleus of that element's atoms. Because elements are electrically neutral, the atomic number also represents the number of electrons in the neutral atom. The peculiar gaps occur in the table because Mendeleev grouped elements with similar reactivities in the same column. For example, the elements hydrogen (H), lithium (Li), sodium (Na), potassium (K), rubidium (Rb), and cesium (Cs) possess many similar properties in their reactivities and form the first column on the lefthand side of the table. The elements below hydrogen in the table are called the alkali metals. When this table was proposed in 1869, several elements found in the current table were unknown. Mendeleev recognized this fact and left a few blanks in his table, believing that the appropriate element had not yet been discovered. Chemists then searched for each missing element, whose properties could be anticipated from its position in the periodic table. When a new element was discovered, with properties similar to those predicted, it was a triumph for Mendeleev. Of the 110 elements that are now known, only 90 occur on Earth in appreciable amounts. We show 103 elements in the table.

The periodic table gives the names and symbols of the elements. The chemical symbol stands for the atoms that make up each element. For example, hydrogen has the symbol H in the table, but the common form of the element is H_2. The names and

s¹ — alkali metals
s² — alkaline earth metals

(d-block) — Transition Metals

(p-block)
s^2p^1 s^2p^2 s^2p^3 s^2p^4 s^2p^5 s^2p^6 — noble gases
halogens
chalcogens

Boundary line between metals and nonmetals

(f-block) — Rare Earth or Lanthanide Metals / Actinide Metals

1 **H** hydrogen							
3 **Li** lithium	4 **Be** beryllium
11 **Na** sodium	12 **Mg** magnesium
19 **K** potassium	20 **Ca** calcium	21 **Sc** scandium	22 **Ti** titanium	23 **V** vanadium	24 **Cr** chromium	25 **Mn** manganese	26 **Fe** iron	27 **Co** cobalt	28 **Ni** nickel	29 **Cu** copper	30 **Zn** zinc	5 **B** boron	6 **C** carbon	7 **N** nitrogen	8 **O** oxygen	9 **F** fluorine	10 **Ne** neon
37 **Rb** rubidium	38 **Sr** strontium	39 **Y** yttrium	40 **Zr** zirconium	41 **Nb** niobium	42 **Mo** molybdenum	43 **Tc** technetium	44 **Ru** ruthenium	45 **Rh** rhodium	46 **Pd** palladium	47 **Ag** silver	48 **Cd** cadmium	13 **Al** aluminum	14 **Si** silicon	15 **P** phosphorus	16 **S** sulfur	17 **Cl** chlorine	18 **Ar** argon
55 **Cs** cesium	56 **Ba** barium	57 † **La** lanthanum	72 **Hf** hafnium	73 **Ta** tantalum	74 **W** tungsten	75 **Re** rhenium	76 **Os** osmium	77 **Ir** iridium	78 **Pt** platinum	79 **Au** gold	80 **Hg** mercury	31 **Ga** gallium	32 **Ge** germanium	33 **As** arsenic	34 **Se** selenium	35 **Br** bromine	36 **Kr** krypton
87 **Fr** francium	88 **Ra** radium	89 †† **Ac** actinium										49 **In** indium	50 **Sn** tin	51 **Sb** antimony	52 **Te** tellurium	53 **I** iodine	54 **Xe** xenon
												81 **Tl** thallium	82 **Pb** lead	83 **Bi** bismuth	84 **Po** polonium	85 **At** astatine	86 **Rn** radon

† | 58 **Ce** cerium | 59 **Pr** praseodymium | 60 **Nd** neodymium | 61 **Pm** promethium | 62 **Sm** samarium | 63 **Eu** europium | 64 **Gd** gadolinium | 65 **Tb** terbium | 66 **Dy** dysprosium | 67 **Ho** holmium | 68 **Er** erbium | 69 **Tm** thulium | 70 **Yb** ytterbium | 71 **Lu** lutetium |

†† | 90 **Th** thorium | 91 **Pa** protactinium | 92 **U** uranium | 93 **Np** neptunium | 94 **Pu** plutonium | 95 **Am** americium | 96 **Cm** curium | 97 **Bk** berkelium | 98 **Cf** californium | 99 **Es** einsteinium | 100 **Fm** fermium | 101 **Md** mendelevium | 102 **No** nobelium | 103 **Lr** lawrencium |

Figure 1.4. *Periodic table of the elements. The number in the upper left corner of the box before each symbol is the atomic number. It represents the number of protons in the nucleus, which is the same as the number of electrons in the neutral atom. Elements with italicized symbols are gases in their normal state at room temperature, and those with outlined symbols are liquids. The rest are solids. Elements to the left of the bold line are metals, except for hydrogen.*

symbols of elements have interesting historical origins. For example, the name **h**ydrogen, symbol H, derives from the two Greek roots *hydro* and *genes*, which together mean "water-forming." This name was given to the element because it burns in air to make water. **I**odine (I) exists as violet crystals that contain I_2 molecules. Its name stems from the Greek *iodes*, "violet." **C**arbon (C) derives its name from the Latin *carbo*, "charcoal," and **ca**lcium (Ca) from the Latin word (*calx*) for "limestone", a mineral that contains calcium.

Sometimes an element's symbol, such as Fe for iron and Au for gold, bears no apparent relation to its modern name. The symbol Fe is derived from the Latin word for "iron," *ferrum*. *Aurum* in Latin means "shining dawn," which refers to the characteristic brilliant yellow color of gold metal. The symbol Pb for lead comes from the Latin word *plumbum*, which means "heavy." This name reflects the high density of metallic lead. The abbreviation Cu is from the Latin word for "copper," *cuprum*.

Cobalt (Co) is the English version of the German word *kobold*, which means "goblin or evil spirits." Cobalt occurred as an unwanted metal in what were thought to be copper ores by German miners of the 1800s. **Ars**enic (As), which was also found in these ores, poisoned the workers in the smelting process. No wonder they thought cobalt contained evil spirits!

Other elements, such as **Curiu**m, **No**belium, **M**endelevium, **La**wrencium, **Ga**dolinium, and **Ei**nsteinium, honor famous scientists. The elements **F**rancium, **G**ermanium, **A**mericium, **Po**lonium, **Ru**thenium (Latin *Ruthenia* = Russia), **Ga**llium (Latin *Gallia* = France), **Haf**nium (Latin *Hafnia* = Copenhagen), **H**olmium (Latin *Holmia* = Stockholm), **Lu**tetium (an ancient name for Paris), **Sc**andium, **Y**tterbium (towns in Sweden), **C**alifornium, and **B**erkelium all stand for a country or city beloved by the element's discoverer. Three of the heaviest known elements, **U**ranium, **N**eptunium, and **Plu**tonium, are named for the outermost planets in the solar system. And the names of some elements reflect their historical uses or the imaginations of their discoverers. The discoverer of the element vanadium (V), impressed by the spectacular colors of its compounds, named it after **V**anadis, the Norse goddess of beauty.

Electronic Structures of Atoms

Rutherford's experiments, performed in 1909 and 1910, showed that rapidly moving electrons occupy most of the space in an atom. The atomic nucleus, which contains the protons and neutrons, occupies less than one-trillionth of an atom's volume but contains nearly all the atom's mass. The chemical properties of the elements depend on the outermost (valence) electrons, and it is important to know how they occupy the space around the atomic nucleus. Are electrons randomly distributed? Do they move around the atom in circular orbits? Are electrons confined to specific regions around the atom? Finding answers to these questions challenged the best scientific minds at the start of the twentieth century.

In 1913, Niels Bohr proposed that electrons move in circular orbits around the nucleus. The number of electrons within a given orbit increases with increasing distance from the nucleus. Bohr further postulated that electrons could not have any arbitrary amount of energy. He proposed that the energy values of electrons in the atom are quantized. Each electron orbit (or orbital) corresponds to an energy level of the atom, and electrons contained within a specific orbital can possess only a specific quantity of energy. Electrons surround the atom in discrete layers, or shells, in Bohr's model. He was the first to recognize that the electrons in atoms determine the atom's chemical reactivity. This creative model of atomic structure earned Bohr the Nobel Prize for Physics in 1922.

In parallel with the struggle to develop an atomic theory, many eminent scientists were perplexed by the nature of light. Throughout the nineteenth century, light was modeled as a wave. By the end of the century, however, several experimental paradoxes had arisen to challenge this description. A monumental advance occured in 1900, when Max Planck postulated that energy comes in packets, or *quanta*. Then, in 1905, Einstein proposed that light also consists of a series of energy packets, which he called *photons*. These bold new views about the nature of light and energy successfully explained the experimental paradoxes of the day. They also provided the inspiration for Bohr's quantized model of electrons in atoms.

When it comes to atoms, language can be used only as in poetry. The poet, too, is not nearly so concerned with describing facts as with creating images.

Niels Bohr (1885–1962) quoted in J. Bronowski, *The Ascent of Man*

The electron is not as simple as it looks.... The important thing in science is not so much to obtain new facts as to discover new ways of thinking about them.

Sir William Lawrence Bragg (1890–1971)

The discoveries of Planck and Einstein showed that in some circumstances light displays wave properties, whereas in others it behaves like a particle. In 1924 Louis de Broglie, a French physics Ph.D. student, suggested that matter should also be regarded as having both wave and particle properties. Clinton J. Davisson and Lester H. Germer at Bell Laboratories validated this idea in 1927 by showing that a beam of electrons (particles) and a beam of X-rays (photons) scatter off a piece of gold foil in a similar manner. The modern electron microscopes found in scientific laboratories and hospitals exploit the wave-like behavior of electrons. The fact that electrons exhibit both wave and particle properties meant that scientists needed to reconsider the theories of atomic structure.

Shortly after de Broglie advanced his proposal, new mathematical treatments appeared that described the wave-like properties of electrons in atoms. These mathematical models, collectively known as *quantum mechanics*, form the foundation of the modern view of the atom. In 1927 Werner Heisenberg introduced the *uncertainty principle*. Heisenberg postulated that it is impossible to examine an atom and definitively locate an electron and determine where it is going. As a result, any model of the atom must treat the locations of electrons in terms of probabilities. Although this idea is accepted today, it generated much controversy and met with considerable opposition when it was originally proposed. Albert Einstein never accepted the idea that absolute determinism of the universe was forbidden.

Erwin Schrödinger unified the ideas put forth by de Broglie and Heisenberg into a comprehensive quantum-mechanical description of the atom. In this model, as in Bohr's, the energy of an electron is quantized. In accordance with Heisenberg's ideas, Max Born, a German physicist, demonstrated that Schrödinger's treatment provided probability maps showing the electron distribution around the nucleus. These maps of the distribution of electrons in space, called the atomic orbitals, vary in shape and size for the different allowed energy levels. The different types of atomic orbitals, in order of increasing complexity, are known as *s*, *p*, *d*, and *f* orbitals. The single electron in the hydrogen atom distributes itself around the nucleus in a pattern represented by the spherical 1*s* orbital sketched in Figure 1.5. Unique

God does not throw dice.

Albert Einstein (1879–1955)

I am now convinced that theoretical physics is actually philosophy.

Max Born (1882–1970)

1*s*

Figure 1.5. *The spherical 1s orbital of the electron in the hydrogen atom. This orbital represents a region where there is a 95% probability of finding the electron.*

properties of the atomic orbitals are intimately tied to the chemical properties of the elements.

Electrons and Chemical Reactivity

The electrons in an atom's outermost orbitals are the ones that participate in chemical reactions. Atomic orbitals of similar energy are grouped into shells, and the outermost layer is called the *valence shell*. The valence shell electrons determine chemical bonding. The first electronic shell of an atom contains up to two electrons in the first atomic orbital, called the $1s$ orbital. The hydrogen atom has one electron in its first shell. Helium atoms have two electrons in this shell, the maximum allowed. Helium is unreactive, because the first shell is full. Hydrogen atoms, which only have a half-filled shell, are very reactive. What about lithium? A lithium atom has three electrons. Two of these fill the first shell and therefore cannot participate in chemical reactions. The third electron occupies the valence shell. Lithium is similar to hydrogen because it has only one electron in its valence shell; this electron occupies a $2s$ atomic orbital. That is why lithium lies in the same column as hydrogen in the periodic table.

The second shell is larger than the first and can hold eight electrons. Besides the $2s$ orbital, the second shell contains the $2p$ orbitals. Each orbital can contain at most two electrons (this statement is called the Pauli Principle); however, the p orbitals occur in sets of three, which can accommodate a total of six electrons. The second shell becomes filled in the neon atom. A neon atom contains a total of ten electrons: two in the $1s$ orbital, two in the $2s$ orbital, and six in the set of three $2p$ orbitals (this structure is abbreviated as the electron configuration $1s^2 2s^2 2p^6$). Like helium, neon is inert; it does not take part in chemical reactions because its valence electronic shell is full. The unreactive elements helium (He), neon (Ne), argon (Ar), krypton (Kr), xenon (Xe), and radon (Rn) make up the rightmost column of the periodic table. These six elements are unreactive because their valence shells of electrons are filled completely. Collectively called the noble gases, they signal the end to chemical reactivity trends across the table and the need to begin a new row.

The next element after neon is the reactive metal sodium, which begins a new row of the periodic table, as electrons go into a new (third) shell. Sodium atoms resemble lithium and hydrogen atoms in that all these atoms have one electron in their outermost valence shell in an *s*-type orbital. The seven reactive elements in the first column of the periodic table—hydrogen, lithium, sodium, potassium, rubidium, cesium, and francium—all have an s^1 valence electronic configuration.

For the first three rows of the periodic table, the number of electrons each atom has in its valence shell is calculated by counting across from the first column of the periodic table. For example, the element carbon is the fourth from the left in its row of the periodic table, and the carbon atom has four electrons in its outermost shell (s^2p^2).

For the fourth row of the periodic table and beyond, the situation is more complicated. Ten new columns appear in the middle of the periodic table. The elements listed in these columns are called the transition metals. Many useful metals, such as iron (Fe), chromium (Cr), and zinc (Zn), occupy this part of the table. The first transition metal is scandium (Sc). An atom of scandium contains one more electron than an atom of calcium (Ca). Within the transition metals, the additional electrons beyond the two required to fill the *s* orbital are placed in a new type of orbital, a *d* orbital. There are five equivalent *d* orbitals in the fourth row (and beyond), which may contain up to ten electrons. When the *d* orbitals are filled (d^{10}), they behave as an inert shell.

After the transition metals, the electrons enter the set of three *p* orbitals just like the elements above them in the periodic table. Consider the atoms of the element germanium (Ge). Counting from the left hand column at potassium, a germanium atom contains 14 electrons in its valence shell; however, ten of them fill the *d* orbitals and behave as an inert set. This leaves four electrons in the valence shell (s^2p^2), the same as in the outermost shell of carbon. This is why the element germanium is chemically like carbon and occupies a place below it in the periodic table. The valence electronic configurations appear above some columns in the periodic table shown in Figure 1.4. Atomic electronic configurations for most of the elements are given in Appendix I.

The elements fluorine, chlorine, bromine, iodine, and astatine are called the halogens. They occur in the column of the periodic table next to the noble gases. Their physical states change from gaseous (F_2 and Cl_2) to liquid (Br_2) to volatile crystals (I_2) as the molecular weight increases. All the halogens exist as diatomic (two-atom) molecules whether they are solids, liquids, or gases. This similarity arises because the seven electrons in their outermost shell (s^2p^5) are one short of the eight in the noble gases (s^2p^6). By sharing an electron with each other in a diatomic molecule, they effectively fill the valence shell of electrons around each atom. This series illustrates the value of the periodic table in showing the relationship among elements with similar chemical properties.

The rare earth or lanthanide metals, cerium (Ce) through lutetium (Lu), and the actinide metals, thorium (Th) through lawrencium (Lr), are placed in two special rows in the periodic table (see Figure 1.4). The rare earth elements exhibit virtually the same chemical properties. Their atomic numbers follow the element lanthanum (La); hence the term *lanthanide metals*. The actinide elements follow the element actinium (Ac); hence the term *actinide metals*. The actinides feature natural radioactive elements, such as uranium (U) and thorium (Th), as well as short-lived radioactive elements, such as plutonium (Pu), that can be made in nuclear reactions. The lanthanides and actinides are built by filling a set of 7 new orbitals, called *f* orbitals, with electrons. Since each orbital can accommodate two electrons, the rows for the lanthanide and actinide elements are 14 elements long.

Gases, Liquids, and Solids

Gases, liquids, and solids are three forms of matter that differ in the microscopic order and motion of their constituent atoms and molecules. For example, helium atoms in helium gas move freely within the entire volume of the container they occupy. Only the noble gases are unreactive as isolated atoms: helium (He), neon (Ne), argon (Ar), krypton, (Kr), xenon (Xe), and radon

(Rn). Some elements exist as gases made up of diatomic molecules: hydrogen (H_2), oxygen (O_2), nitrogen (N_2), chlorine (Cl_2), and fluorine (F_2).

Several elements exist as liquids at room temperature. A liquid, unlike a gas, need not fill its container. Gravity pulls a liquid to the bottom of its container. Even if there were no gravity, a liquid would exist as droplets, held together by weak attractive forces between its atoms or molecules. The weak character of the forces in a liquid allows the atoms or molecules to slide freely by one another, although their motion is much more constrained than in the gaseous state. Only three elements, bromine (Br_2), francium (Fr), and mercury (Hg), are liquids at room temperature. Gallium (Ga) and cesium (Cs) melt just above room temperature. Bromine is a brown liquid with brown swirls of Br_2 vapor above. Mercury is a silver liquid metal with much less vapor above it. Mercury is highly toxic, so the small amount of Hg vapor above an open container poses a serious health hazard. Most gases liquefy if cooled sufficiently. For example, gaseous water vapor in air, called humidity, condenses to form liquid water droplets on a cold surface.

Most elements are solids under normal conditions. Solids make up the third common form of matter. Solids are usually slightly denser than liquids. They retain their shapes when moved or shaken. Unlike the atoms in liquids, the atoms in solids may adopt a regular three-dimensional stacking pattern. Only small vibrational motions of the atoms occur in solids; however, the forces between atoms may vary greatly. Solid iron or diamond (carbon) is hard and strong. Other solids, such as wax and table sugar, deform easily because only weak attractive forces hold individual molecules in their stacking patterns. In solid iodine (I_2), the forces between the molecules are so weak that I_2 vaporizes to a purple gas without forming a liquid. Dry ice (solid carbon dioxide) behaves similarly and evaporates without forming a liquid. Liquid carbon dioxide is stable only when confined under pressure, such as inside a carbon dioxide fire extinguisher. Figure 1.6 depicts the differences in the atomic arrangements of the three states of matter for the element nitrogen. It exists as a solid at low temperature, as a liquid upon warming,

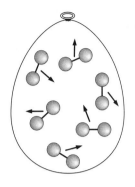

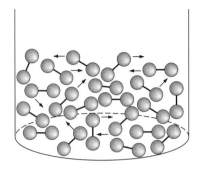

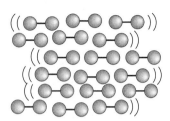

Gaseous state

Molecules in a balloon of gaseous nitrogen move about the container randomly.

Liquid state

The disordered nitrogen molecules in the liquid state conform to the shape of the container and are free to slip by one another.

Solid state

Heat causes molecules of nitrogen (N_2) to jiggle slightly about their ordered positions in the crystal.

Figure 1.6. *Differences in the closeness of packing and the atomic motions in the three states of matter, using nitrogen as an example.*

and as a gas at room temperature. Note the presence of the diatomic N_2 molecule in each state.

Mass and Density

Mass is a measure of the amount of matter, based on an object's resistance to being moved when a standard force acts on it. This resistance is called the inertia of an object. Weight is a similar measurement, based on the force of gravity. Mass and weight have the same value at the Earth's surface, so the terms are often used interchangeably. However, because gravitational forces vary from one planet to another, the weights of objects also vary. The mass of a particular substance remains the same wherever it is measured in the universe.

A property related to mass is density. The density of a substance is the mass contained in a given volume. Gases are not very dense; compared to solids and liquids, a small mass of gas occupies a relatively large volume. Under atmospheric pressure at 25°C, 1 mole of a gas occupies 22.4 liters. If the gas were water vapor, this volume of gas would weigh only 18 grams (less than $1/20$ of a pound). Liquids and solids are much denser than gases. The corresponding 22.4 liters of liquid water would weigh 22,400 grams (about 50 lb), and would contain 1240 moles of water molecules. Liquid water is over a thousand times denser than gaseous water. Water is unusual in that it expands on freezing. Whereas the densities of most substances increase in going from the liquid to the solid state, the density of ice is slightly less than that of liquid water. (The volumes of 1 mole of ice, of liquid water, and of water vapor are shown in Figure 1.7.) Ice floats on water thereby enabling life to exist through the winter in ponds. This unusual property of water also makes the winter sport of ice skating possible. Compression of ice beneath the thin blades of ice skates forces the water molecules closer together and transforms ice into the higher density liquid state. So, an ice skater glides on a thin film of liquid. Ice skating is not like skiing. Skis, with a larger surface area, exert a weaker force on the ice crystals in snow, so melting does not occur. That is why the surfaces of skis are often coated with wax to make them slide more readily across the ice crystals in snow.

The different densities of solids, liquids, and gases account for the buoyancy of objects. For example, a piece of Styrofoam rises when it is released underwater. Because a volume of water weighs more than an equal volume of Styrofoam, gravity pulls the denser water toward the Earth with a greater force than it exerts on the less dense Styrofoam. Styrofoam floats to the surface because the Earth exerts a weaker gravitational attractive force on it than on an equal volume of water.

Helium is used to fill airships such as the Goodyear blimp. A balloon, filled with a gas lighter than air, floats atop the heavier oxygen and nitrogen molecules of the atmosphere. Hot-air balloons work by a related principle. Heat causes molecules and atoms to move more rapidly. In gases, heat causes molecules to move about faster, striking the walls of their container at greater speeds. This exerts a greater pressure and expands the balloon. Once the balloon is expanded, the density of hot air inside it is less than that in the surrounding cool atmosphere. Thus a balloon filled with hot air rises.

24

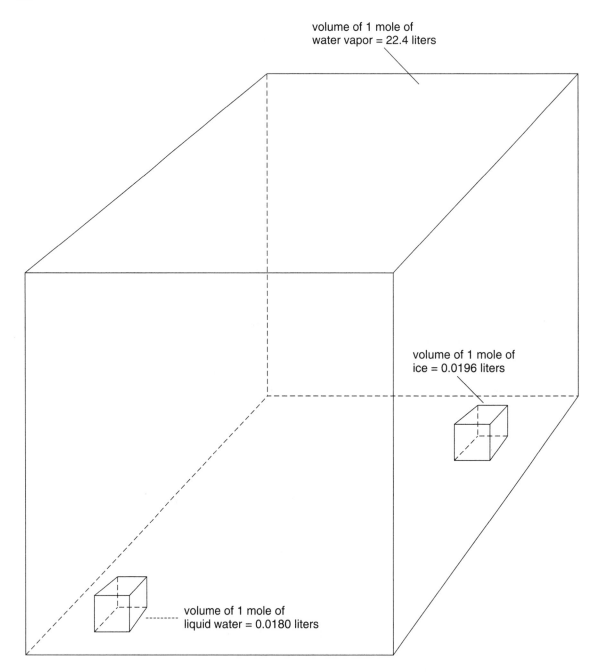

Figure 1.7. *The relative cubic volumes occupied by a mole (18 grams) of ice, a mole of liquid water, and a mole of water vapor (steam).*

Solid Forms of Elements

Metals.   Most elements (82%) in the periodic table are metals. Metals exhibit a characteristic luster, conduct heat and electricity efficiently, and have the ability to bend without breaking. The alkali metals (the elements in the first column of the periodic table) are soft and low melting. Alkali metals can react explosively with water! The metals in the center portion of the periodic table, called the transition metals, are stronger and less reactive. Structurally important transition metals with high strengths and melting points include titanium (Ti), chromium (Cr), iron (Fe), and nickel (Ni). The excellent electrical conductors copper (Cu), silver (Ag), and gold (Au) also are transition metals.

Aluminum (Al) lies in a column next to the transition metals in the periodic table. Though aluminum is a highly reactive metal, its ability to form a tough surface coating of aluminum oxide protects the metal underneath from chemical reactions with air and water. This makes aluminum suitable for a variety of structural applications. Even so, the inherent reactivity of aluminum sometimes leads to problems. Ship builders have favored aluminum in construction of modern warships, because it is so much lighter than steel. Aluminum metal burns in air (reacts with oxygen) if heated enough, and it will even continue burning under water by ripping oxygen atoms out of water (H_2O) molecules. Aluminum also melts much more readily than steel and deforms in a severe fire. During the Falklands War of 1982, the British destroyer H.M.S. Sheffield burned furiously after it was hit by a single missile (Figure 1.8). The intense heat caused failure of the aluminum superstructure. This catastrophic event led naval designers to rethink the use of aluminum in warships. A similar problem occurred on the U.S. frigate Belknap after an accidental collision with the aircraft carrier John F. Kennedy (November 22, 1975) in the Ionian sea. Fire-resistant coatings and automatic fire extinguishers are being used aboard existing warships to mitigate the problem.

Figure 1.8.   *Destroyer H.M.S. Sheffield ablaze. (UPI/Bettman.)*

Nonmetals. A boundary line (see Figure 1.4) separates the metallic from the nonmetallic elements in the periodic table. Elements near this boundary line may possess both metallic and nonmetallic properties. The semiconductors used in the electronics industry (silicon, germanium, and arsenic) lie along this boundary. The boundary elements are often referred to as metalloids.

Carbon is an element adjacent to this region. Elemental carbon can exist as graphite, diamond, or one of a series of cage structures best exemplified by C_{60}. (Different forms of the same element are called allotropes.) Diamond and graphite are solids that differ only in the way the carbon atoms are bound to one another (Figure 1.9). In diamond the carbon atoms are in a rigid three-dimensional network; in graphite they form hexagonal-patterned sheets. Only weak forces bind the different sheets of graphite together, so they slide easily by one another. Graphite is a glossy black solid used in pencil "lead." It crumbles readily when rubbed across a piece of paper, leaving a trace of black powder in the paper fibers. Diamond, one of the hardest substances known, will scratch most surfaces it scrapes. Graphite conducts electricity fairly well, but diamond is an insulator. Graphite is a solid lubricant, but diamond is an abrasive agent. Diamond and graphite were known long before anyone suspected their common elemental origin.

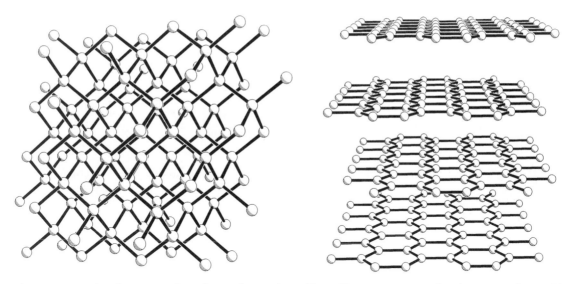

Interconnected carbon atoms in a diamond crystal.   Sheet-like arrangement of carbon atoms in graphite.

Figure 1.9. *Structures of diamond and graphite.*

In the last decade, molecular forms of elemental carbon have been identified. When carbon is vaporized in an electric arc under certain conditions, the soot that is produced contains discrete C_{60} molecules (Figure 1.10). The C_{60} molecule was named buckminsterfullerene after Buckminster Fuller, the creator of the geodesic dome, because the C_{60} structure resembles one of his domes. Fuller also postulated a correspondence between his geometric structural principles and those operative at a molecular level. Each C_{60} molecule weakly attracts neighboring C_{60} molecules in solid buckminsterfullerene, and it has the solid lubricating properties of graphite. Much research is underway to find technological uses of this newly recognized form of carbon.

Like carbon, phosphorus exists in allotropic forms. One of them, called white phosphorus, is a waxy solid that bursts into flame when exposed to air (it is used in tracer bullets), whereas a red, solid form of phosphorus is inert in air. The difference in reactivity arises from the way the phosphorus atoms bind to one another in the solids. White phosphorus consists of small molecules containing four phosphorus atoms, P_4. Pyramid-shaped molecules of P_4 are loosely stacked together in the solid. It melts readily and spontaneously ignites in air. Red phosphorus is an

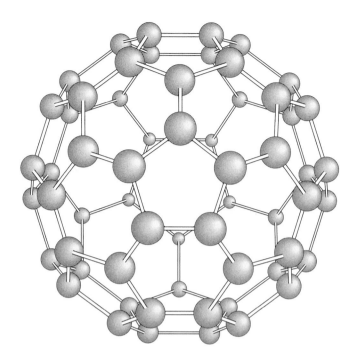

Figure 1.10. *Soccer ball cage structure of the C_{60} molecule, a component of soot.*

infinite network of tightly bound phosphorus atoms. It does not liquefy readily or burn spontaneously in air.

Temperature and Phase Changes

The state of an element can often be changed (this is called a phase change) by changing the temperature. Scientists usually measure temperature in degrees Celsius (°C). The Celsius temperature was standardized to water's melting point, 0°C (32°F), and boiling point, 100°C (212°F). The temperatures at which changes in state occur are unique to each substance. Pressure can change the temperature of a phase change; for example, water boils at a higher temperature inside a pressure cooker. The nearly constant pressure at the Earth's surface is usually assumed in quoting the temperatures of phase changes. For example, nitrogen is a gas at room temperature, but it liquefies when cooled to –196°C (–321°F). With further cooling, nitrogen solidifies at –210°C (–346°F). On the other hand, iron is a solid

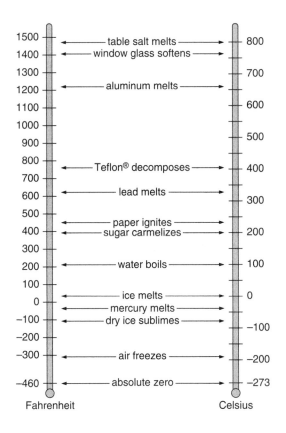

Figure 1.11.
Comparison of the Fahrenheit and Celsius temperature scales.

at room temperature. It melts in a blast furnace when heated to 1535°C (2795°F). Iron vaporizes near 3000°C (5432°F) or when a high-powered laser is focused on it. On a hot star like our Sun, which has a surface temperature of about 6000°C (10800°F), all the known elements are gases.

Absolute zero, −273.15°C (−460°F), is a temperature below which matter cannot be cooled further. It represents the temperature at which all atomic motion ceases. This temperature has not been achieved in the laboratory. The coldest temperature at which matter has been studied is about 1/10,000 of a Celsius degree above absolute zero. In contrast, no upper limit for temperature exists. The center of the Sun is about 15 million Celsius degrees. Figure 1.11 compares the two temperature scales and gives the boiling and melting points for some common substances. The two scales can be converted by applying these formulas:

$$°F = [(9/5)°C] + 32°$$

$$°C = (5/9)(°F - 32°)$$

In the Kelvin temperature scale, which is also used in scientific literature, absolute zero is defined as 0 K. Temperatures above this point are measured in kelvin (K), which are equivalent to Celsius degrees. (Ice melts at 273.15 K.)

On to Nuclear Chemistry

Pure elements are the simplest building blocks of matter. Although solid matter can be made of an element such as graphite or diamond, it also can be made of compounds. A compound is a substance with a constant elemental composition. Solid quartz consists of pure silicon dioxide, a compound that contains silicon and oxygen atoms in a 1:2 ratio. Even a microscopic chip of quartz contains this 1:2 ratio of Si and O atoms. Ice is another example of a pure solid compound. It contains many molecules of water (H_2O) packed together in the solid. Most familiar substances consist of compounds or mixtures of compounds. Chemistry focuses on understanding these substances on an atomic and molecular level.

The familiar reactions of chemistry involve rearrangements of atoms in compounds. Although the atoms are shuffled around in normal chemical reactions, their identities are not changed. A nitrogen atom can be bonded to a carbon atom or to an oxygen atom, but it is still a nitrogen atom because its nucleus is intact. The reactions in which the nuclei of atoms are transformed to other nuclei are called nuclear reactions. We discuss *nuclear chemistry*—the alchemist's dream of changing one element into another—in the next chapter.

CHAPTER 2

The Alchemist's Dream

A Historic Experiment

The sign said "To Squash Courts." As I headed down the hallway, I thought about how long it had been since anyone had played squash down here. Outside, it was a typical Chicago winter day—cold and about 10°F. The effect of the war on our lives was evident everywhere. It was the second day of gas rationing in the city. The streetcars and trains had been jammed on the way to work. This morning's *Tribune* described the Allied assault on Tunisia and told how our soldiers were battling the Japanese for control of New Guinea. Unbeknownst to the general population, today would be historic. In the subterranean squash courts, with oblivious pedestrians overhead, we would attempt something that could change the course of history.

I hadn't slept well. I had kept running through the numbers over and over. What if the calculations were wrong? What if the experiment got out of hand? I had spent the whole night wondering what we might have overlooked. Scientists suffer constantly from this problem. A bit of second guessing could never do any harm, especially when it came to this project. What we were about to attempt wasn't an ordinary laboratory experiment. It had the potential of making the *Hindenburg* seem like a firecracker. I didn't feel comfortable with the shared responsibility. We probably should have told the University of Chicago officials exactly what was happening down in their abandoned squash courts. I remember Arthur Compton, the project leader, telling us why he hadn't explained the dangers to the university's president. "The only answer he could have given would have been no. And this answer would have been wrong."

At the end of the hall, I stopped to show my ID to the guard. Security checks didn't bother me anymore. They seemed a normal part of life now. My breathing quickened as I opened the door and entered the laboratory. The air smelled dusty. Black dust lightly covered the walls, floors, and hallways. But after all, we had moved 80,590 pounds of refined uranium oxide, 12,400 pounds of uranium metal, 771,000 pounds of graphite, and countless pieces of electronic equipment into the area. Who would have imagined that less than a decade ago, college students slammed little balls against these walls to test their agility and deviousness? Today the room was a sea of dark suits. Several colleagues stood on the platform that overlooked the pile of dark metallic ore in the center of the room. There was something uneasy about the tone of their voices. It was little comfort to know that I wasn't alone in my anxiety.

I lingered at the edge of the platform for a minute to gaze at the pile. It seemed so simple. To me, it was a beautiful structure, a work of art. The pile consisted of alternating layers of solid graphite bricks and bricks drilled out so they could contain two short, rounded cylinders of compressed uranium oxide powder. Interspersed were rods of cadmium supported on wood, the ends of which stuck out of the pile with attached handles. Then there were the buckets of cadmium sulfide solution. These were our neutron-absorbing fire extinguishers. They were held in ready by three courageous scientists positioned above the pile. The technicians behind me were running various tests, throwing switches, reading meters, and scribbling notes into lab books. Their instruments included redundant boron trifluoride counters for measuring low neutron intensities, as well as ionization chambers for high-intensity readings. All these sounds blended into a steady whine. I stared at the pile and thought about the events that had led up to the experiment we would perform today.

It was still hard to believe what three simple solids could do. At the blackboard the idea looked so simple, so elegant. To me, much of science seems like that. But putting theory into practice was something different. We had all worked hard and long, not to mention the money that this work had required. It would all begin when one uranium nucleus absorbed one neutron. This new nucleus would be unstable, splitting into barium and krypton, which would release three high-energy neutrons and heat. On average, one neutron in produces two to three neutrons out. Simply amazing! I still remember the feeling of excitement the first time I contemplated the possibility of a sustained and controlled nuclear chain reaction. In practice, natural uranium oxide ore (U_2O_3) could not be used. I had overseen the development of chemical techniques needed to purify the metal and produce the refined oxide (UO_2) that was being used today. The speed of the neutrons also had to be moderated so they could be captured by uranium nuclei. This was the purpose of graphite. Who would ever have thought that pencil lead would be used in the core of a nuclear reactor? The last component, on which our lives depended, was the cadmium. A runaway chain reaction would produce more heat than that needed to transform the pile into a molten mass. Our converted squash courts could become a raging inferno and shower the South Side of Chicago with radioactive fallout. All hope of surviving the experiment relied on the cadmium rods. Laboratory experiments showed that cadmium was an excellent neutron absorber. The cadmium rods should control the neutron flux in the pile. They were our brake pedal. But just to be sure, we were ready to flood the monolith with solutions of cadmium sulfide on a moment's notice.

After an aborted morning test, the moment of truth was at hand. Tension mounted in the room, and I felt both queasy and excited. Eating lunch hadn't been a wise decision. I glanced around the room. Everyone was in his designated place. Enrico Fermi, our scientific leader, gave the command to proceed with the test of his brainchild. Slowly, the cadmium control rods were withdrawn from the pile. My ears strained to hear the clicks that the scalars would make when the detectors in the pile began registering neutrons. These small devices would tell us whether the work was a success or a failure. Nervously, my friends with the buckets of cadmium sulfide readied themselves for the worst. The scalars increased their rates of clicking for a short while, and then the clicking sound became steady. The pile was undergoing a large-scale nuclear reaction. We slid the cadmium rods further out of the pile. The clicking increased to a roar. We hoped that we were in complete control of a nuclear reaction! I watched the pen climb on the chart recorder for what seemed like an eternity, at the end of which Enrico Fermi ordered the control rods inserted to quench the reaction. At 3:53 P.M. central time on December 2, 1942, the road to nuclear power was paved, open, and ready for traffic.

An artist's rendition of the first controlled nuclear fission reaction, The Chicago Pile. (The Chicago Historical Society. Painting by Gary Sheehan (1964. 521.))

Modern Alchemy

Through the Middle Ages, the alchemists embraced Empedocles and Aristotle's theory of matter, which included the possibility of transforming one element into another. Much of their effort was directed to attempts to synthesize the precious metal gold from an inexpensive metal. The alchemists frequently used a metal of similar density, such as lead, for their experiments. Dalton's atomic theory, which emerged much later, in the 1800s, stated that atoms were not altered during chemical reactions. The bearing out of predictions based on Dalton's theory suggested that the alchemist's dream of transforming one element into another could never be realized. But this changed again with the discovery of radioactive elements around 1900. With the realization that changes could occur in the very nuclei of atoms, it became clear that elements *could* be transformed.

At the beginning of the twentieth century, radioactivity was a scientific curiosity whose main application was in the manufacture of luminescent paint. With the discovery of nuclear fission in 1938, however, the race for atomic weapons and nuclear reactors began. The development of the hydrogen bomb followed, as humans attempted to harness the same energy source that stars use. In addition to the high-profile weapons and reactors, the chemistry of radioactive elements has led to less widely known advances in technology. The high sensitivity with which radioactivity can be detected led to important analytical applications. Small samples of paint and hair samples from crime scenes can be characterized by a nuclear chemistry technique called neutron activation analysis. The need to sequence small amounts of DNA for biotechnology applications (including DNA fingerprinting) relies on radioactive chemical "tags." Chemicals that contain small numbers of radioactive atoms are also given to patients in many medical imaging techniques. This enables physicians to obtain pictures of suspected diseased organs or tumors without invasive surgical procedures. The decay of radioactive elements can even be exploited to estimate the age of human relics or dinosaur bones.

Isotopes of Elements

The nucleus of an atom contains protons and neutrons. The neutrons act as glue to hold the positively charged protons together. If the number of neutrons in the nucleus varies while the number of protons remains the same, there is no change in the chemical properties of the atom. However, the mass of the atom is altered. Atoms of a particular element that differ only in mass are called isotopes. The number of neutrons in the nucleus also determines the stability of nuclei toward radioactive decay. Different isotopes of a particular element may exhibit different nuclear stabilities. This difference in stability gives rise to radioactivity.

Superscripts and subscripts are added before the symbol for an element to distinguish isotopes. The superscript, or mass number, is the total number of protons and neutrons in the nucleus. The subscript, or atomic number, specifies the number of protons in the nucleus. The difference between the mass number and the atomic number is the number of neutrons in the nucleus. The symbol ^1_1H stands for a hydrogen atom of atomic number 1 and mass number 1. The symbol ^4_2He represents a helium atom with an atomic number of 2 (two protons) and a mass number of 4 (two protons plus two neutrons). The isotope ^1_1H is the most abundant form of hydrogen; however, 0.016% of atoms in a naturally occurring sample of hydrogen exist as the heavier isotope ^2_1H. This heavy isotope of hydrogen is called deuterium. It is often given the special symbol D to eliminate the need for superscripts. A deuterium isotope of mass number 2 weighs approximately twice as much as the more abundant hydrogen isotope. Like hydrogen, deuterium exists as the gas D_2, and it explodes (equation 1) when ignited in the presence of oxygen to form D_2O, called heavy water.

$$2D_{2(g)} + O_{2(g)} \rightarrow 2D_2O_{(g)} \qquad (1)$$

A deuterium–oxygen gas mixture explodes
when ignited to form heavy water.

Although we include the superscripts and subscripts for the atomic weights and numbers when describing nuclear reactions, these superscripts and subscripts are usually omitted in the equations for ordinary chemical reactions. When no specific isotopic information is given, each element's atoms are understood to be distributed among the natural mixture of isotopes. In natural water, for example, about 0.03% of the molecules occur as HDO. Natural sources of elements often contain a mixture of isotopes of different masses. Therefore, the atomic weights used by chemists, given in Table 2.1, are average masses for each element's natural mixture of isotopes.

Most elements consist of several isotopes. For example, there are five known isotopes of the gas helium: $^{4}_{2}He$, $^{3}_{2}He$, $^{5}_{2}He$, $^{6}_{2}He$, and $^{8}_{2}He$. The $^{4}_{2}He$ isotope is the most stable. The others are radioactive and undergo decay processes. In general, elements with more than one proton in the nucleus usually need an equal or greater number of neutrons to stabilize the nucleus. Cobalt-60, $^{60}_{27}Co$, is a dangerous gamma ray (γ–ray)-emitting isotope used to provide a radiation source for cancer therapy. However, the most abundant isotope of cobalt, $^{59}_{27}Co$, is not radioactive. Natural uranium also exists in two isotopic forms, $^{235}_{92}U$ (0.72 %) and $^{238}_{92}U$ (99.28 %). Only the rare $^{235}_{92}U$ undergoes nuclear fission easily (the splitting of the nucleus), which makes nuclear reactors and atomic bombs possible.

One of the great challenges during World War II was to find a way to separate uranium isotopes. This task was very difficult because the isotopes have virtually the same chemical properties. A dedicated group of scientists and engineers working at Oak Ridge, Tennessee, solved this separation problem in the early 1940s. These individuals were part of the Manhattan Project, the name given to the effort in the United States to develop and build atomic weapons. The project culminated in the successful test of the first atomic bomb at Alamogordo, New Mexico, in 1945. The first bomb used $^{239}_{94}Pu$ (plutonium) as the fissionable fuel ($^{239}_{94}Pu$ is made from uranium in a nuclear reactor). The difficulties associated with obtaining high-purity samples of fissionable plutonium and uranium limited the initial production to enough material for only two atomic bombs, referred to as Little Boy and Fat Man. Photographs of Little Boy and the mushroom cloud made upon its explosion at Hiroshima are shown in Figure 2.1.

Table 2.1. *Atomic Weights of the Elements*

Name	Symbol	Atomic Weight (grams/mole)	Name	Symbol	Atomic Weight (grams/mole)
Actinium	Ac	227.028	Mercury	Hg	200.59
Aluminum	Al	26.982	Molybdenum	Mo	95.94
Americium	Am	243.061	Neodymium	Nd	144.24
Antimony	Sb	121.757	Neon	Ne	20.180
Argon	Ar	39.948	Neptunium	Np	237.048
Arsenic	As	74.922	Nickel	Ni	58.693
Astatine	At	209.987	Niobium	Nb	92.906
Barium	Ba	137.327	Nitrogen	N	14.007
Berkelium	Bk	247.070	Nobelium	No	259.101
Beryllium	Be	9.012	Osmium	Os	190.2
Bismuth	Bi	208.980	Oxygen	O	15.999
Boron	B	10.811	Palladium	Pd	106.42
Bromine	Br	79.904	Phosphorus	P	30.974
Cadmium	Cd	112.411	Platinum	Pt	195.08
Calcium	Ca	40.078	Plutonium	Pu	244.064
Californium	Cf	251.080	Polonium	Po	208.982
Carbon	C	12.011	Potassium	K	39.098
Cerium	Ce	140.115	Praseodymium	Pr	140.908
Cesium	Cs	132.905	Promethium	Pm	144.913
Chlorine	Cl	35.453	Protactinium	Pa	231.036
Chromium	Cr	51.996	Radium	Ra	226.025
Cobalt	Co	58.933	Radon	Rn	222.018
Copper	Cu	63.546	Rhenium	Re	186.207
Curium	Cm	247.07	Rhodium	Rh	102.906
Dysprosium	Dy	162.50	Rubidium	Rb	85.468
Einsteinium	Es	252.083	Ruthenium	Ru	101.07
Erbium	Er	167.26	Samarium	Sm	150.36
Europium	Eu	151.965	Scandium	Sc	44.956
Fermium	Fm	257.095	Selenium	Se	78.96
Fluorine	F	18.998	Silicon	Si	28.086
Francium	Fr	223.020	Silver	Ag	107.868
Gadolinium	Gd	157.25	Sodium	Na	22.990
Gallium	Ga	69.723	Strontium	Sr	87.62
Germanium	Ge	72.61	Sulfur	S	32.066
Gold	Au	196.966	Tantalum	Ta	180.948
Hafnium	Hf	178.49	Technetium	Tc	97.907
Helium	He	4.003	Tellurium	Te	127.60
Holmium	Ho	164.930	Terbium	Tb	158.925
Hydrogen	H	1.008	Thallium	Tl	204.383
Indium	In	114.82	Thorium	Th	232.038
Iodine	I	126.904	Thulium	Tm	168.934
Iridium	Ir	192.22	Tin	Sn	118.710
Iron	Fe	55.847	Titanium	Ti	47.88
Krypton	Kr	83.80	Tungsten	W	183.85
Lanthanum	La	138.906	Uranium	U	238.029
Lawrencium	Lr	262.11	Vanadium	V	50.942
Lead	Pb	207.2	Xenon	Xe	131.29
Lithium	Li	6.941	Ytterbium	Yb	173.04
Lutetium	Lu	174.967	Yttrium	Y	88.906
Magnesium	Mg	24.305	Zinc	Zn	65.39
Manganese	Mn	54.9380	Zirconium	Zr	91.224
Mendelevium	Md	258.1			

Figure 2.1. *The atomic bomb, Little Boy (left), and the mushroom cloud generated when it was exploded over Hiroshima (right). (Left, courtesy of Los Alamos National Laboratory; right, UPI/Bettman.)*

It is ironic that the shortage and questionable reliability of such weapons contributed to the political decision to use them. President Harry Truman decided that because only two atomic bombs had been produced, neither could be wasted in an uninhabited area for a "show of force." Even today the complex technology required to separate uranium isotopes makes it difficult for terrorists to obtain nuclear weapons. Construction of a bomb from uranium or plutonium enriched in the fissionable isotope is an easier task. This is why stockpiles of enriched uranium and plutonium must be guarded closely and why international trade in these materials must be regulated.

Nuclear Chemistry and Radioactive Decay

The first experimental evidence for radioactivity was reported in 1896 by the French scientist Henri Becquerel. Becquerel showed that uranium gave off radiation that could expose protected photographic film. Soon thereafter, several kinds of radioactive decay were discovered. One type of radioactive decay was termed alpha decay (α-decay). This process occurs when an unstable nucleus of a heavy element spits out a cluster of two protons and two neutrons to form a lighter element with an atomic number decreased by 2 and a mass number decreased by 4. The small fragment emitted is called an α-particle. Examining the periodic table (See Figure 2.2) reveals that an α-particle is simply a helium nucleus, $_2^4\text{He}^{2+}$. In fact, the main source of helium in our atmosphere comes from α-decay processes of elements in the Earth. Helium is so light that it eventually escapes from the Earth's atmosphere into outer space. Because helium is so useful in modern technology, it is fortunate that it is constantly replenished by radioactive decay.

Often a gamma ray (γ-ray) accompanies the emission of an α-particle. γ-rays are a form of electromagnetic energy like radio waves, visible light, and X-rays; however, γ-rays have higher energies than X-rays. They can penetrate several feet of concrete. In contrast, a thin piece of paper stops an α-particle. The energies of γ-rays depend on the specific nuclear decay event that produces them. Equations (2) and (3) show the main initial nuclear decay processes that occur in uranium ores. The net effect of these nuclear reactions is the transformation of a uranium atom into a thorium atom and a helium atom with the emission of a γ-ray.

$$_{92}^{235}\text{U} \rightarrow \ _{90}^{231}\text{Th} + \ _2^4\text{He} + \gamma \qquad (2)$$

Uranium-235 undergoes α-decay, forming thorium-231 and emitting an α-particle and a γ-ray.

$$_{92}^{238}\text{U} \rightarrow \ _{90}^{234}\text{Th} + \ _2^4\text{He} + \gamma \qquad (3)$$

Uranium-238 undergoes α-decay, forming thorium-234 and emitting an α-particle and a γ-ray.

In Becquerel's experiment, it was the γ-rays produced by the reactions represented by equations (2) and (3) that exposed the photographic film.

In 1898 Marie and Pierre Curie discovered the element radium, $_{88}^{226}\text{Ra}$, in pitchblende, a uranium ore. Radium undergoes α-decay according to equation (4). It emits both an α-particle and

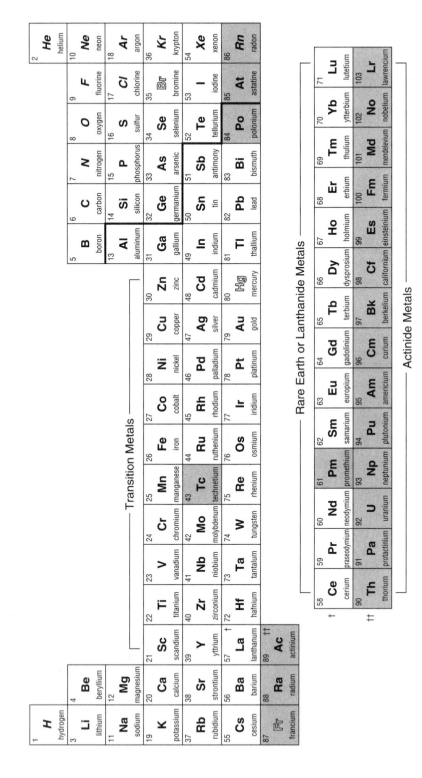

Figure 2.2. *Periodic table of the elements wherein the naturally radioactive elements are shown with a gray background.*

a γ-ray to produce the lighter element radon, Rn. Radium exists as a metallic element, but radon is a noble gas.

$$^{226}_{88}Ra \rightarrow \, ^{222}_{86}Rn + \, ^{4}_{2}He + \gamma \qquad (4)$$

Radium-226 undergoes a-decay, to forming radon-222
and emitting an a-particle and a g-ray.

Radon, $^{222}_{86}Rn$, produced by decay of radium in the Earth, also undergoes α-decay to form polonium, as shown in equation (5).

$$^{222}_{86}Rn \rightarrow \, ^{218}_{84}Po + \, ^{4}_{2}He + \gamma \qquad (5)$$

Radon-222 undergoes α-decay, forming polonium-218
and emitting an α-particle and a γ-ray.

Another decay process that occurs naturally is termed beta decay (β-decay). A small amount of carbon dioxide in the Earth's atmosphere contains radioactive $^{14}_{6}C$. This isotope of carbon originates in the upper atmosphere from the bombardment of nitrogen nuclei with cosmic rays. The $^{14}_{6}C$ undergoes a nuclear decay process, which involves the splitting of a neutron in the nucleus into a proton and an electron. The electron, $_{-1}^{0}e$, leaves the nucleus. Conservation of charge (a law of nature) requires the nucleus left behind (with a new proton) to increase its positive charge, or atomic number, by 1. The mass number is unchanged. With the new atomic number of 7, the atom is no longer carbon (which has an atomic number of 6), but $^{14}_{7}N$, as shown in equation (6).

$$^{14}_{6}C \rightarrow \, ^{14}_{7}N + \, ^{0}_{-1}e \qquad (6)$$

Carbon-14 undergoes β-decay, forming
nitrogen-14 and emitting an electron.

The emitted electrons, called β-particles, can penetrate the piece of paper that stops α-particles, but a piece of aluminum foil stops them. Isotopes such as ^{11}C and ^{18}F undergo a form of β-decay wherein a "positive electron" instead of a negative electron is emitted. The positive electron, $^{0}_{1}e$, called a positron, is an example of antimatter. A positron exists briefly before it encounters an atom with normal electrons. When an electron and a positron collide, they destroy each other, producing two γ-rays that go in opposite directions.

Half-lives of Radioisotopes

Nuclear decay processes always occur at a specific rate described by a half-life for decay. The *half-life* is the time it takes for half the atoms in a sample of a specific isotope to undergo nuclear decay. Each isotope exhibits a characteristic time of radioactive decay. The decay of $^{238}_{92}U$ has a measured half-life of 4.5 billion years; other examples are $^{14}_{6}C$, 5730 years; $^{226}_{88}Ra$, 1600 years; $^{60}_{27}Co$, 5.3 years; $^{234}_{90}Th$, 24 days; $^{218}_{84}Po$, 3 minutes; and $^{214}_{84}Po$, less than 1/1000 of a second. The long time it takes for several radioactive elements to decay creates a problem in the disposal of nuclear waste. For some of the radioactive isotopes in nuclear waste, the decay takes thousands of years. Disposal sites need to be selected with a commitment that corresponds to 6 to 10 times the half-life of the radioisotopes being discarded!

Radiocarbon dating, for which Williard Libby received the 1960 Nobel Prize for Chemistry, is extensively used in archeology. Radiocarbon dating is based on the decay of $^{14}_{6}C$. This isotope of carbon is constantly being generated in the upper atmosphere as the result of a nuclear reaction between nitrogen and cosmic rays. The abundance of atmospheric nitrogen ensures a constant source of $^{14}_{6}C$. As a result, the abundance of $^{14}_{6}C$ is nearly constant, varying slightly with the cosmic ray flux from the Sun. The generation and decay of this carbon isotope constitute a balanced process. Living plants acquire a small amount of radioactive $^{14}_{6}C$ from the CO_2 that they take in from the atmosphere when they make carbohydrates by photosynthesis. After death, of course, they no longer consume atmospheric $^{14}_{6}C$, and the amount of radioactive $^{14}_{6}C$ in the plant then begins to decrease. Animals ingest the same fraction of radioactive carbon when they eat plants. When an animal or plant dies, the $^{14}_{6}C$ it accumulated while alive begins to decay with a half-life of 5730 years. That means that a tree chopped down 5730 years ago contains only half as much of $^{14}_{6}C$ (for an equivalent mass of carbon) as one cut recently. Therefore, if we unearth a wooden spear with half as much of $^{14}_{6}C$ in its carbon as a modern spear, then

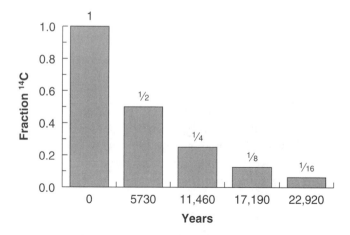

Figure 2.3. *Bar graph showing the decay of ^{14}C in an artifact over a time period corresponding to four half-lives. The amount of ^{14}C remaining successively decreases by half with each interval of 5730 years.*

the relic must be 5730 years old. If the remains of a primitive wooden boat have only one fourth $^{14}_{6}C$ per gram of carbon as freshly cut wood, then two half-lives, or 11,460 years, have elapsed. (It took 5730 years for half the $^{14}_{6}C$ to decay, and after another 5730 years it was halved again to one-fourth the original amount.) A graph of this decay process is shown in Figure 2.3.

Modern forgeries of wooden relics are easily detected because they possess the telltale $^{14}_{6}C$ content of the modern wood used to construct them. With this technique archeologists estimate the dates of objects made from living plants or animals (bones, paper, clubs) over the past 400–15,000 years. Minor corrections need to be included for the variability in cosmic ray flux during this time period.

The technique of radiocarbon dating had a major impact on estimates of the age of the European Ice Man that hikers found in the Alps in 1991. The mummified body appeared stuck in the edge of a receding glacier. An axe, thought to be bronze, lay near the body. On the basis of this evidence, archeologists estimated that the Ice Man lived around 2000 B.C., during the Bronze Age. (The use of bronze, a hard alloy of copper and tin, in tools came after the use of copper, which is a much softer metal.) However, subsequent analysis of the carbon-14 content in protein samples extracted from bones of the corpse indicated that the skeleton was more than 1000 years older. The results were checked by independent laboratories. The radiocarbon date corresponds to the Neolithic Age, or Stone Age, that preceded the Bronze and Copper Ages. Elemental analysis of the metal in the axe head showed that it was not bronze but copper. These results established the Ice Man as the oldest preserved human body and caused archeologists to reconsider when early Europeans first began to use metal tools.

Geologists estimate the age of rocks by using a dating procedure based on the relative amounts of $^{238}_{92}U$ (which has a half-life of 4.5 billion years) and of its decay product, $^{206}_{82}Pb$. The oldest rocks appeared 3 to 4.5 billion years ago, at the time the Earth formed.

Other Reactions Involving Nuclei

Nuclear Fission

By 1930 radioactive decay reactions were the only ones believed to be possible for nuclei, but that soon changed. In 1934 Enrico Fermi claimed to have made a new element by bombarding uranium nuclei with neutrons. Fermi, however, did not have good chemical evidence for the new element. In 1938 two German chemists, Otto Hahn and Fritz Strassman, tried to identify the element (or elements) produced in Fermi's experiment. To their astonishment, they found that the collision of neutrons with uranium produced the lighter element barium. The other products were later found to be krypton and neutrons, as shown in equation (7).

$$^{235}_{92}U + ^{1}_{0}n \rightarrow ^{141}_{56}Ba + ^{92}_{36}Kr + 3^{1}_{0}n \tag{7}$$

Fission of the uranium-235 nucleus by a neutron to form barium-141, krypton-92, and three neutrons

Hahn reported this striking result to his friend and former co-worker on the uranium project, the physicist Lise Meitner. As an Austrian Protestant with Jewish ancestors, she had been forced to flee Nazi Germany in 1933. Meitner deduced that a tremendous amount of energy must have been liberated because of the mass loss in the fission process. The products of equation (7) have slightly less mass than the reactants. A calculation of the energy liberated, via the famous Einstein relationship $E = mc^2$, showed that *nuclear fission* liberates about a million times more energy per mole of substance than any known chemical reaction. The other unusual aspect of this reaction is that three neutrons are released for each neutron absorbed, thereby causing a sustained chain reaction if the density of $^{235}_{92}U$ nuclei in the sample is high enough to capture the secondary neutrons. For example, the first fission event in equation (7) generates three neutrons. These can react with three more $^{235}_{92}U$ nuclei, releasing nine more neutrons, as shown in Figure 2.4. And these neutrons, in turn, collide with nine more $^{235}_{92}U$ nuclei to release 27 neutrons. The reaction cascade continues until the $^{235}_{92}U$ nuclei are depleted.

A chain reaction begins only when there are enough $^{235}_{92}U$ nuclei in the sample to absorb the neutrons. Otherwise, nothing happens. By varying the amount of the rare $^{235}_{92}U$ isotope in the more abundant $^{238}_{92}U$, it is possible to make the chain reaction occur relatively slowly, as in a nuclear reactor; if the chain reaction occurs rapidly, it causes an explosion, as in an atomic fission bomb. The reaction of equation (7) is not the only process that occurs. The nucleus splits in many ways upon fission of uranium. Three different radioactive barium species were identified by Hahn and Strassman in their original report [the long-lived (12.8 days) isotope $^{140}_{56}Ba$ makes up about 6% of the fission products]. Additional experiments have shown that more than 35 different elements occur among the fission products of $^{235}_{92}U$. Most of the $^{235}_{92}U$ fission paths yield two or three neutrons. On average, 2.5 neutrons are produced per $^{235}_{92}U$ fission event. Hahn received the 1944 Nobel Prize for Chemistry for the discovery of nuclear fission.

Fission bombs use a piece of uranium enriched in $^{235}_{92}U$ that contains almost the critical density of nuclei needed to sustain an explosive chain reaction. A chemical explosion is used to

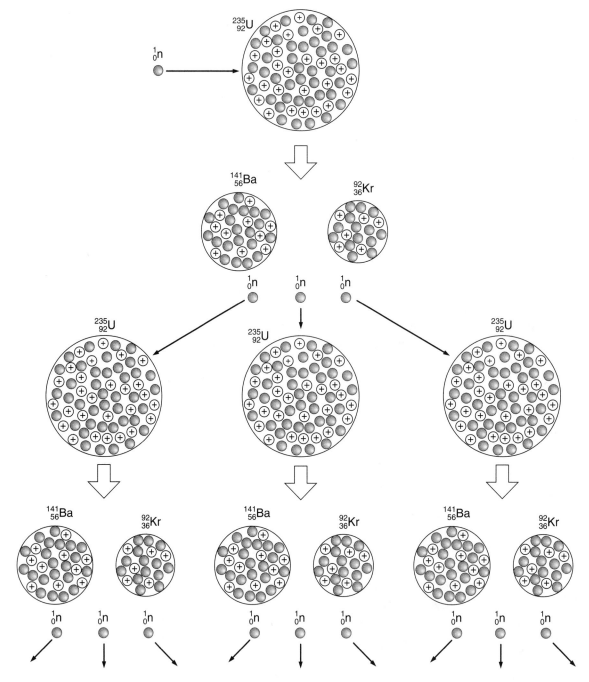

Figure 2.4. *Diagram of the fission of a $^{235}_{92}U$ nucleus by a neutron and the chain reaction that can ensue.*

compress the atoms into a smaller volume, which increases the concentration of $^{235}_{92}U$ to the point (called the critical mass) where the cascade shown in Figure 2.4 begins. At the urging of the physicist Leo Szilard, Albert Einstein outlined the possible significance of such a nuclear fission chain explosion in a letter to President Franklin Roosevelt. Einstein's letter impressed Roosevelt, and Roosevelt launched the Manhattan Project.

Nuclear Fusion

Of all elements, iron-56 ($^{56}_{26}Fe$) has the most stable nucleus. Elements heavier than iron tend to decay to it. The high stability of the iron nucleus makes it one of the most abundant metals in the universe. Fission is not energetically favorable for elements lighter than iron, but fusion may occur under the right conditions. *Nuclear fusion* is the process whereby two light nuclei join together to make a larger, more stable nucleus. Repulsive forces between positively charged nuclei must be overcome before their fusion can occur. Only when the separation between nuclei becomes very small can the short-range, strong, attractive nuclear forces bind the nuclei together. In normal collisions between atoms or molecules, the nuclei never get close enough for fusion to commence. The filled electron shells surrounding nuclei repel atoms if they get too close to one another. Thus, to start nuclear fusion, a collection of light nuclei (for which fusion is favorable) must first be given a tremendous amount of energy. Very hot atoms move so fast and collide so violently that the filled electron shells are stripped away and the nuclei collide. This occurs in the Sun and in other stars, where extreme temperatures and huge forces from gravitational compression slam nuclei together in sustained fusion (thermonuclear) reactions. At the center of the Sun, the temperature is about 15 million degrees Celsius, and the pressure is about 200 billion times that of the Earth's atmosphere. Under these conditions, hydrogen gas is compressed to the point where it is denser than uranium metal.

On Earth, two deuterium nuclei (^2_1H, or D) have been fused to form ^3_2He and a neutron as major products. Detonation of an

atomic fission bomb provides the high temperatures and pressures necessary for fusion to occur. The sequential fission–fusion
nuclear explosion results in the so-called hydrogen bomb (although "deuterium bomb" would be a more accurate name). On
a per-gram basis, a hydrogen bomb puts out twice as much energy as a uranium fission bomb. The more important practical
aspect is that the deuterium fuel costs less than the enriched uranium required in a fission bomb. Tritium ($^3_1$H) nuclei (tritium is
another isotope of hydrogen) may be injected with the deuterium fuel to form a more easily detonated fusion device. The
fusion of $^2_1$H and $^3_1$H produces a nonradioactive α-particle and a
neutron. The barrier to fusing a tritium nucleus and a deuterium
nucleus is lower than the barrier to fusing two deuterium nuclei.
Tritium is highly radioactive and expensive.

Nuclear fusion explosions also can occur on a cosmic scale
when, for example, a giant star becomes a supernova (Figure
2.5). A star begins as a hydrogen fusion energy source. Several
fusion reactions, some of which are shown in equations (8)
through (10), make helium fuel for the next stage of a star's life.

$$^1_1H + {}^1_1H \rightarrow {}^2_1H + {}^0_1e + \text{energy} \tag{8}$$

Fusion of hydrogen in a star yields deuterium, a positron, and energy.

$$^1_1H + {}^2_1H \rightarrow {}^3_2He + \text{energy} \tag{9}$$

Fusion of hydrogen and deuterium in a star yields helium-3 and energy.

$$^3_2He + {}^3_2He \rightarrow {}^4_2He + 2{}^1_1H + \text{energy} \tag{10}$$

Fusion of helium-3 in a star yields helium-4, hydrogen, and energy.

After the hydrogen supply in a star is exhausted, helium nuclear
fusion begins. Fusion reactions involving helium nuclei, some of
which are shown below, make heavier elements such as beryllium (equation 11) and carbon (equation 12). If a star is too small,
the gravitational field and inner temperatures may not be sufficient to cause these large nuclei to fuse. As a result, fusion may
never progress beyond this stage.

Figure 2.5. *Supernova 1987A: (Left) Photograph of the Large Magellanic Cloud taken with a 1.5 meter telescope before February 23, 1987. (Right) Image of the same area of sky after that date. The bright image at right-center of the photograph shows the supernova detected in this nearby galaxy. (Courtesy of the National Optical Astronomy Observatories.)*

$$\,_{2}^{4}\text{He} + \,_{2}^{4}\text{He} \rightarrow \,_{4}^{8}\text{Be} + \text{energy} \tag{11}$$

Fusion of helium-4 in a star yields beryllium-8 and energy.

$$\,_{4}^{8}\text{Be} + \,_{2}^{4}\text{He} \rightarrow \,_{6}^{12}\text{C} + \text{energy} \tag{12}$$

Fusion of beryllium-8 and helium-4 in a star yields carbon-12 and energy.

In more massive stars, carbon and helium nuclei may further fuse to produce oxygen and neon, as shown in equations (13) and (14).

$$\,_{6}^{12}\text{C} + \,_{2}^{4}\text{He} \rightarrow \,_{8}^{16}\text{O} + \text{energy} \tag{13}$$

Fusion of carbon-12 and helium-4 in a star yields oxygen-16 and energy.

$$\,_{8}^{16}\text{O} + \,_{2}^{4}\text{He} \rightarrow \,_{10}^{20}\text{Ne} + \text{energy} \tag{14}$$

Fusion of oxygen-16 and helium-4 in a star yields neon-20 and energy.

Continued fusion reactions generate even heavier elements. Fusion reactions in a star ultimately stop with the production of iron-56, the most stable nucleus. Iron nuclei do not undergo further fusion in stable stars. If the formation of iron occurs too rapidly in a large star, then the heavy nuclei may collapse under their own gravitational force. This collapse compresses the remaining light nuclei, which suddenly undergo fusion and release a tremendous burst of energy—an event known as creating a supernova. A dense neutron star, or black hole, remains, and it contains the bulk of the original mass. Small amounts of nuclei heavier than iron are also generated in the explosive process, even though they are less stable. This explains the origin of heavy metals such as iridium, lead, platinum, gold, and uranium. It also explains why elements much heavier than iron are relatively rare in the universe.

There is much interest in the peaceful use of nuclear fusion. Radioactive waste products from fusion pose less of a disposal problem than those made in fission reactors. A confined high-temperature plasma (a collection of electrons and nuclei at high temperatures) that contains tritium and deuterium nuclei shows promise as a controlled fusion energy source. Tokamak reactors, developed by the Russian physicists Andrei Sakharov and Igor Tamm in the early 1950s, appear to be the best candidates for fusion reactors in the near term. They offer the advantage that because of the small amount of fuel in the plasma at any given time, an accidental runaway reaction cannot occur. The problem to date has been the containment of a plasma that must operate at a temperature of about 100,000,000 K to sustain fusion. Figure 2.6 shows a design sketch for the International Thermonuclear Experimental Reactor known as the ITER Tokamak, currently scheduled for construction between 1997 and 2004. It uses an array of magnets to create a field that serves as a magnetic "bottle" for the hot plasma.

The Synthesis of New Elements

Before 1940, 92 elements had been identified in the solar system. Chemists Edwin McMillian and Glenn T. Seaborg at the University of California at Berkeley were able to synthesize sev-

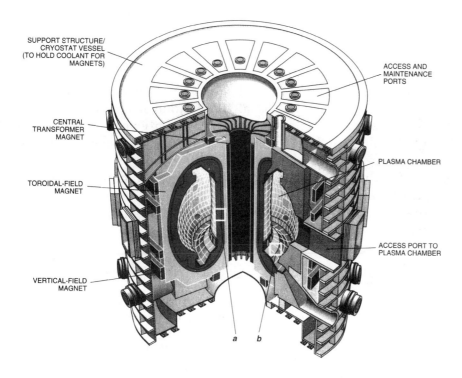

SUPPORT STRUCTURE/ CRYOSTAT VESSEL (TO HOLD COOLANT FOR MAGNETS)

ACCESS AND MAINTENANCE PORTS

CENTRAL TRANSFORMER MAGNET

TOROIDAL-FIELD MAGNET

PLASMA CHAMBER

ACCESS PORT TO PLASMA CHAMBER

VERTICAL-FIELD MAGNET

a b

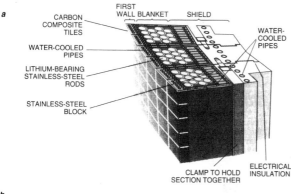

a

FIRST WALL BLANKET   SHIELD

CARBON COMPOSITE TILES

WATER-COOLED PIPES

WATER-COOLED PIPES

LITHIUM-BEARING STAINLESS-STEEL RODS

STAINLESS-STEEL BLOCK

CLAMP TO HOLD SECTION TOGETHER

ELECTRICAL INSULATION

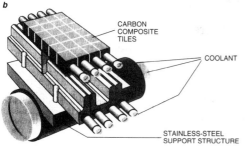

b

CARBON COMPOSITE TILES

COOLANT

STAINLESS-STEEL SUPPORT STRUCTURE

Figure 2.6. *Sketch of the International Thermonuclear Experimental Reactor the ITER Tokamak, currently planned for construction at the end of the decade. (Illustration by Ian Warpole from "The International Thermonuclear Experimental Reactor," by Robert W. Conn, Valery A. Chuyanov, Nobuyuki Inoue and Donald R. Sweetman, April 1992. Copyright © 1992 by Scientific American, Inc. All rights reserved.)*

eral new elements by neutron bombardment of the most massive known element, uranium. These and other transuranium elements appear in the row at the bottom of the periodic table. Some have a very short existence, and only very small amounts of the element have been synthesized. For instance, only ten atoms of $^{256}_{103}\text{Lr}$ have ever been made.

Synthesis of the transuranium elements requires the neutron flux produced by fission of uranium-235 in a nuclear reactor. For example, the nonfissionable isotope of uranium, $^{238}_{92}\text{U}$, reacts with neutrons by equation (15) to produce a new element, neptunium.

$$^{238}_{92}\text{U} + {}^{1}_{0}\text{n} \rightarrow {}^{239}_{93}\text{Np} + {}^{0}_{-1}\text{e} \qquad (15)$$

Uranium-238 collides with a neutron to form
neptunium-239 and emit an electron.

Neptunium, in turn, undergoes β-decay with a half-life of 2.4 days, according to equation (16), to produce plutonium.

$$^{239}_{93}\text{Np} \rightarrow {}^{239}_{94}\text{Pu} + {}^{0}_{-1}\text{e} \qquad (16)$$

Neptunium-239 undergoes β-decay to form
plutonium-239 and emit an electron.

Like uranium-235, plutonium-239 can undergo nuclear fission, and it is more convenient to use for fission weapons. Figure 2.7 shows a diagram of a plutonium triggering device employed in hydrogen bombs. The Tom Clancy novel *The Sum of All Fears* gives a detailed description of a plutonium triggering device.

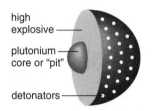

high explosive
plutonium core or "pit"
detonators

Figure 2.7. *Diagram of a plutonium triggering device for a hydrogen bomb. (Copyright 1991 The Washington Post. Redrawn with permission.)*

Radiation and Health

Besides making it possible to prepare new elements, the development of nuclear chemical techniques has helped chemists generate radioactive isotopes of some of the 92 "natural" elements. These radioisotopes are used today in the treatment of cancer, and they also play a key role in many of the imaging techniques of modern medicine.

The Effects of Radiation on Living Tissue

For many years the main medical uses of radioactive elements, such as radium and cobalt-60, were as convenient sources of γ-radiation for the control of cancerous tumors. Cobalt-60 is produced by exposing $^{59}_{27}\text{Co}$ to neutrons in the core of a fission reactor. The new cobalt-60 isotope decays with a half-life of 5.3 years to produce two γ-rays of different energies (equation 17).

$$^{60}_{27}\text{Co} \rightarrow \,^{60}_{28}\text{Ni} + \,^{0}_{-1}\text{e} + 2\gamma \tag{17}$$

*Cobalt-60 undergoes β-decay, forming nickel-60
and emitting an electron and two γ-rays.*

γ-radiation penetrates the human body and is stopped only by several feet of concrete (or 6 inches of lead). On passing through the human body, a γ-ray occasionally kicks an electron out of an atom. The high-energy electron and positive ion produced may collide with nearby molecules and break them apart to create free radicals. This molecular destruction can lead to cell death, mutations, and even cancer of the organism. Accordingly, cobalt-60 must be stored and shipped in lead containers. Plants and animals are constantly exposed to γ-radiation from the cosmic rays that bombard the Earth. Mutations caused by cosmic rays are thought to play an important role in evolutionary changes.

Although radiation can cause cancer, it also helps in cancer control treatment. In radiation therapy, a technician directs a lethal dose of γ-rays selectively toward the tumor. Unfortunately, radiation damage to the surrounding tissue may occur, and the process involves some risk to the patient.

Cobalt-60 γ-radiation can also help preserve food (Figure 2.8). This method is in practice but is not widely accepted by the public. Irradiation of a piece of meat in a sealed container kills all the bacteria present. The sterilized food remains fresh for years when kept in its original sealed package. Treatment with γ-radiation prevents stored potatoes from sprouting, reduces microorganism contaminants in herbs and spices, and controls

Figure 2.8.   *A cobalt-60 γ-radiation machine used to sterilize food.*
(Courtesy of U.S. Army Natick, MA Research, Development and
Engineering Center.)

trichinosis in pork. Some have expressed fear that such food
might be tainted by radioactivity. This fear is unfounded because
γ-rays, which are pure energy, don't change nuclei in the food.
The irradiation process does not make the food radioactive.
However, in addition to killing bacteria, radiation causes a very
small amount of chemical destruction in the irradiated food.
Proponents argue that the trace amounts of chemical decomposition products produced by free radicals in irradiated food are
less objectionable than the present practice of adding much larger quantities of preservatives. On the other hand, food preservatives have been tested for toxicity, and the decomposition
products produced by γ-ray irradiation have not. Such testing
would be difficult because of the low concentrations involved.
In concluding that irradiated foods are not a toxic hazard, a committee of the United States Food and Drug Administration reasoned that people eating irradiated food would ingest only a few
micrograms (1 microgram = 0.000001 gram) of foreign chemicals
each day.

In contrast to γ-rays, the electrons produced by β-decay of
radioactive elements penetrate only half an inch of plant or animal tissue, and massive α-particles go through less than a fraction of an inch of human tissue before they are stopped. An
α-emitter therefore presents a minor hazard when outside the
body, because the α-particles collide with the oxygen and nitrogen molecules of air before they have a chance to enter the skin.

On the other hand, if an α-emitter is inhaled or ingested, it can create serious problems in the tissue it enters. An internal α-emitter does much more damage per decay event than an internal γ-emitter or β-emitter.

The health hazards of radioactive elements were not appreciated at first. Some years ago it was common to find radium, an α-emitter, and zinc sulfide, a phosphor, painted on watch dials so the dials would glow in the dark. This practice continued for years until factory workers began to die in large numbers from leukemia and lung cancer. A dentist first noticed that rare cancers of the mouth occurred in workers who licked their radium-tipped paint brushes. Marie Curie, who received the 1911 Nobel Prize for Chemistry for the discovery and isolation of radium, died from radiation-induced leukemia. Occupational ingestion and inhalation of small amounts of radium dust and radioactive decay products (radon gas) were soon recognized as the cause of these problems. Radium (Ra) is in the same column of the periodic table as calcium (Ca). Because it exhibits similar chemical properties, radium can substitute for calcium in the bone—and the concentrated radiation damage to bone marrow produces cancers such as leukemia. Marie Curie's daughter, Irène, also died from leukemia, and her husband, Frédéric Joliot, suffered from radiation sickness. Irène and Frédéric received the 1935 Nobel Prize for Chemistry for their work on the synthesis of radioactive elements.

The radioactive gas radon (Rn), an α-emitter, occurs naturally during the radioactive decay of elements in the soil. When the gas is inhaled, some of the α-particles bombard lung tissues. What makes the problem more serious is that the decay of gaseous $^{222}_{84}$Rn produces a solid radioactive element, polonium-218 ($^{218}_{84}$Po), which can cling to fine aerosol particles that are inhaled. The $^{218}_{84}$Po is deposited on the inner walls of the lungs, where it undergoes a series of decay events that emit α-particles and γ-radiation. The radiation damage to the lungs caused by the deposited polonium-218 increases the risk of developing lung cancer. As radon gas is generated in the soil, it enters homes through cracks in the basement or foundation. Tightly insulated homes, which restrict ventilation in the name of energy efficiency, exacerbate this problem. Kits are available to test for radon gas. If a problem exists, contractors can seal entry points for the gas and provide air circulation so that dangerous radon concentrations won't build up. Many of the kits are not very reliable, however, so it is best to try several types and see whether they yield consistent measurements. The United States Environmental Protection Agency *estimates* that environmental radon may be second only to smoking as a cause of lung cancer.

Radioisotopes in Medicine

X-ray cameras "see" bones and teeth well but do not resolve details in soft organs, such as the heart. For this reason, there has long been interest in finding ways to image the organs that X-rays cannot see. Because heavy-metal nuclei scatter X-rays well, one early approach was to inject solutions that contain X-ray–visible nuclei into patients. Heavy-metal toxicity limits this to a few applications, such as the "barium enema" for colon imaging. Barium sulfate particles slurried in water adsorb on the walls of the intestine, and the heavy barium atoms make the surface they coat show up in an X-ray picture. Only nontoxic compounds should be used for X-ray imaging of this type. The water-insolubility of barium sulfate prevents its uptake by the body. In contrast, soluble barium salts are extremely poisonous. To image other body organs, very clever techniques are needed. Some recently devised approaches take advantage of the radioactive isotopes that are synthesized in nuclear reactors. Several radioactive nuclei used in diagnostic medicine are listed and described in Table 2.2.

The element technetium (Tc) occurs only in trace amounts on Earth. The irradiation of $^{98}_{42}\text{Mo}$ with neutrons in a nuclear reactor produces $^{99}_{42}\text{Mo}$. This unstable isotope of molybdenum undergoes β-decay with a half-life of 67 hours to produce $^{99\text{m}}_{43}\text{Tc}$. The superscript m means that $^{99}_{43}\text{Tc}$ forms in a metastable (briefly stable) excited nuclear state. The excited atom, $^{99\text{m}}_{43}\text{Tc}$, is called metastable because it decays with a half-life of 6 hours to form $^{99}_{43}\text{Tc}$ by the emission of two γ-rays. The $^{99}_{43}\text{Tc}$ then decays slowly (it has a half-life of 210,000 years). This new element, $^{99\text{m}}_{43}\text{Tc}$, which was made initially out of scientific curiosity, is immense-

Table 2.2. *Radioisotopes Used in Medical Diagnostics*

Element	Half-life	Mode of Administration	Uses
^{51}Cr	27 days	i.v. (intravenous)	Determination of serum protein loss into the gastrointestinal tract; determination of red cell volume; spleen imaging; placenta localization
^{47}Ca	4.5 days	i.v.	Calcium metabolism studies
^{60}Co	5.27 years	orally, as radioactive vitamin B_{12}	External, intracavity, or interstitial irradiation
^{18}F	1.8 hours	i.v.	Brain imaging: bone scan
^{67}Ga	3.2 days	i.v.	Tumor-seeking agent: detection of inflammatory lesions
^{198}Au	2.6 days	i.v.,	Liver imaging
^{111}In	2.8 days	i.v.	Tumor detection; bone marrow imaging; detection of gastrointestinal bleeding, detection of deep vein thrombosis
^{113m}In	1.6 hours	i.v.	Liver and spleen imaging; blood pool imaging; determining cardiac output
^{131}I	8 days	i.v.	Brain imaging; liver function
^{125}I	60 days	i.v.	Pancreatic function; thyroid imaging; liver function
^{197}Hg	2.7 days	i.v.	Brain scan; renal imaging
^{32}P	14.2 days	orally or i.v.	Localization of ocular, brain, and skin tumors
^{43}K	22.3 hours	i.v.	Myocardial scan
^{42}K	12.3 hours	orally or i.v.	Localization of brain tumors; determination of intracellular space
^{75}Se	119.7 days	i.v.	Imaging of pancreas and parathyoid glands
^{85}Sr	64.8 days	i.v.	Bone imaging
^{99m}Tc	6 hours	orally or i.v.	Brain imaging; thyroid imaging; salivary gland imaging; placenta localization; blood pool imaging; gastric mucosa imaging; cardiac function studies; urinary bladder imaging; liver imaging; renal imaging; bone and bone marrow imaging
^{24}Na	14.7 hours	i.v.	Determination of blood circulation times

ly valuable for imaging the heart (to locate blockages) and heart muscle (to locate damaged areas). It is also useful for brain, lung, spleen, liver, bone marrow, colon, and kidney imaging. The γ-radiation from the minute amounts of radioactive $^{99m}_{43}Tc$ consumed in the procedure lights up all areas in the patient's body in which it is localized. Destructive α-particles and β-particles are not emitted when the $^{99m}_{43}Tc$ decays, and the dose of γ-radiation is below the level that is harmful. Most of the γ-radiation passes through the body without causing damage. γ-ray detectors placed around the patient fix the direction of the emanating radiation. In the process called tomography, computer analysis of this geometric information generates a three-dimensional image of the organ in the body where the radiation originated. Alternatively, simple two-dimensional camera pictures, which resemble an X-ray image, are produced. Much research is being done to synthesize technetium compounds that localize in specific organs. If the body concentrates a certain $^{99m}_{43}Tc$ compound in the heart muscle, then the γ-ray picture shows the heart most clearly (and, as a bonus, the level of $^{99m}_{43}Tc$ in the rest of the body is reduced). Figure 2.9 shows an image of several organs (heart, liver, and kidneys) obtained with $^{99m}_{43}Tc$. As time passes, the imaging agent is cleared from the body through the kidneys and liver and concentrates in the bladder.

The half-life requirements for the radioisotopes that are used in medicine differ widely. When a physician is employing cobalt-60, a γ-ray source used to irradiate tumors in patients, a high intensity of γ-rays is necessary but the source shouldn't decay rapidly; if it did, it would need to be replaced too often. For an internal imaging agent, such as $^{99m}_{43}Tc$, a physician should use the lowest dose possible. Radioisotopes with a short half-life, such as $^{99m}_{43}Tc$, provide a brief burst of γ-rays for an intense image. Furthermore, a short half-life means that any remaining radioisotope in the body decays within a few days.

Another common radioisotope used in medical imaging is iodine-131 $\left(^{131}_{53}I\right)$. This isotope of iodine has a half-life of 8 days; it decays by emitting a β-particle and a γ-ray. The β-emission makes $^{131}_{53}I$ less desirable than $^{99m}_{43}Tc$, because β particles can cause tissue damage. However, owing to the low price and ready

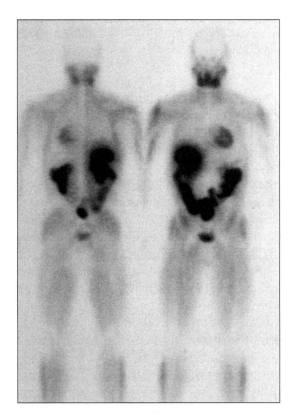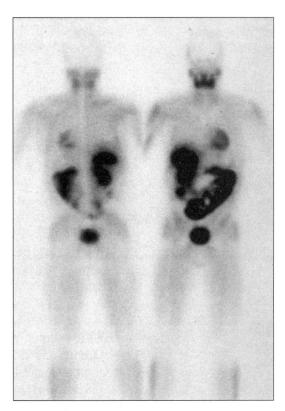

Figure 2.9. *A 29-year-old normal male volunteer was injected with the technetium-99m compound TECHNESCAN Q12, a new myocardial imaging agent for the detection of coronary artery disease. The images shown were acquired 1 hour (left pair) and 2 hours (right pair) after injection. The images on the left and right of each pair are the views from the front and back of the body, respectively. The heart is observed as a dark horseshoe-shaped object in the upper chest region (the dark areas show intense uptake of the radioisotope). The agent also shows uptake in the salivary glands of the neck. The imaging agent is rapidly cleared from the body through the kidneys and liver (via the gall bladder and intestines). Hospital San Raffaele, Milan, Italy. (Courtesy of Mallinckrodt Medical.)*

availability of $^{131}_{53}\text{I}$, this isotope is often used to image tumors of the thyroid. A different isotope of iodine, $^{123}_{53}\text{I}$, has a half-life of only 13 hours, which makes it better for thyroid imaging than $^{131}_{53}\text{I}$. But the $^{123}_{53}\text{I}$ isotope is difficult to obtain and is more expensive than $^{131}_{53}\text{I}$.

Nuclei that undergo positron emission are employed in a relatively new medical imaging technique called positron emission tomography (PET). First, a patient ingests a substance containing some $^{11}_{6}C$ or $^{18}_{9}F$ atoms. Once inside the body, the $^{11}_{6}C$ or $^{18}_{9}F$ nuclei emit positrons, which combine with nearby electrons to produce two γ-rays that go in opposite directions. Two detectors on opposite sides of the patient's body detect the simultaneously emitted γ-rays. Because the point of origin of the two detected γ-rays must lie on a line between these detectors, the difference in their arrival times can be computer-processed in such a way as to image organs. Medicinal chemists have increased the effectiveness of this technique by synthesizing chemicals that concentrate in certain organs or tumors. The incorporation of $^{11}_{6}C$ or $^{18}_{9}F$ atoms greatly enhances the PET images of the organs that take up these chemicals. Figure 2.10 shows a PET machine and an image of the brain obtained by this device.

Research on the use of radioisotopes in medicine may someday lead to "magic bullets" for the cure of cancer. If α-emitters could be attached to molecules that recognize cancer cells, then it might be possible to kill cancer cells selectively. Current surgical and radiation therapies suffer the drawback that single cancer cells, which escape to unaffected regions of the body, spread tumors. Both chemists and biologists are trying to find ways to attach deadly radioisotopes specifically to cancer cells.

Neutron Activation Analysis

The nuclei of most nonradioactive elements absorb neutrons to generate unstable radioisotopes. Thus, when a sample containing a mixture of atoms is bombarded with neutrons from a nuclear reactor, a great many unstable radioisotopes are produced. These unstable radioisotopes emit γ-rays, the energy of which differs for different elements. In neutron activation analysis, a sample of unknown composition is exposed to neutrons. The γ-rays emitted by the radioisotopes made in the sample are analyzed, and their intensities and energies reveal what quantities and what kinds of nuclei must have been in the original sample. For example, natural arsenic atoms all consist of the stable isotope $^{75}_{33}As$. Irradiation of a material containing natural

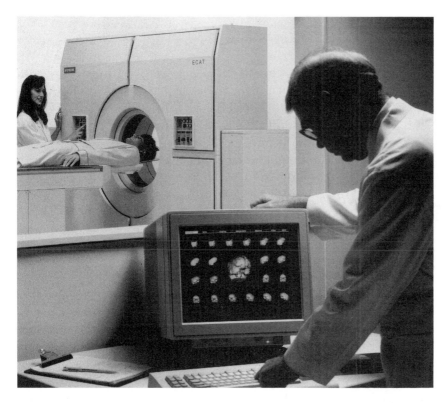

Figure 2.10. *A PET imaging machine. (Siemens Medical Systems, Inc. PET scanner. Courtesy of the Society of Nuclear Medicine.)*

arsenic with neutrons produces $^{76}_{33}$As, an isotope that undergoes simultaneous β-decay and γ-emission. The specific energy of the emitted γ-ray is characteristic of arsenic. The intensity of the emission establishes how much arsenic is present in the sample. With this method, as little as 5 billionths of a gram (5 nanograms) of arsenic can be detected in 1.0 gram of material.

Neutron activation analysis has been an asset in archeology and forensic science. It was used, for example, to debunk the old rumor that Zachary Taylor, the twelfth president of the United States, died from arsenic poisoning. Taylor opposed admitting new states that endorsed slavery, and this was claimed to be the motive for his "assassination." His body was exhumed in 1990, and samples of hair, nails, and bone were subjected to neutron activation analysis. The small amounts of arsenic found did not justify the poisoning hypothesis.

Because it is an elemental fingerprinting technique, neutron activation analysis is useful in tracking down criminals. A precise match between the paint, soil samples, gunshot residue, shell casings, or other physical evidence found on a suspect, and those at the scene of a crime, usually is of interest to a jury. And art forgeries can be detected easily, because the specific composition of trace elements in a paint pigment from an authentic masterpiece invariably differs from that in the modern paint used in a forgery.

Nuclear Power

As of January 1, 1989, there were 429 commercial nuclear reactors around the world; there also were 105 commercial reactors under construction and 324 research reactors in operation. Together these facilities were producing over 310,000 megawatts of energy. With the dramatic increase in nuclear power facilities (there were only 66 commercial reactors in 1970), substantial spent-fuel inventories have accumulated. Since 1970 the United Kingdom, the United States, France, and Japan have generated more than 65,000 tons of heavy-metal nuclear wastes. The storage and disposal of these toxic materials raise serious safety issues.

A diagram of a nuclear power plant is shown in Figure 2.11. Nuclear power is generated by the fission of the $^{235}_{92}U$ nucleus in the reactor core. The core contains fuel rods, control rods, and a coolant such as water. For safety reasons, the core is located inside a containment building constructed of thick, steel-reinforced concrete walls. Nuclear reactions in the fuel rods generate heat, which is used to produce steam. The steam pressure turns electrical turbines. Because of the hazards associated with the nuclear materials, several steps are involved in using the heat of the nuclear reaction to generate electricity. The fission of 1 gram of uranium produces energy equivalent to the burning of 6 tons of coal. The fission of $^{235}_{92}U$ generates more neutrons than are consumed, so a runaway chain reaction can occur rapidly, exhausting the entire fuel supply. To harness this energy in a controlled fashion, the number of energetic neutrons is regulat-

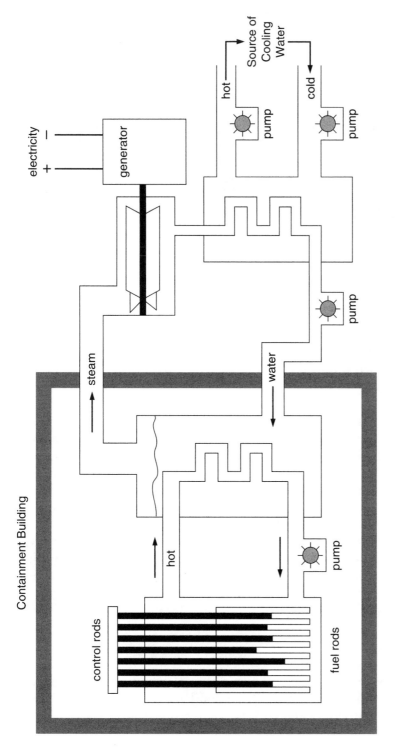

Figure 2.11. *Diagram of a nuclear power plant. The heat generated in the core is transferred to a secondary closed water system, where the water is vaporized to generate steam. The steam drives the turbine, which generates electricity. The steam is recondensed using external cooling water from a nearby river or lake.*

ed by placing control rods made of neutron-absorbing materials
such as cadmium between the uranium fuel rods. The lowering
or raising of the control rods modulates the flux of neutrons in
the reactor. The fission process releases radioactive nuclei, many
of which dissolve in the water. For this reason, the coolant water
is in a closed-cycle system. The energy released by the fission
reaction heats the water, but it is under high pressure so that it
does not boil at the operating temperature of about 350°C (the
boiling point of water at atmospheric pressure at sea level is
100°C). It is important that the coolant not vaporize at the operating temperature of the reactor, because formation of gas bubbles in the reactor core could result in insufficient cooling,
melting the fuel rods. Steam is needed to drive the turbines that
power the electrical generators, so the hot coolant water in the
core is used to heat a secondary closed water system via a
process called heat exchange that involves running the two
water systems through pipes that are in contact. Using piping
materials that efficiently conduct heat, the hot coolant water
from the core heats the secondary water system. The secondary
water system is at a much lower pressure than that in the reactor core; the heat exchange process causes the water in the secondary system to boil, which generates steam. The steam drives
the turbine, which generates electricity. In this two-step process,
the energy released by the nuclear reaction generates electricity.
After flowing through the turbine, the steam needs to be recondensed—and this is also done by heat exchange. Generally, cooling water from a nearby river or lake is used.

In addition to $^{141}_{56}$Ba and $^{92}_{36}$Kr, the fission of $^{235}_{92}$U produces
over 40 different radioactive isotopes, many of which build up
to significant concentrations in the reactor core. The nuclear
accidents at Three Mile Island and Chernobyl have taught us a
lot about the hazards of these nuclear materials.

Three Mile Island

In March 1979, there was a partial loss of reactor core coolant
water from the Three Mile Island Nuclear Reactor in Harrisburg,
Pennsylvania (Figure 2.12). After the loss of water, the reactor

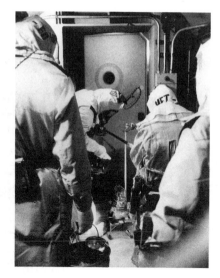

Figure 2.12. *The intact containment domes and the inside of the damaged reactor of the Three Mile Island Nuclear Reactor in Harrisburg, Pennsylvania, after the accident in 1979. (Left, UPI/Bettman; right, AP/Wide World Photos.)*

core overheated, which led to its partial disintegration. The core became so overheated that its zirconium metal casing reacted with water to generate over 450 kilograms of hydrogen gas.

The hydrogen gas created a 30-cubic-meter bubble inside the core vessel. This exposed portions of the core and further prevented the cooling water from functioning properly. For nearly 14 hours after the incident began, the reactor core was partially exposed. Many times during the next few days, radioactive xenon $\left(^{131}_{54}Xe\right)$ and krypton $\left(^{85}_{36}Kr\right)$ gases were released into the air. Water containing radioactive contaminants was dumped into the river, and among the pollutants were radioactive isotopes of iodine $\left(^{125}_{53}I \text{ and } ^{131}_{53}I\right)$. Because iodine localizes in the thyroid gland, these radioactive isotopes posed a serious health risk. Government officials recommended that people exposed to the radiation take potassium iodide (KI) pills to reduce their need for iodine.

The major problem was the enormous volume (2.4 million liters) of radioactive water that spilled into the containment building of the reactor. Fortunately, the concrete shell contained most of the radioactivity. Over half a million curies (radioactivity is measured in curies) of $^{90}_{38}Sr$ and $^{137}_{55}Cs$ were in the water, exceeding the Public Health Services "safe water" standard (0.00000000001 curies per liter) by a factor of 10 billion! The water was decontaminated by removing the radioactive Sr and Cs isotopes with ion exchange materials. By 1983 the only radioactive element left in the water was tritium, $^{3}_{1}H$.

Chernobyl

"Three Mile Island" was a serious reactor accident, but at least it was contained. In April of 1986, a greater disaster occurred; it started with an explosion at the nuclear reactor in Chernobyl, Ukraine. What happened at this site demonstrates that nuclear accidents can result from poor human judgment. According to the report released by the Soviet government, plant operators were conducting various tests at the facility at the time. (To approximate an emergency, the operators had deactivated many of the safety systems.) During these tests, the level of the coolant water in the reactor core fell rapidly (as it did at Three Mile Island). However, without the safety systems, the reactor temperature soared to 5000°C, twice the temperature at which steel melts. The water in the core vaporized instantly, blowing the roof off the building and shooting flames from the materials burning in the reactor core over 30 meters into the air. On the advice of West German and Swedish nuclear experts, Soviet authorities had helicopters drop sand, lead (a γ-ray absorber), and boron (a neutron absorber) on the molten mass of graphite control rods and uranium fuel. Heroic workers tunneled under the molten core to deposit additional cement to make sure that radioactive materials would not contaminate the ground water. Eventually, the entire reactor was encased in concrete. Workers also sprayed a plasticizing agent on the nearby soil to prevent the wind from spreading radioactive particles. Figure 2.13 shows the reactor containment building after the explosion.

The effects of the Chernobyl tragedy extended far beyond Soviet soil. Scientific studies conducted in the years since the disaster have revealed that the fallout from the reactor spread throughout Europe. The main cloud of radioactive contaminants

Figure 2.13. *The remains of the Chernobyl nuclear reactor in Ukraine after the explosion in 1986. (ITAR-Tass/Sovfoto.)*

passed over the United Kingdom on May 2 and 3, six days after the explosion. While the cloud was over Cumbria, in the English Lake District, heavy rains carried many of the radioactive particles down onto the soil, precipitating toxic levels of radioactive cesium, $^{137}_{55}$Cs. On June 20, measurements of radioactivity in parts of Cumbria were so high that a region containing over 4 million sheep (20% of the total sheep population in the United Kingdom) and nearly 7000 farms was declared restricted. Four days later, part of Scotland was added to the restricted region. By the end of September 1986, the restricted area was cut back to 150 farms. However, high levels of radioactivity were present in many of the animals. This had a significant impact on both the food supply and the economic health of the region.

Breeder Reactors

Scientists estimate that at the current rate of use, the world's supply of $^{235}_{92}U$ will last about a century. Many countries are supporting active research to develop alternative nuclear reactor designs. One such device is the "breeder reactor," which uses plutonium-239 $\left(^{239}_{94}Pu\right)$ as the fuel source. Neutrons produced by plutonium-239 fission collide with $^{238}_{92}U$, which makes up 80% of the fuel, and produces plutonium-239 (see equations 15 and 16). Since the initial fission of plutonium-239 generates more than 2 neutrons, there is a net production of plutonium-239 from uranium-238. $^{238}_{92}U$ is the most abundant isotope of uranium; in the United States alone, the supply of $^{238}_{92}U$ is sufficient to run breeder reactors for many centuries. A major way in which breeder reactors differ from conventional reactors is that the cooling is accomplished using materials such as liquid sodium. The potentially increased hazard of operation, as well the concern that plutonium from such reactors might find its way into terrorists' bombs, has limited the development of breeder reactors.

CHAPTER 3

Chemical Bonding

Only the elements called the noble gases, which fall in the extreme righthand column of the periodic table, "prefer" to exist as isolated atoms. The atoms in a pure element usually bond with one another to form a solid (carbon atoms in diamond), a liquid (mercury atoms in liquid mercury), or a gas (hydrogen molecules, H_2, in hydrogen gas). Different elements may also react to make compounds that contain different types of atoms bonded to each other. A chemical bond arises when the forces between the electrons and nuclei of two atoms pull them together. Chemical reactions involve the making and breaking of chemical bonds.

Chemical Bonds and Molecular Properties

Two hydrogen atoms bond together readily and form H_2 gas. This gas is stable by itself; when oxygen (O_2) is present, however, it may react violently and form water (H_2O). Combustion of hydrogen occurs because the chemical bonds in water are stronger than those in the hydrogen and oxygen molecules. Spontaneous chemical changes always proceed in such a way as to yield products that are more stable than the reactants.

$$2H_2 \quad + \quad O_2 \quad \rightarrow \quad 2H_2O \qquad (1)$$

Two molecules of hydrogen react with one molecule of oxygen $\rightarrow$ *to make two molecules of water.*

This reaction may be written in chemical shorthand as in equation (1). The actual bonds present before and after the reaction are depicted with ball-and-stick models below equation (1). In this reaction, the chemical bonds of both reactants (H–H and O=O) are broken, and new bonds are formed between the hydrogen and oxygen atoms to create two water molecules as products. Reactants always appear on the lefthand side of a reaction arrow and products on the righthand side. The total number of oxygen atoms (two) and hydrogen atoms (four) is the same on both sides of the arrow, because atoms are neither created nor destroyed in chemical reactions (conservation of mass). The reaction of equation (1) generates a tremendous amount of heat, so the water is produced as vapor. This hot gas expands so rapidly that the reaction occurs with explosive force, as in the *Hindenburg* disaster.

Covalent Bonds

The bonds in H_2 and O_2 are called *covalent bonds.* In a covalent bond, one electron each from the valence shell of two bonded atoms contributes to a bond. These two-electron bonds appear as linear connections between atoms in structural formulas for molecules. For H_2, the two bonded atoms are equivalent, so they share equally both electrons in the bond. The oxygen molecule contains a double bond (O=O). Two electrons in the outermost electronic shell of each oxygen are shared to make 2 two-electron bonds.

The entire crystal of diamond consists of a cross–connected network of pure covalent carbon–carbon (C–C) bonds. Each carbon atom binds to four neighboring carbon atoms in the diamond crystal. A solid interconnected with covalent bonds is called a network solid. Because the outermost electrons are constrained in the covalent bonds and cannot wander around the crystal freely, network solids behave as thermal and electrical insulators or as semiconductors. Solids held together by a network of covalent bonds tend to be brittle, because rigid bonds must be broken to deform the solid structure.

Covalent bonds can also form between two different atoms (C and H, for example) if both have a similar power to attract electrons. Hydrogen and carbon atoms bond together to make hydrocarbons, such as methane, ethane, ethylene, acetylene, benzene, and isooctane. Hydrocarbon molecules contain only covalent C–C and C–H bonds. Solid hydrocarbons (wax) are soft materials, because there are no bonds connecting adjacent molecules.

Polar Covalent Bonds

Some atoms are better at attracting electrons than others, thereby causing a lopsided distribution of electron charge in the bonds they form. Bonds of this type are called *polar covalent bonds*. In the O–H bonds in the water molecule, oxygen attracts electrons much better than does hydrogen. Each hydrogen atom acquires a partial positive charge, and the oxygen has a partial negative charge. Because the water molecule is neutral, the partial negative charge on the oxygen atom $(2\delta^-)$ must be twice as large as the positive charges (δ^+) on each of the hydrogen atoms. The separation of charge in a polar bond is indicated by the symbols δ^- and δ^+ in Figure 3.1. The sum of all the charges equals zero $(2\delta^- + \delta^+ + \delta^+ = 0)$, as required by the overall neutrality of the molecule.

Figure 3.1. *Charge separation resulting from unequal sharing of the bonding electrons in the water molecule.*

Weak forces exist between the electric charges on different polar molecules. These forces are attractive when the molecules align as shown in Figure 3.2. The attractive forces between two polar molecules tend to be greater than for comparable nonpolar molecules. Water exists as a liquid at room temperature, but the similar-sized nonpolar molecule methane exists as a gas. Stronger forces between water molecules help stabilize the liquid state. Microwave ovens use the charge separation in water to heat food. The oscillating electric field of microwave radiation interacts with the charge separation in water molecules, which forces the molecules to oscillate rapidly. The friction generated by this molecular motion in foods that contain water heats the food.

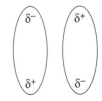

Figure 3.2. *Two polar molecules arranged in such a way as to maximize the attractive electric forces between the opposite charges.*

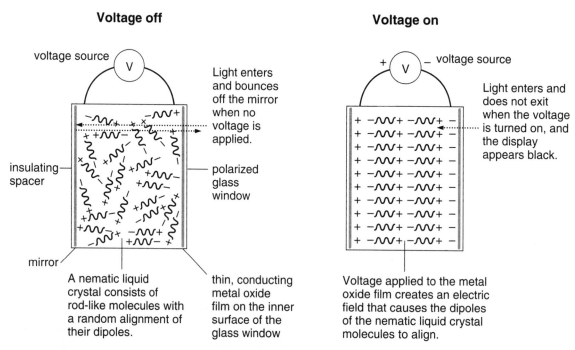

Voltage off

voltage source

Light enters and bounces off the mirror when no voltage is applied.

insulating spacer

polarized glass window

mirror

A nematic liquid crystal consists of rod-like molecules with a random alignment of their dipoles.

thin, conducting metal oxide film on the inner surface of the glass window

Voltage on

voltage source

Light enters and does not exit when the voltage is turned on, and the display appears black.

Voltage applied to the metal oxide film creates an electric field that causes the dipoles of the nematic liquid crystal molecules to align.

Figure 3.3. *A liquid crystal display works by using an electric field to orient molecular dipoles.*

Liquid crystal displays (LCDs) in watches, clocks, calculators, and portable computer screens also rely on manipulating polar molecules with electric fields. One type of liquid crystal, the nematic liquid crystal, contains long, rod-shaped molecules with a charge separation lengthwise along the rod. The transparency of the liquid differs considerably between random and aligned (called liquid crystal) orientations. In the absence of any applied voltage, the liquid crystal exists in a disordered state. But in the presence of an electric field, all the molecules in the liquid stack in a regular fashion, as though they were in a crystal.

In one type of LCD, a liquid crystal solution is placed between a polarizing glass plate and a mirrored glass plate held apart by an insulating spacer less than one-thousandth of an inch thick (Figure 3.3). The inner surfaces of the two plates are coated with a thin, transparent film of tin oxide (SnO_2) doped with the element indium. This film acts like a metal and conducts electricity. In the absence of an applied voltage, the solution is clear. When a voltage is applied across the two plates in the sandwich, the polar liquid crystal molecules respond by aligning their positive ends toward the negatively charged plate and their negative ends toward the positively charged plate. This crystal-like alignment of the molecules can make the solution between the glass plates appear opaque (black). Research has advanced liquid crystal technology to the point where image color can be controlled. Color LCDs have applications in thin-screen color televisions and in color screens for lap-top computers and video games.

Molecules and Structures

A molecule is a discrete collection of atoms held together by covalent and polar covalent bonds. The chemical properties of drugs, materials, food additives, gasoline, and proteins all depend on the structures of their molecules. Many atoms form fixed numbers of bonds in making molecules; for any given atom this number is called the atom's *valence.*

A carbon atom generally forms four chemical bonds. This seemingly mysterious property can be understood by examining the periodic table. Because each atom bonded to carbon shares one of its electrons with the carbon atom, four chemical bonds would give carbon the same number of electrons as the noble gas neon, which has eight valence electrons. The carbon atom uses its four valence shell electrons to make four covalent bonds in methane. The four valence electrons of carbon and the four electrons from the four bound hydrogens fill the valence shell of carbon with eight electrons. The tendency for atoms to have the same number of valence electrons as noble gases explains many aspects of chemical bonding. Hydrogen atoms normally form only one chemical bond. The sharing of an electron from one other atom provides a hydrogen atom with the same number of electrons (two) as the noble gas helium. Each hydrogen atom in methane gains the one electron it needs by forming a single bond to carbon, while carbon gains four electrons by forming four single bonds to different hydrogen atoms.

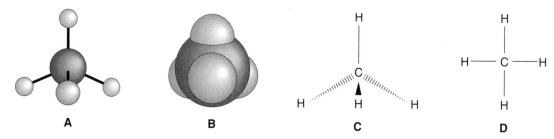

Figure 3.4. *Four ways of drawing the methane molecule (CH$_4$).*
A: *A ball-and-stick representation of the four hydrogen atoms bound
around carbon in a tetrahedral geometry.* ***B:*** *A space-filling model
that represents an estimate of the relative sizes of the atoms.* ***C:*** *A
sketch that chemists use to suggest three-dimensional structure.* ***D:*** *A
two-dimensional representation, which shows the connectivity of
bonds, but not the three-dimensional structure. This representation is
often used for simplicity.*

Chemists use formulas as a shorthand to describe molecules.
In a molecular formula, each type of atom in the molecule is represented by the elemental symbol, and a subscript is used to
indicate how many of that type of atom appear in the molecular
structure. Thus the molecular formula for methane is CH$_4$; the
subscript 1 is assumed for an element, carbon (C) in this case, if
none is given. This formula provides no information about the
three-dimensional structure.

Molecules exhibit a variety of three-dimensional structures.
Figure 3.4 illustrates several ways in which chemists draw the
molecular structure of methane. The carbon atom of methane
lies in the center of a pyramid with a triangular base. Four
hydrogen atoms occupy the vertices of the pyramid. Structure A
depicts each atom as a sphere with lines indicating the chemical
bonds. Structure B shows the space occupied by the rapidly
moving electrons (the "electron cloud") for each atom. The central carbon, with a total of six electrons in the free atom, has a
larger electron cloud than any one of the four hydrogen atoms.
Although structure B may be the most realistic, it is very difficult to draw and comprehend for molecules that contain large
numbers of atoms. Structure C shows an approach that chemists
use to keep the drawing simple and still provide three-dimensional perspective. A solid wedge represents a chemical bond to
an atom that lies above the plane of the drawing, and a dashed
wedge represents a bond to an atom that lies below the plane. An

even simpler representation, such as structure D, shows only the chemical bonds connecting atoms.

More than one pair of electrons may combine to form a multiple bond between two atoms. The common examples are limited to single, double, and triple bonds wherein the two atoms share two, four, and six electrons, respectively. A doubled line indicates a double bond, as found in O_2, and three lines denote a triple bond. A triple bond is formed by two nitrogen atoms: Each N shares three of its electrons with the other N atom in the diatomic molecule N_2. Thus each nitrogen atom in N_2 has access to eight valence electrons (five in the valence shell of a N atom and three shared from the triple bond to the other N), the same number present in the nearest noble gas, neon. In the top structure shown in Figure 3.5, the three lines between the nitrogen atoms denote the triple bond, which consists of three shared electron pairs. The molecule ethylene (also shown in Figure 3.5) bonds so that each carbon atom shares one electron each with two hydrogen atoms and two electrons with one carbon atom. Every line between atoms represents the sharing of two electrons, one from each atom, to form a chemical bond.

Single bonds are weaker than double bonds, and triple bonds are the strongest of all. In ethane (molecular formula C_2H_6), which is represented in Figure 3.6, the two carbon atoms bind with a weaker force than in ethylene (C_2H_4), which contains a double bond. Acetylene (C_2H_2), also shown in Figure 3.6, contains a carbon–carbon triple bond, the strongest carbon–carbon bond of the three. The separation between the carbon atoms becomes smaller as the bond strength increases. In this group, the carbon atoms are closest together in acetylene and farthest apart in ethane. But in every case, carbon satisfies its valence of four by making four bonds.

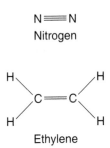

Figure 3.5. *Structures of the nitrogen (N_2) and ethylene (C_2H_4) molecules.*

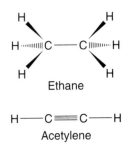

Figure 3.6. *Structures of the ethane (C_2H_6) and acetylene (C_2H_2) molecules.*

Simplified Structures for Organic Molecules

Chemical compounds are usually differentiated as organic or inorganic. The organic molecules that make up organic chemicals contain carbon–hydrogen bonds, and they often contain bonds with other elements as well. Ethylene (see Figure 3.5), ethane, and acetylene (see Figure 3.6) are examples of organic molecules.

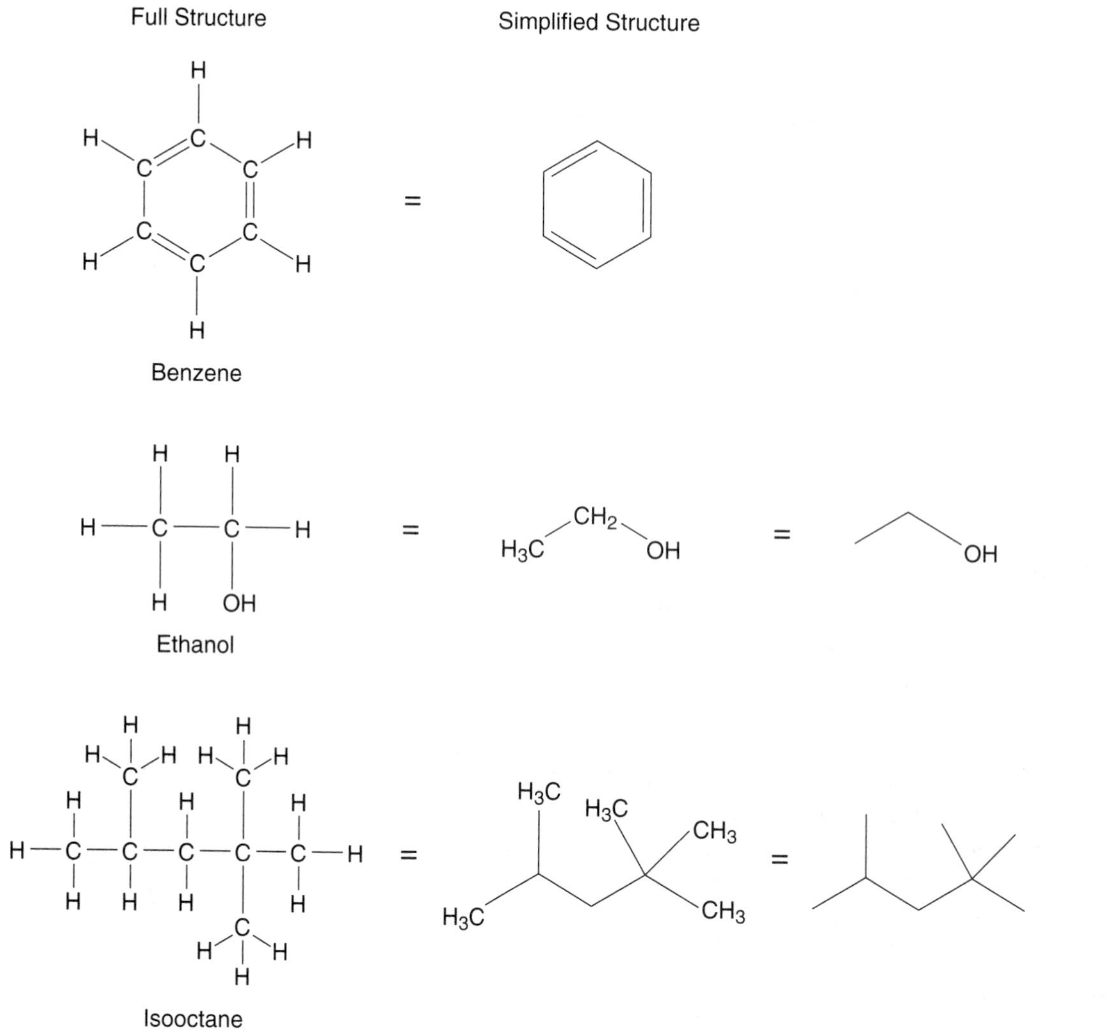

Full Structure   Simplified Structure

Benzene

Ethanol

Isooctane

Figure 3.7. *Simplified representations of the molecular structures of benzene (C_6H_6), ethanol (C_2H_5OH), and isooctane (C_8H_{18})*

Compounds that do not contain any C–H bonds are inorganic. Oxygen (O_2), nitrogen (N_2), carbon dioxide (CO_2) and water (H_2O) are examples of inorganic molecules. Thus the chemist's definition of *organic* differs from that of the popular term, which refers to pesticide-free foods produced with natural fertilizers.

The structures of complex organic molecules can be represented even more simply, as shown in Figure 3.7. In streamlined stick figures, only the bonds to nonhydrogen atoms are shown. Symbols for carbon atoms, and for the hydrogens bound to them, are not drawn. Elemental symbols for atoms other than carbon appear in the drawing. Convention dictates that carbon atoms lie at the end of each bond unless another atom, such as oxygen, is specified. Benzene (C_6H_6), shown in detail on the lefthand side of Figure 3.7, is a very stable organic molecule found in gasoline. In this molecule, carbon attains its valence of four by forming a closed, six-membered ring structure. The full structure can be shortened even further by dropping the hydrogen atoms, as shown in the simplified structure on the righthand side of the figure. Any carbon atom short of four bonds must have the right number of hydrogens there to complete the missing bonds. In the simplified structure for ethanol, the intoxicating component in alcoholic drinks, the oxygen atom must be specified, along with the hydrogen bound to it. Each bend or kink in the stick figure denotes a carbon with attached hydrogens, because it represents a point where one bond to carbon ends and another starts. In the abbreviated formula for isooctane, a hydrocarbon found in gasoline, the end point of each line, and all points where lines intersect correspond to a carbon atom with the necessary number of hydrogen atoms attached.

Bonds in Metals

The unique properties of metals arise because the atoms pack closely together and share their outermost electrons equally with all their neighbors. At the atomic level, a metal can be viewed as containing nuclei suspended in a cloud of highly mobile electrons. These mobile electrons conduct heat and electricity through metallic structures. Metal atoms bond equally well in all directions. This permits metals to bend, because strong directional bonds don't have to break. All elements (except H) on the lefthand side of the periodic table (see Figure 1.5), in addition to those in the bottom two rows, exist as metals when pure. Eighty-two percent of the known elements are metals!

Often it is possible to mix different metallic elements by melting them together. Upon cooling, the mixture produces a new mixed metal called an *alloy*. Typical compositions for several common alloys appear in Table 3.1. The percentages of the component elements can be varied somewhat to "customize" the properties of the alloys in the table. For example, brass alloys may contain 67–90% copper and 10–33% zinc. Alloys are used in many ways. Wood's metal, an alloy of bismuth (Bi), lead (Pb), tin (Sn), and cadmium (Cd), melts below the temperature of boiling water. It is used in electrical fuses and as the heat-sensitive plug in fire sprinkler systems. Copper is too soft to be useful as a pure metal, except as conducting wire. Brass and bronze alloys of copper have mechanical properties superior to those of the metals from which they are made. The ancient Greeks used bronze to cast statues. Unfortunately, many of the early bronze statues were melted down so that the strong, workable metal could be reused in weapons. Modern gun barrels are still made of a bronze alloy.

Gold jewelry and watches usually consist of an alloy of gold, silver, and copper. The alloy is harder and more scratch-resistant than pure gold. The proportion of gold in alloys is expressed in carats. Pure gold has a 24-carat designation. "One-carat gold" means the alloy is 1 part gold and 23 parts another metal by weight. The common 14-carat alloy consists of 14 parts gold and 10 parts of silver and copper. White gold jewelry is an alloy of gold that contains about 10% added nickel or palladium. This gives the metal the appearance of platinum. Sterling silver is an alloy of silver that contains 7.5% copper.

Pure iron, which is called wrought iron, is softer and weaker than steel. Adding carbon to iron increases its hardness and strength, but too much carbon makes it brittle and difficult to work. Cast iron has a carbon content above 2% but below that of brittle pig iron (3.5–4.5% carbon). Oxygen is blown over molten pig iron, the crude product from a blast furnace, to burn away some carbon and form steel (Figure 3.8). Steels are classified as mild, medium, or high-carbon as the carbon content varies from 0.15% to 1.4% by weight. Softer mild steels make useful cables and nails, and hard high-carbon steels are better for tools and springs. Structural steels for use in skyscrapers and rails generally consist of medium steels. Steel alloys are further tailored for

Table 3.1. *Typical Elemental Composition of Common Metallic Alloys*

Alloy	Approximate Elemental Composition (% weight)
Stainless steel (cutlery)	Fe, 74%; C, 0.15%; Cr, 15%; Ni, 8%; Mn, 2%; Si, 1%
High-carbon steel	Fe, 97.1%; C, 1.4%, Mn, 1.2%; Si, 0.3%
Structural steel	Fe, 98.3%; C,0.08%; Mn, 1.25%; P, 0.01%; S, 0.01%; Si, 0.28%; V, 0.06%
Monel	Ni, 68%; Cu, 32%
Nichrome	Ni, 60%; Cr, 40%
Aluminum (2000 series)	Al, 95%; Cu, 5%
Aluminum (3000 series)	Al, 98.8%; Mn, 1.2%
Aluminum (5000 series)	Al, 97%; Mg, 3%
Aluminum (7000 series)	Al, 94%; Zn, 5%; Mn, 1%
Magnesium alloy	Mg, 89%, Al, 9%; Zn, 2%
Yellow brass	Cu, 67%; Zn, 33%
Bell metal	Cu, 77%; Sn, 23%
Bronze	Cu, 80%; Sn, 20%
Bronze gunmetal	Cu, 88%; Sn, 10%; Zn, 2%
Aluminum bronze	Cu, 90%; Al, 10%
Coinage, U.S. 5¢, 10¢, ...	Cu, 75%; Ni, 25%
Common solder	Pb, 50%; Sn, 50%
Plumber's solder	Pb, 66%; Sn, 34%
Silver solder	Ag, 63%; Cu, 30%; Zn, 7%
Pewter	Sn, 92%; Sb, 5%; Pb, 3%
Dental amalgam (traditional)	Hg, 50%; Ag, 35%; Sn, 13%; Cu, 1.5% ; Zn, 0.5%
Dental amalgam (high-Cu)	Hg, 2%; Ag, 70%; Sn, 18%; Cu, 10%;

increased strength and hardness by the addition of manganese (Mn), nickel (Ni), copper (Cu), vanadium (V), tungsten (W), and molybdenum (Mo). Stainless steels incorporate chromium (Cr) for corrosion resistance. The properties of carbon steels can be further modified by heat treatment. Tempered steel is made by heating steel below 1000°C. Slow cooling produces a ductile steel, whereas rapid cooling makes the steel more brittle.

Aluminum, another important structural metal, requires added Mn, Cu, silicon (Si), or zinc (Zn) for strength. The 3000-

Figure 3.8. *Oxygen is blown over steel to remove carbon and other impurities and samples are withdrawn periodically to monitor its composition. (Bethlehem Steel Corporation.)*

series aluminum alloy (see Table 3.1) is used to make cooking utensils, lawn furniture, and highway signs. Corrosion-resistant 5000-series alloys work best in boats and streetlamp poles. Aircraft structures and skins require a 7000-series or a 2000-series alloy. Magnesium can be alloyed with aluminum and zinc for use in lightweight wheels for automobiles and the air-cooled engine block of the classic "VW Beetle."

Bonds in Ionic Solids

Sodium metal will catch fire if exposed to chlorine gas, and the reaction produces white crystals of sodium chloride—common table salt. In this reaction, each sodium atom (Na) transfers its outermost electron to a chlorine atom (Cl), forming a positive ion (+1), Na^+, and a negative ion (–1), Cl^-. Atoms or molecules that possess an overall electric charge are called *ions*. A positively charged ion is called a cation, and a negatively charged ion is called an anion. Ions differ from polar compounds in the amount of charge transfer, which results in integral positive or negative

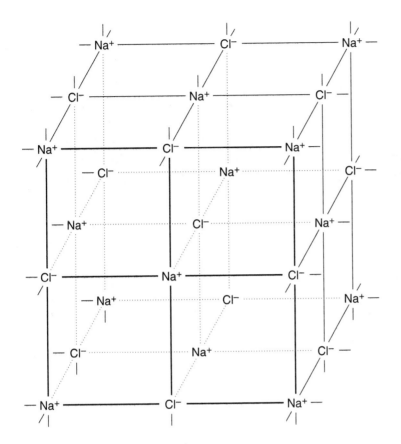

Figure 3.9. *The alternating arrangement of chloride anions (Cl⁻)and sodium cations (Na⁺) in a crystal of salt (NaCl).*

charges (rather than fractional unit charges) on atoms or molecular fragments. When sodium loses an electron, its total number of electrons equals that of neon, a noble gas. Chlorine gains an electron to increase its number of electrons to that of argon, another noble gas. The pioneering X-ray diffraction studies of the Braggs (see Chapter 1) showed that the three-dimensional sodium chloride crystal contains alternating sodium and chloride ions (Figure 3.9).

The sodium chloride crystal is uncharged overall. It contains an equal number of sodium cations and chloride anions. The oppositely charged sodium and chloride ions attract each other, and this attractive force is referred to as an *ionic bond*. Ionic compounds normally are solids, which allows the positive ions to be surrounded by negative ions and the negative ions to be surrounded by positive ions. Each ion in the solid is stabilized by attractive forces to its oppositely charged neighbors.

Metallic elements in the first three columns of the periodic table (excluding the transition metals) lose one (Li^+, Na^+, K^+, ...), two (Be^{2+}, Mg^{2+}, Ca^{2+}, ...), or three (B^{3+}, Al^{3+}, Ga^{3+}, ...) valence electrons to produce positive ions. These ions have the same number of electrons as a noble gas and are especially stable. Cations with a charge greater than +3 are rare. Ionic solids usually form when metallic elements from the lefthand side of the periodic table combine with elements on the right that need to gain one (F and Cl) or two (O and S) electrons to have the same number of electrons as a noble gas. Solid sodium (Na) combines with chlorine gas (Cl_2) to form solid sodium chloride (NaCl). Solid magnesium (Mg), on the other hand, combines with chlorine gas (Cl_2) to form solid magnesium dichloride ($MgCl_2$). The difference in the number of chlorine atoms in the NaCl and $MgCl_2$ molecules reflects the different positive charges of the stable sodium (Na^+) and magnesium (Mg^{2+}) cations. Different numbers of chloride anions (Cl^-) are needed to make the molecule electrically neutral. These reactions involve reactants that are in different states (gases and solids).

In writing a chemical equation to describe such reactions, we must indicate the state of matter for each reactant and product. This is done by using the subscripts (s), (l), and (g) to denote solids, liquids, and gases. Using these subscripts, we write the reactions for forming NaCl and $MgCl_2$ as shown in equations (2) and (3).

$$2Na_{(s)} + Cl_{2(g)} \rightarrow 2NaCl_{(s)} \tag{2}$$

Sodium metal reacts with chlorine gas to form sodium chloride (table salt).

$$Mg_{(s)} + Cl_{2(g)} \rightarrow MgCl_{2(s)} \tag{3}$$

Magnesium metal reacts with chlorine gas
to form the salt magnesium dichloride.

The number 2 in front of the symbol for sodium in equation (2) is needed to balance the chemical equation. A balanced chemical equation has the same number of atoms of each type on each side of the arrow. In equation (2), two atoms of sodium metal react with two atoms of chlorine (in Cl_2) to form two sodium ions and two chloride ions in the solid state. In equation (3), one atom of magnesium reacts with two atoms of chlorine to form one magnesium ion and two chloride ions.

Transition metals also form ionic solids. Silver metal forms technologically useful halide salts, such as silver chloride (AgCl) and silver bromide (AgBr). Unlike sodium salts, silver salts are sensitive to light and insoluble in water. Ionic compounds may also contain ions that consist of more than one atom. The nitrate (NO_3^-), sulfate (SO_4^{2-}), ammonium (NH_4^+), carbonate (CO_3^{2-}), and phosphate (PO_4^{3-}) ions occur frequently in compounds. In writing molecular formulas that involve cationic and anionic groups, we use parentheses to define the charged constituents, and we write the positive ion first. For example, ammonium sulfate is written $(NH_4)_2(SO_4)$ to show that it contains ammonium cations and sulfate anions in a 2:1 ratio.

Ionic Solutions

Ionic compounds tend to be hard and brittle, with high melting points. As solids they behave as electrical insulators, but when melted they conduct electricity. They can also conduct electricity in solutions. Pure water is a poor conductor of electricity. Soluble ionic compounds, such as table salt and baking soda, dissolve in water to make electrically conducting solutions. The dissolution of table salt is described by equation (4). Writing the water symbol above the arrow indicates that the solid NaCl is dissolved in a large container of water. If the water symbol appeared as a reactant, the equation would need to be balanced, depicting a specific reaction between the salt and water molecules. The subscript (aq) stands for *aqueous*, which means that the compound has been dissolved in water.

$$NaCl_{(s)} \xrightarrow{\quad H_2O \quad} Na^+_{(aq)} + Cl^-_{(aq)} \qquad (4)$$

Sodium chloride dissolves in water to form
aqueous sodium and chloride ions.

The separated positive and negative ions in aqueous salt solutions are mobile, which provides a way for electricity to be conducted through the solution. This conductivity is exploited by modern technologies. Steam vaporizers available in drug stores work by passing an electric current through a water solution of mineral salts. Because the solution conducts electricity and also possesses an electrical resistance, the water is heated and then vaporized. If you live in a part of the country where the water contains few dissolved minerals, or if you try to use distilled water in a steam vaporizer, then you must add a gram or so of salt. But if too much salt is added, the solution can become too conducting. The high current that then flows can lead to excess steam or even trip a circuit breaker.

Electrical conduction through ionic media differs from that in metals. Electrons move freely through metals and provide the mechanism for electrical conduction. For ionic substances, conduction of charge can occur only once the ions themselves, such as Na^+ and Cl^-, are able to move about freely. Unrestricted movement cannot occur in the solid state, but it can occur in the molten state or when the ions are dissolved in water. The human body relies on the conductivity of ionic solutions (salt solutions in cells) to relay electrical signals, that cause, among other things, the movements of muscles. The absence of spontaneous electrical activity in the brain is what defines "brain dead." Because multicellular organisms rely on small electrical signals for their activity, they are susceptible to electric shock. The disruption of the body's electrical signals by an electric shock can cause the heart to stop beating. On the other hand, electric shock may be used as a last resort to "jump-start" the nervous system of a person who has gone into cardiac arrest.

Sodium, potassium, and chloride ions are essential conductors in the body's electrical system, yet too much Na^+ in the diet aggravates hypertension in certain individuals. Sea water contains so much salt that it is toxic to drink. The optimal concentration for potassium ions in the body is even smaller than that for sodium. Potassium chloride (KCl) looks like common table salt and occurs as an ingredient in many sports electrolyte drinks. Yet absorption of too much potassium chloride into the blood may cause death. The imbalance in electrical activity created by a sudden influx of potassium ions can cause the heart to stop beating. The solution used to cause death by "lethal injection" often contains KCl.

Solvents and Solubility

Many substances dissolve in liquids to make solutions. The substance dissolved is called the *solute*, and the liquid is referred to as the *solvent*. The term *solubility* is used to describe the extent to which a solute dissolves in a solvent. A general rule is that "like dissolves like." Polar solvents consist of polar molecules, such as water. These solvents may dissolve polar substances and ionic solids. In contrast, low-polarity solvents (liquids of nonpolar molecules) such as paint thinner and gasoline dissolve nonpolar substances that are insoluble in water. Three common nonpolar solvents are shown in Figure 3.10.

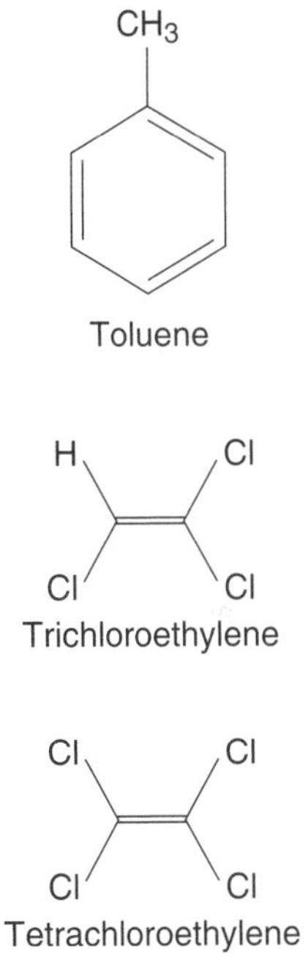

The general rule of solubility is typified by the procedure used in dry-cleaning clothes. Many garments cannot be cleaned with water–based soap solutions and must be dry-cleaned. Dry cleaning does not mean that liquids are not used. The name indicates that the process does not involve water. Instead, volatile nonpolar chlorinated solvents such as tetrachloroethylene are used. Before 1925 gasoline was used as a nonpolar solvent for dry-cleaning; however, the nonflammable chlorinated solvents are much safer to use. When a garment becomes soaked with dry-cleaning solvent, the greasy component in the dirt dissolves to leave solid particles that flake away. The odor of freshly dry-cleaned clothes comes from the evaporation of residual solvent left in the fibers.

For some materials there are drawbacks to dry-cleaning. The feathers in down-filled garments contain natural oils that are necessary if the material is to keep its loft. Dry-cleaning solvents remove these natural oils and lower the material's insulating ability. For that reason, down garments usually bear a "Do Not Dry Clean" warning. A similar problem occurs in the wild when waterfowl and aquatic animals encounter oil spills. Many birds

Figure 3.10.
Molecular structures of three common nonpolar solvents.

and water mammals (such as otters) have, in their feathers or fur, nonpolar oils that repel polar water molecules. This keeps their skin dry and helps insulate their bodies. The hydrocarbons in oil spills can dissolve these natural oils or adhere to them in the feathers or fur of the animal, destroying the natural oils' insulating role.

Whether the molecules in a drug are polar or nonpolar plays an important role in physiological action. A polar water-soluble drug, such as aspirin, can enter the blood stream readily and exert its effect. Many drugs are also soluble in nonpolar substances. Biological lipid membranes form a barrier at the cell wall and behave as nonpolar media. Drugs often must cross such membrane barriers before they can alter cell chemistry. This behavior is accentuated at the blood/brain barrier, which protects the central nervous system from foreign substances. The nonpolar character of anesthetics, sedatives, and opiates enables them to cross this barrier and alter brain chemistry. Fatty tissues, another nonpolar oily medium, can also take up nonpolar drugs and act as a storage depot in the body. THC, for example, the active component in marijuana, remains dissolved in body fat long after its use. That is why drug tests can detect whether a person has used a narcotic or other drug, even after the overt effects have disappeared.

CHAPTER 4

Newsworthy Molecules

Miracle plastics, vitamin supplements, anticancer drugs, acid rain, plastique explosives, food preservatives, pesticides, birth control pills, steroid abuse, water purification, water pollution, cancer, and drugs for AIDS—all involve chemicals. Chemistry is an inescapable science! The Earth and all living things are made of atoms, molecules, and chemicals. Even the thoughts and actions of humans can be broken down into a complex series of biochemical events.

Molecules in the Atmosphere

A few simple molecules play a major role in the Earth's atmosphere (Figure 4.1). All occur naturally, but problems may arise when human activity alters their concentrations from those established by nature. Oxygen (O_2) is essential for the oxidation reactions (respiration) that provide energy for life processes in mammals. Nitrogen (N_2), an inert molecule, takes part in no useful chemical reactions in the body. As so often happens, these molecules can be beneficial or harmful, depending on the dose; both pure oxygen and pure nitrogen are deadly, but mammals readily breathe a mixture of the two. Other molecules necessary for life must be present in the proper concentration. Carbon dioxide (CO_2), a minor gas in the atmosphere, provides the essential carbon source for the photosynthesis of carbohydrates by plants. Without carbon dioxide, all life on Earth would perish

Oxygen, a reactive gas necessary for combustion and human life. It makes up 21% of the atmosphere.

Nitrogen, an unreactive gas that makes up 78% of the atmosphere.

Carbon dioxide, a gas that humans exhale and plants inhale. A major greenhouse gas, it makes up 0.04% (400 millionths) of the atmosphere.

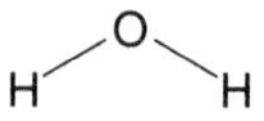

Water occurs in its gaseous state in amounts up to 4%. The vaporization and condensation of water largely determine weather. It is the largest con-tributor to the greenhouse effect.

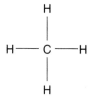

Methane, the primary component of natural gas. It makes up 2 millionths of the Earth's atmosphere and is a major greenhouse gas.

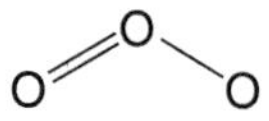

Ozone, a toxic pollutant at ground level, is an essential component of the stratosphere. It absorbs harmful ultraviolet radiation from the Sun, although it makes up less than 10 millionths of the atmosphere. Stratospheric ozone depletion in recent years has been a cause for world-wide concern.

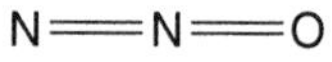

Nitrous oxide, or laughing gas, is produced by bacterial and human activity. It makes up less than 1 millionth of the atmosphere. It causes ozone loss in the stratosphere and is a greenhouse gas.

Figure 4.1. *Major natural components of the Earth's atmosphere.*

from starvation. Carbon dioxide, water, and methane also absorb infrared radiation (heat) in the atmosphere, which helps keep the Earth warm. Yet increased amounts of carbon dioxide in the atmosphere from burning fossil fuels threaten to warm the climate too much (global warming).

The two regions of the atmosphere closest to the Earth are called the troposphere and the stratosphere. Plants and animals breathe air from the troposphere, which consists of the gas layer that blankets the Earth from sea level to 15 km high (0–9 miles). The stratosphere extends 15–50 km (9–30 miles) above the Earth's surface. The same molecule can play different roles in these two regions of the atmosphere. For example, ozone (O_3), a component of photochemical smog, represents a summertime health hazard in cities like Los Angeles, Mexico City, and Tokyo. Ozone damages the cells that line the nose, mouth, and lungs. In the stratosphere, however, ozone is essential. It screens out harmful ultraviolet light from the Sun, which causes skin cancer in humans and stunts growth in plants. One modern environmental threat is depletion of the Earth's protective sheath of ozone.

Nitrous oxide (N_2O) is a by-product of nitrogen assimilation in bacteria and algae. It ultimately causes ozone depletion in the stratosphere and is a potent greenhouse gas. Like carbon dioxide, it has been present in the atmosphere for millions of years. And like that of carbon dioxide, its atmospheric concentration is increasing.

One group of problematic atmospheric substances arises because of human activity. Several such molecules are shown in Figure 4.2. Gaseous sulfur dioxide (SO_2) and nitrogen dioxide (NO_2) pollutants form in the troposphere from burning fuels. These gases exist briefly before they react with atmospheric oxygen and water to form sulfuric acid (H_2SO_4) and nitric acid (HNO_3) either dissolved in rain water or present as a fine aerosol or mist. In the early part of the twentieth century, sulfuric acid formed routinely in the atmosphere over London because large amounts of high-sulfur coal were used to heat homes and power industries. When combined with soot particles, this acid caused several "killer fogs." In December 1952, one such fog resulted in 2000–3000 deaths, spurring passage of the British Clean Air Act. Some noxious atmospheric gases have natural sources as well. Volcanoes emit large amounts of sulfur dioxide. High temperatures in lightning strikes form natural atmospheric nitric acid from reactions involving water, nitrogen, and oxygen.

Carbon monoxide (CO) gas is a tropospheric pollutant that arises from the combustion of fuels, such as coal, oil, and natural gas. It also has natural biological sources. Carbon dioxide is the primary product of combustion, but small amounts of carbon monoxide form when there isn't enough oxygen for complete oxidation. Carbon monoxide binds tightly to hemoglobin molecules in red blood cells and prevents them from transporting oxygen to cells. It is a deadly component in automobile exhaust and is a by-product of wood, oil, and coal fires. Because carbon monoxide has no odor, its presence is not evident.

The chlorofluorocarbon molecules found in the atmosphere are a result of human activities. Freon gases, such as CFC-12, are

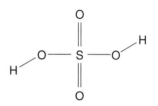

Sulfur dioxide is a corrosive pollutant formed during the burning of coal or oil that has high sulfur content. It reacts in the air to form sulfuric acid.

Nitrogen dioxide is a major pollutant from high-temperature combustion. This gas imparts the yellow-brown color to smog in cities.

Sulfuric acid forms in the atmosphere by the reaction of water, O_2, and SO_2. It is a major constituent of acid rain. The increased acidity of soil and water can lead to the death of plants and of fresh-water life. Concentrations are high near coal-fired power plants and volcanoes.

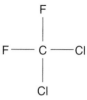

Nitric acid, another component of acid rain, forms from the reaction of NO_2 with OH radical in the atmosphere. Automobile pollution in cities is a major source, as is lightning.

Carbon monoxide is a highly toxic gas that (fortunately) makes up less than 1 millionth of the atmosphere. Locally higher concentrations can occur in cities. It is produced both by the combustion of fuels and by the decay of organic matter. Carbon monoxide poisoning may occur in poorly ventilated garages and when a gas, wood, or fuel heater is used with inadequate ventilation. It also occurs in cigarette smoke.

CFC-12 belongs to a class of molecules called freons or chlorofluorocarbons. Developed as nontoxic gases for use in refrigerators and air conditioners, they have been found to cause a serious loss of ozone in the stratosphere. Their use is being phased out in some countries. They make up less than 1 billionth of the atmosphere.

Figure 4.2.   *Newsworthy air pollutants.*

nontoxic chemicals designed for air conditioner and refrigerator compressors. They also are used as nontoxic cleaning solvents in the electronics industry and as blowing agents for plastic foam. The term CFC is an acronym for chlorofluorocarbon, which means a molecule that contains only chlorine, fluorine, and carbon atoms. The high stability of CFCs permits their vapor to diffuse up into the stratosphere, a process that takes several years. In the stratosphere, the chemical bonds in CFCs are broken by high-energy radiation from the Sun to produce chlorine atoms.

Chlorine atoms are highly reactive and cause ozone depletion. The severity of the ozone depletion problem has led to world-wide agreements on the phasing out of CFCs.

Compounds that contain only hydrogen and carbon atoms are called hydrocarbons. They occur as pollutants from incompletely burned fuel and the evaporative loss of gasoline. Hydrocarbons and carbon monoxide have both been targeted, along with nitrogen dioxide, for elimination from automobile exhaust. In the Los Angeles basin, there are plans to ban the use of charcoal lighter fluid, because the high frequency of backyard barbecues significantly contributes to air pollution in this region. Gaseous hydrocarbons also occur in nature. Methane (see Figure 4.1) is the primary chemical (about 85%) in natural gas for use in furnaces, boilers, and kitchen stoves. Natural gas often occurs trapped in underground deposits near oil fields, and in coal mines. If the atmosphere in a mine contains more than 5.5% methane, a deadly explosion may occur. Methane is also the primary constituent of marsh gas, which forms in swamps from the decay of vegetable matter.

Many other trace molecules generated by human activity are in the atmosphere but do not affect the global climate. They may create local air quality problems, as with cigarette smoke. Tobacco belongs to the poisonous nightshade family of plants and contains a variety of toxic organic compounds. Over 4000 different molecules have been detected in cigarette smoke. Some, such as carbon dioxide and water, are harmless, but many others (carbon monoxide, ammonia, hydrogen cyanide, acrolein, acetaldehyde, phenol, dimethylnitrosoamine, and benzopyrene) are toxic, carcinogenic, or both. The exhaust from many chemical manufacturing plants is safer to breathe than cigarette smoke! The structures of several components of cigarette smoke are shown in Figure 4.3. It is a proven fact that chronic exposure to the chemicals in cigarette smoke increases the probability of developing lung cancer. Even so, the latency period of 10–30 years gives many young people a false sense of security.

Although smokers are at greatest risk from these toxins, non-smokers also suffer. A U.S. Environmental Protection Agency draft report in mid-1992 places the health effects of passive

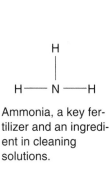

Ammonia, a key fertilizer and an ingredient in cleaning solutions.

Hydrogen cyanide, a rodenticide, an insecticide, and a highly toxic gas.

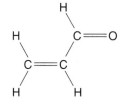

Acrolein, an industrial chemical used to make plastics, has also been employed in poison gas. Industrial emissions are banned.

Acetaldehyde, an industrial chemical of low toxicity that is also a waste product in human biochemistry. It irritates the respiratory pathways.

Phenol, a disinfectant liquid that is moderately toxic. It is an aerosol component of cigarette smoke.

N-nitrosodimethylamine, a potent human carcinogen. This liquid is an aerosol component in cigarette smoke.

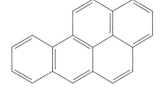

Benzo[a]pyrene, found in coal tar as well as in the "tar" content of cigarettes. This solid is an aerosol component of cigarette smoke and a potent human carcinogen.

Figure 4.3. *A few of the 4000 molecules found in cigarette smoke.*

smoking in the same category of carcinogens as benzene, asbestos, and radon. Environmental tobacco smoke is estimated to cause 2500–3000 lung cancer deaths annually in nonsmokers. Nonsmokers who live with a smoking spouse are about 30% more likely to die of lung cancer than other nonsmokers. Secondary smoke has also been blamed for an increased incidence of lower respiratory tract infections in infants of parents who smoke.

Chemicals in the Body

The human body consists of an incredibly complex array of chemicals. These range from organic molecules, such as fats and hormones, to the inorganic calcium phosphate minerals found in bones and teeth. Maintaining the chemical balance among proteins, fats, carbohydrates, essential minerals, and vitamins is the cornerstone of nutrition.

Vitamins

Vitamins are chemicals needed in small amounts for normal cell function. Elmer McCollum, working in the Chemistry Department at Johns Hopkins University Medical School, separated vitamins A and D in the first quantitative nutrition experiments. In 1917 he showed that xerophthalmia, a dry, thickened condition of the eyeball, occurred in rats if their diets did not include vitamin A. These studies led him to suggest that milk and green leafy vegetables are important in the human diet, a nutritional principle now taken for granted.

Figure 4.4 presents the molecular structures of a few vitamins important for human nutrition. Some, such as vitamin C, are an essential part of the diet, because the body lacks the machinery to make these compounds. Others, such as vitamin D, can either be consumed or be produced by the body from ingested chemicals (food). Vitamins vary in their toxicity. Massive doses of vitamin C cause little harm, because any excess is eliminated in urine. Others, such as vitamin D, are poisonous when taken in excessive amounts.

Advertising plays an important role in marketing vitamins. For example, there is no intrinsic difference in the chemical you purchase whether you buy "name brand" vitamin C, a generic, or "natural" vitamin C in a health food store. Chemists have developed methods for synthesizing vitamins, which provide an alternative source for their production. The biological activity of synthetic vitamins is identical with that of those obtained from natural sources. Some vitamins, however, may be formulated for more efficient absorption in the body. Natural sources of vitamins may also contain other impurities that could be beneficial, harmless, or harmful. Synthetic vitamins may also contain impurities, depending on the degree of purification in their manufacture. Sometimes health food manufacturers stress the phrase "natural" as implying a better product. The more important fac-

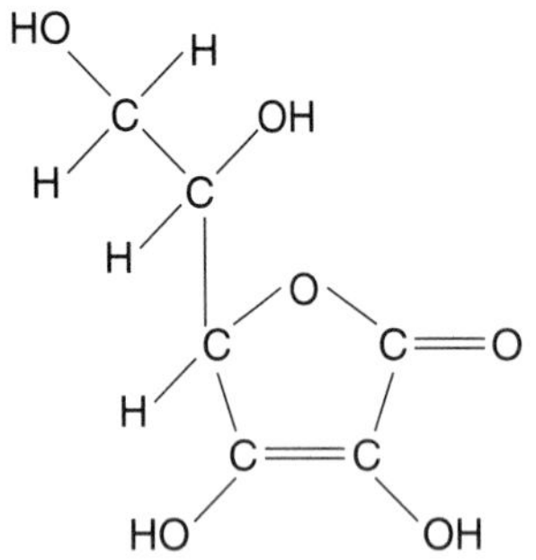

Vitamin C, or ascorbic acid, is essential for humans. A deficiency results in scurvy. The human body does not make it, so foods or vitamin supplements must provide it.

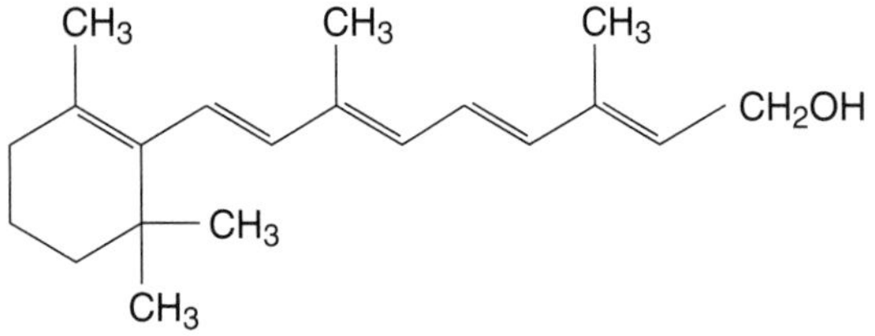

Vitamin A, or retinol, is an essential component for the vision receptors in the eyes. A deficiency leads to a loss in night vision. The body synthesizes vitamin A_1 from carotene, a chemical found in plants such as carrots.

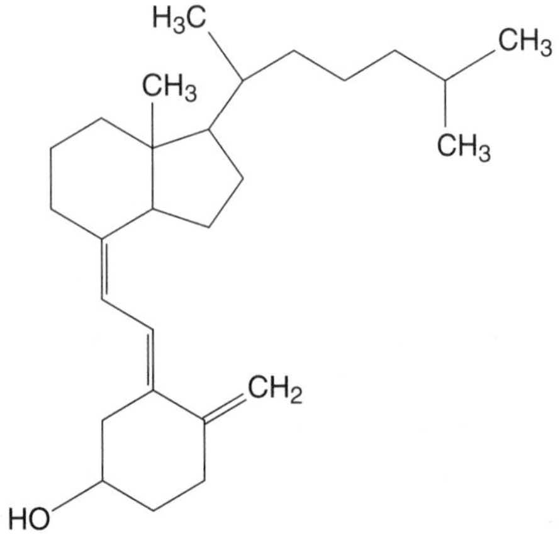

Vitamin D_3 is made in the skin by the action of sunlight on ingested plant sterols. A deficiency causes rickets. Small amounts are beneficial, but an overdose of Vitamin D is toxic to humans. A synthetic form called vitamin D_2 has been used as a rodenticide as well as a vitamin supplement.

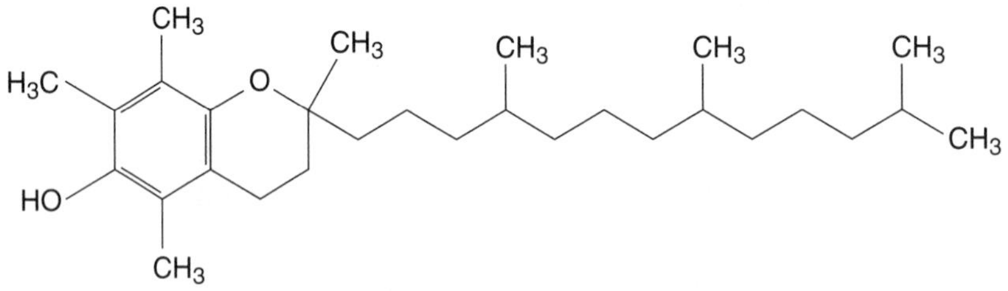

Vitamin E is a fat-soluble vitamin essential for muscle development.

Figure 4.4.   *Some vitamins essential to human development.*

tor for a cost-conscious consumer is the dose (usually given in milligrams, a unit of weight, or in potency units termed I.U.) of essential vitamin or mineral per dollar.

Therapeutic properties of vitamins attract increasing interest. Recent clinical studies have shown that people infected with HIV are prone to deficiencies in vitamins A, B, and C. A study of male intravenous drug users has shown that they are five times as likely to develop AIDS sooner and to die sooner than men without these vitamin deficiencies. Furthermore, researchers at Johns Hopkins University found that large doses of vitamins A, C, B_1, and niacin delay the onset of severe AIDS symptoms by as much as two years in HIV infected individuals. These effects have been partially attributed to the ability of vitamins A and B to stimulate the immune system and vitamin C's ability to protect against damage from an activated immune system. Indiscriminately boosting all vitamins and minerals would be a mistake, however. In these same clinical studies, an increased intake of dietary zinc led to an increased risk in progression to AIDS.

Cholesterol and Its Derivatives

Molecules important to the human diet may also cause problems. One example is cholesterol. Cholesterol is a steroid found in the tissues of mammals but not of plants. It is an essential component of mammalian cell membranes, and the body can make it if needed. Cholesterol is an important chemical precursor to a number of steroid hormones, such as the sex hormones testosterone (male) and estradiol (female). Hormones are chemicals that bind to cellular receptors and turn on or off essential biochemical processes. The molecular structures of some essential steroids appear in Figure 4.5. This figure also depicts two synthetic analogs of the female hormones estrogen and progesterone, which are produced by the ovaries. Although steroids are essential molecules for life, problems may occur when there is too much cholesterol in the diet. High fat intake compounds the problem, because it stimulates the additional synthesis of cholesterol by the body. Cholesterol, a large waxy molecule, is insol-

steroid ring system

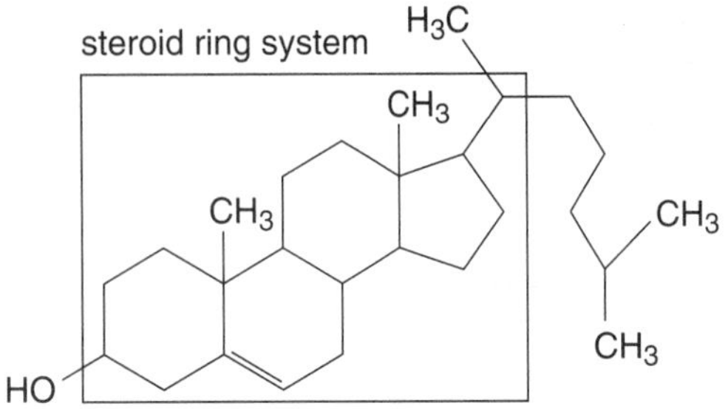

Cholesterol, a component of the tissue of all mammals and a precursor to the human sex hormones testosterone and estradiol.

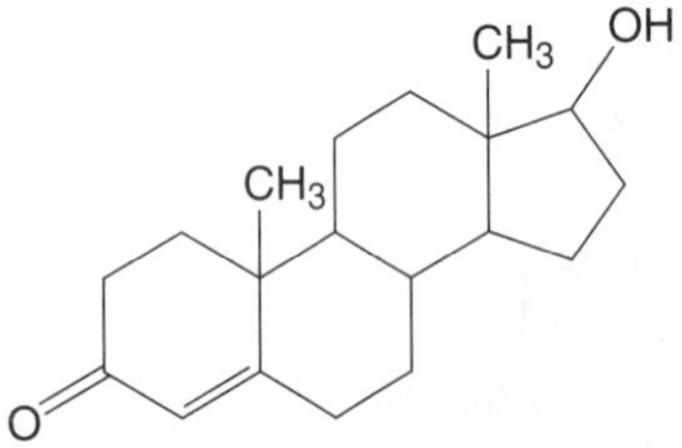

Testosterone, a male sex hormone that is produced by the body from cholesterol. A surge in testosterone levels occurs in males at the onset of puberty to trigger the changes of adolescence.

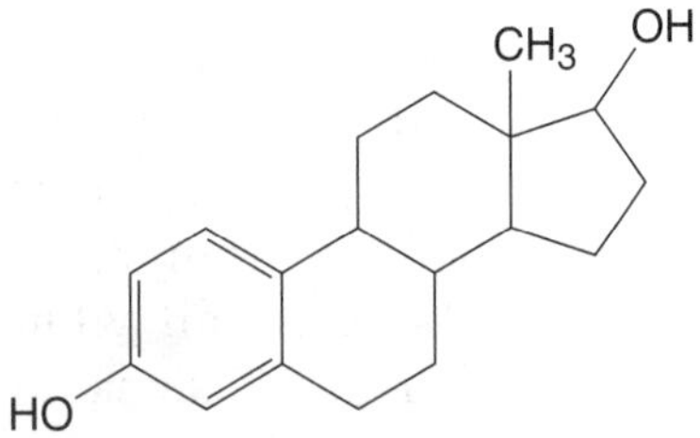

Estradiol, a female sex hormone responsible for the development of breasts and other characteristics of females. Note the similarity to the structure of testosterone. Small chemical differences can lead to a large difference in biological effects.

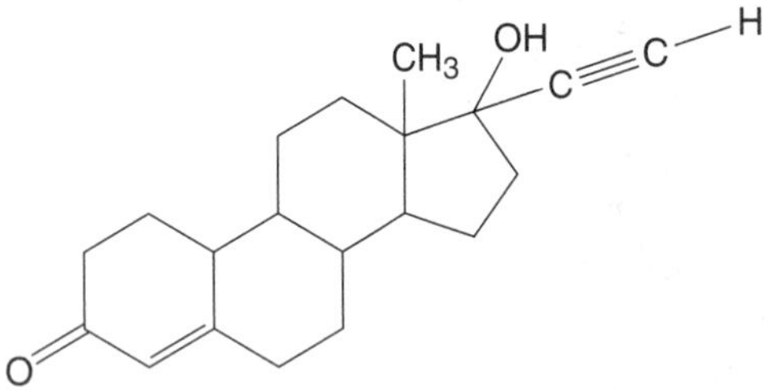

Norethindrone, a synthetic analog of the female hormone progesterone used in birth control.

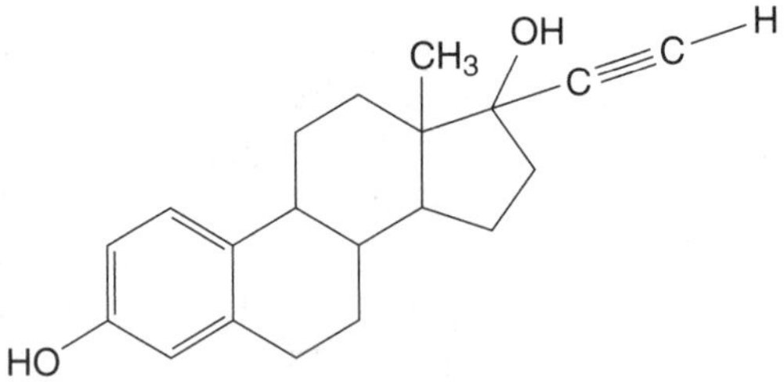

Ethinylestradiol, a synthetic analog of the female estrogen hormones (estradiol and estrone) used in birth control pills.

Figure 4.5.   *Structures of cholesterol, sex hormones, and hormone analogs used for birth control.*

uble in water. It may accumulate on the walls of arteries if present in excess. This restricts blood flow and, if a blockage occurs in a coronary artery, leads to a heart attack.

Anabolic Steroids.   Anabolic steroids are synthetic derivatives of the male hormone testosterone. Their abuse has even tainted Olympic competition. Anabolic steroids are potent tissue-building drugs used in treating anemia, post-surgical care, and enhancing the recovery of burn patients. Because they can also cause acne, calcium deficiency, decreased libido, nausea, suppressed blood-clotting ability, imbalance in blood electrolytes, and other undesirable effects, their use requires medical supervision. In females these drugs may cause male-pattern baldness, a deepening of the voice, and other undesirable changes. Some athletes use these drugs illegally when they want a rapid increase in body tissue mass.

Sex Hormones.   Hormone chemical balance underlies human growth and sexual development. Estrogen or its synthetic analogs may be prescribed during menopause, when a natural decrease in estrogen production occurs. This therapy helps retard osteoporosis, a condition that weakens bones. Perhaps the most widely debated application of estrogen is its use in the birth control pill, either alone or in combination with a progesterone. Estrogen levels drop in a woman's body just before the time when ovulation occurs. This decrease triggers production of follicle-stimulating hormone (FSH) by the pituitary gland. The FSH binds to a receptor that stimulates the ovaries to prepare an egg for conception. High doses of estrogen suppress the formation of FSH and thereby prevent release of an egg.

The role of the female hormone progesterone differs from that of estrogen. It regulates synthesis of another pituitary hormone called luteinizing hormone (LH). This hormone stimulates release of an egg and prepares the uterus for pregnancy. High levels of progesterone inhibit the synthesis of LH, but its synthesis is permitted when the progesterone concentration in the blood drops. This inverse regulation mechanism resembles that for estrogen. An absence of LH helps prevent sperm from reaching the egg and makes the uterus unreceptive if a fertilized egg reaches it. Thus a high progesterone hormone level in the blood also can prevent pregnancy.

Either a progesterone-only or an estrogen-only oral medication is effective in birth control, but the combination of the two is more reliable and is referred to as "the pill." The dual-component pill keeps the levels of estrogen and progesterone in the blood high enough to shut off production of the pituitary hormones FSH and LH. Synthetic derivatives of the natural hormones are often used because of their higher efficacy on oral administration. In spite of the pill's reliability, concern arose because high blood estrogen levels significantly increase blood clotting and associated heart attacks, pulmonary embolisms (blood clots in lungs), and strokes (blood clots in the brain). Other side effects include an increased risk for certain cancers. The amount of estrogen-like hormone in "the pill" now prescribed for 18–35-year-olds has been reduced, and side effects are thought to be minimal.

Recent public debate in the United States has focused on another progesterone receptor antagonist, called RU 486 or the "day-after pill." Its structure, shown in Figure 4.6, resembles that of the birth control agent norethindrone (see Figure 4.5). RU 486 also behaves as an oral contraceptive. Furthermore, it induces abortions in pregnant females and has been prescribed for that purpose in Europe and China.

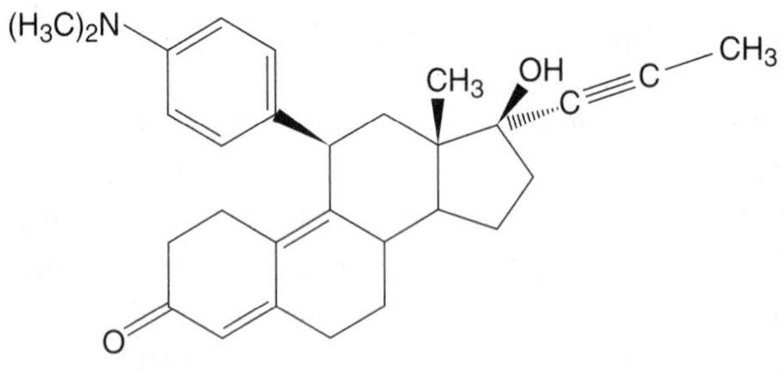

Figure 4.6.   *Mifepristone, or RU 486, an abortifacient (abortion-inducing) drug.*

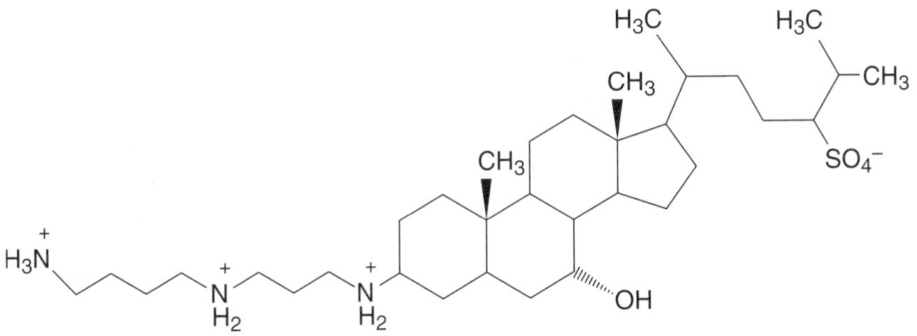

Figure 4.7. *Squalamine, an antimicrobial steroid found in sharks.*

A New Application? An antimicrobial steroid, squalamine (Figure 4.7), has been isolated recently from the stomach and other organs of the dogfish shark. Not only is it effective at killing bacteria in cell cultures, but it also kills a common fungus, *Candida*, that severely infects cancer and AIDS patients. These findings raise the possibility of a new type of application for a steroid drug. The hypothesis that squalamine is a key weapon in the shark's defense against infection is being tested. Its toxicity and effectiveness as an antimicrobial drug are also being studied.

Fats

Fats constitute another class of waxy molecules necessary for life, but they too can cause problems in the diet. Fats possess about twice the energy content of proteins or carbohydrates. They are a calorie-rich food. The general chemical structure of a fat molecule appears in Figure 4.8. It consists of a small molecular fragment derived from glycerol or glycerin coupled to three fatty acid side chains 16 or 18 carbon atoms long. Saturated fats, unsaturated fats, and polyunsaturated fats differ in whether there are zero, one, or several carbon–carbon double bonds in the fatty acid side chains. Double bonds in unsaturated fats provide reactive sites in the molecules where chemicals in the body can attack and break them down. Some evidence suggests that fats with a high percentage of monounsaturated chains may be

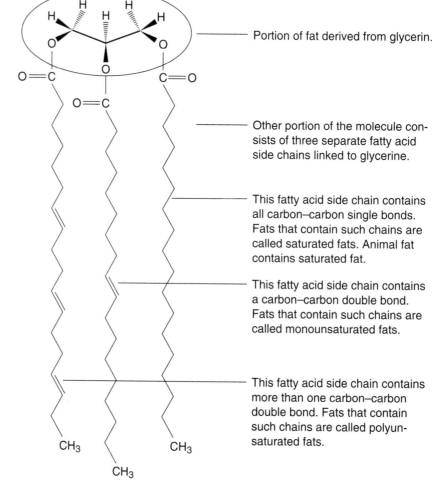

Portion of fat derived from glycerin.

Other portion of the molecule consists of three separate fatty acid side chains linked to glycerine.

This fatty acid side chain contains all carbon–carbon single bonds. Fats that contain such chains are called saturated fats. Animal fat contains saturated fat.

This fatty acid side chain contains a carbon–carbon double bond. Fats that contain such chains are called monounsaturated fats.

This fatty acid side chain contains more than one carbon–carbon double bond. Fats that contain such chains are called polyunsaturated fats.

Figure 4.8. *Sketch of a fat or triglyceride molecule. The classification of a fat as saturated, monounsaturated, or polyunsaturated depends on which fatty acid side chain is present.*

especially beneficial in reducing blood cholesterol. Saturated and unsaturated fats are similar in calorie content; both contain the same number of carbon atoms that the body burns to carbon dioxide to generate energy. But saturated fats are more difficult for the body to process chemically and are less desirable in the diet.

Bulk quantities of fats contain many different fat molecules. In some molecules all three side chains may be unsaturated, whereas others may have completely saturated side chains. All the various mixtures are also possible. Table 4.1 lists the per-

Table 4.1. *Approximate Composition of Some Common Oils and Fats*

Oil or Fat	% Saturated	% Monounsaturated	% Polyunsaturated
Coconut oil	91	8	1
Palm kernel	81	18	1
Butter	59	31	10
Beef tallow	48	50	2
Palm oil	47	43	10
Lard	42	50	8
Cottonseed oil	27	25	48
Peanut oil	14	56	30
Soybean oil	14	29	57
Olive oil	9	84	7
Corn oil	15	51	34
Linseed oil	9	19	72
Safflower oil	7	19	74
Canola oil	6	58	36
Rapeseed oil	1	32	67

centage composition of the side chains in various dietary triglycerides. Natural compositions can also be altered by chemical treatment. Hydrogenation procedures add an H_2 molecule across a carbon–carbon double bond. Partial hydrogenation of vegetable oils removes some of the carbon–carbon double bonds, which gives them a solid consistency like that found in margarine. Complete hydrogenation would produce a solid saturated fat.

Animal fat contains saturated as well as unsaturated and polyunsaturated fats. In addition, animal fat products contain some cholesterol. Plants usually contain unsaturated or polyunsaturated fats; however, some plant oils also contain large amounts of saturated fats. Thus the label "vegetable oil" doesn't guarantee a product free of saturated fat. Coconut and palm oils are high in saturated fat. Nondairy cream substitutes that use coconut oil contain considerable amounts of saturated fats. Plant fat products are usually cholesterol-free.

102

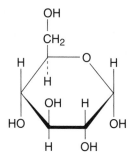

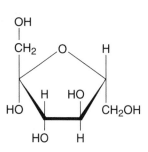

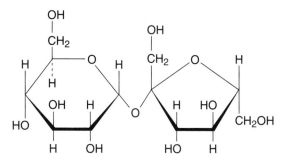

Glucose occurs in fruits and plants as well as in cellulose, starch, and glycogen. It provides an energy source for plants and animals.

Fructose is the chief sugar in most fruits and in honey. It also occurs in human semen as the sole sugar component to provide energy for sperm.

Sucrose (common table sugar) is abundant in sugar cane and sugar beet plants. It consists of glucose and fructose joined together through a bridging oxygen atom.

Figure 4.9. *Molecular structures of some simple sugars.*

Carbohydrates and Starch

Carbohydrates, such as sugars and starch, provide the most important energy source in the diet, because they enter the body's fuel-producing chemical cycle the fastest. Carbohydrates derive their name from their chemical formula, which consists of carbon, C, and water, H_2O, with the general formula $(CH_2O)_n$. The value of n, the number of carbon atoms, lies between 3 and 8 for monosaccharides. Three common sugar carbohydrates, glucose, fructose, and sucrose, are illustrated in Figure 4.9. Sucrose (cane sugar) consists of the two smaller carbohydrates, glucose and fructose, linked by an oxygen atom. Single sugars are often called monosaccharides, and double-component sugars are disaccharides. Maltose (malt sugar) and lactose (milk sugar) are disaccharides similar to sucrose.

Starch, a bland-tasting complex carbohydrate found in potatoes and rice, consists of many glucose molecules linked by bridging oxygen atoms, just as the two sugar components of sucrose are joined. Complex carbohydrates derive their name from their chemical structures, which contain large numbers of individual sugars joined together. Starch in food consists of small, white, hard, tasteless grains. It dissolves in hot water as a sticky substance; an example is the addition of corn starch to hot gravy to make it thicken. If a corn starch solution is treated with acid, it breaks down into its sugar component, dextrose or glucose, which is called corn syrup. Unlike sucrose, dextrose is not a very sweet sugar.

Sugar units even play structural roles in plant cell walls or in plant fibers such as cellulose. Cellulose contains many glucose molecules linked together to form a long fibrous chain. More than 50% of the organic carbon in the biosphere is contained in cellulose. Wood consists of 50% cellulose, and cotton is nearly pure cellulose. Humans cannot digest cellulose found in plant tissue. It serves as a fiber source in the diet. Ruminants (cows, deer, and sheep) are able to digest cellulose and use the released glucose molecules as food. Cellulose can also be chemically processed to make rayon, an artificial silk, and cellophane. Sugars and complex carbohydrates have about half the calorie content of fats by weight.

Fragrances and Flavors

The sense of smell is one of the most intriguing molecule-detector systems in mammals. The scent receptors of some animals and insects can distinguish traces of molecules in the air. Taste is often linked to the sense of smell. In this case, the molecule being detected plays a dual role. Smells can trigger appetite, sexual desire, or fear. Figure 4.10 shows molecules responsible for familiar odors and tastes. Most are complex organic chemicals that are toxic in excessive concentrations; however, in low amounts they serve a valuable function in biology. Life is much enhanced by fragrances and flavors.

Natural flavors and fragrances often involve a complex mixture of naturally occurring chemicals that are difficult to reproduce accurately in a synthetic substitute. Vanillin (see Figure 4.10) is the main, but not the only, component in natural vanilla extract. Synthetic vanillin, available inexpensively in grocery stores, approximates the taste of expensive vanilla extract; but it lacks the rich, full flavor of naturally extracted vanilla.

Thiols (molecules containing S–H groups) are foul-smelling molecules. A few are shown in Figure 4.10. The slow-moving, docile skunk employs two thiols to deter even the most fearsome

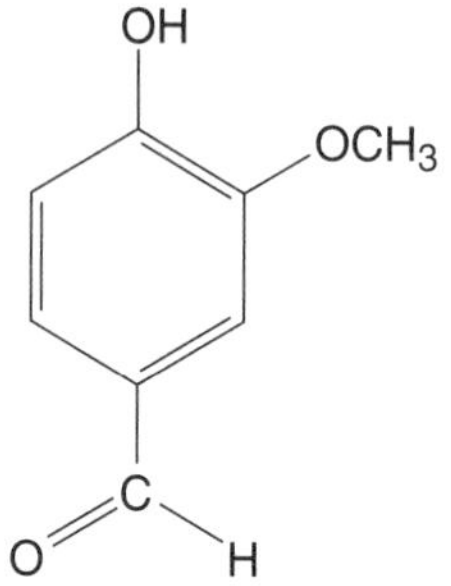

Vanillin, the vanilla taste and smell.

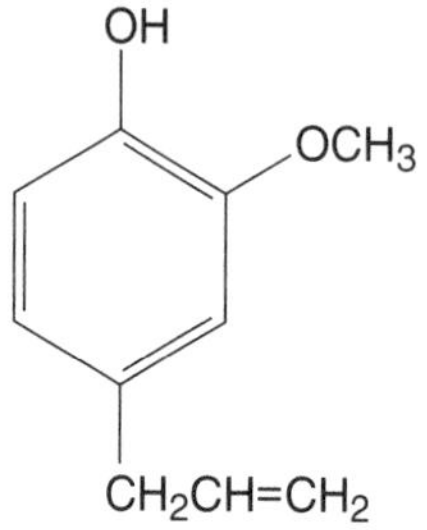

Eugenol, the odor of cloves.

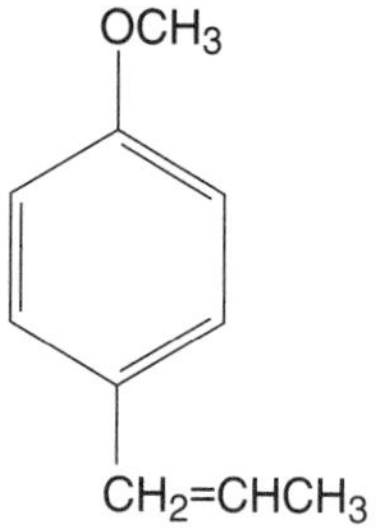

Anethole, the odor of licorice.

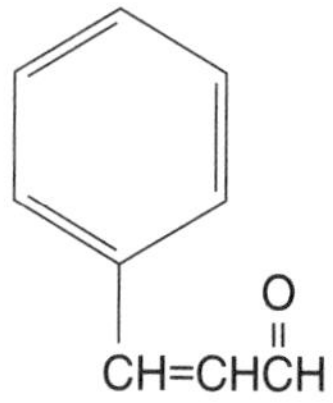

Cinnamaldehyde, the odor of cinnamon.

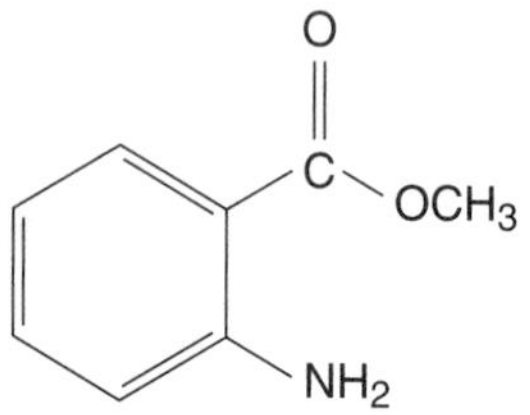

Methyl anthranilate, a component of grape odor.

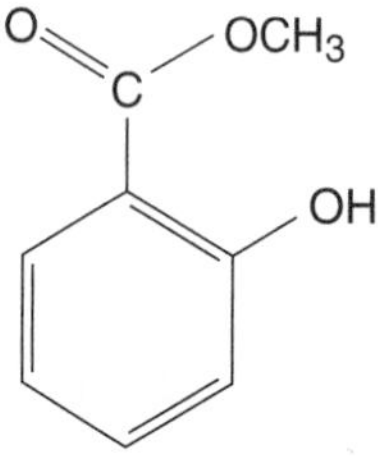

Methyl salicylate, wintergreen odor

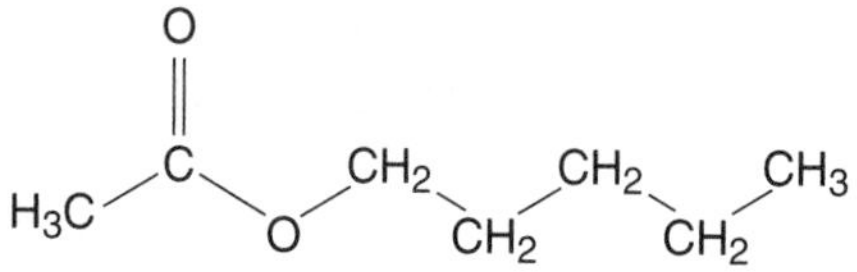

Amyl acetate, banana odor.

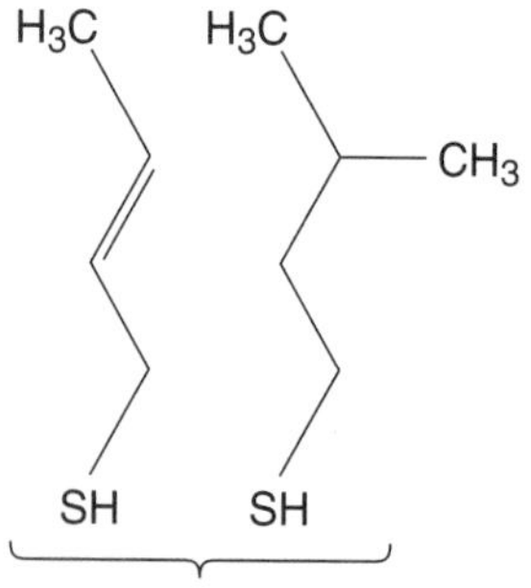

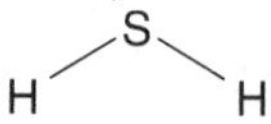

Hydrogen sulfide, the smell of rotten eggs. It is a highly toxic gas.

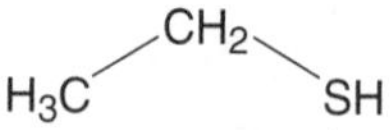

Butenethiol and 3-methyl butanethiol are responsi-ble for the defensive scent of skunks.

Ethanethiol, a foul-smelling compound that is put into natural gas so that leaks can be detected easily. One can smell as little as a single part of the vapor diluted in 50 billion parts of air!

Figure 4.10.   *Structures of some good- and bad-smelling/tasting molecules.*

predators. Hydrogen sulfide, the sulfur analog of water, smells particularly bad. Visitors to geothermal areas such as the Yellowstone and Mount Lassen Volcanic National Parks may inhale enough of this gas to become nauseated from its rotten egg smell while they marvel at the sights of nature. Not many realize that hydrogen sulfide is almost as toxic as the feared chemical hydrogen cyanide. A dangerous aspect of H_2S poisoning stems from a saturation of the sense of smell when too much of it is inhaled. The ability to smell it goes away moments before the onset of unconsciousness!

Utility companies make good use of ethanethiol, another vile-smelling chemical. Very small amounts of it are added to natural gas (as well as to liquefied gas for barbecue and camping stoves) to make a gas leak noticeable by its intense odor. Pure natural gas is itself odorless. The chemist who discovered ethanethiol probably thought it repugnant, but its use as an odorant to provide an early warning of gas leaks has saved numerous lives.

Two foul-smelling chemicals present in rotting flesh are aptly named putrescine and cadaverine. Both are relatively simple molecules (Figure 4.11). Chemists who are unlucky enough to use these compounds in a synthesis often must curtail their social activities for a day or so until the residual trace odor washes away!

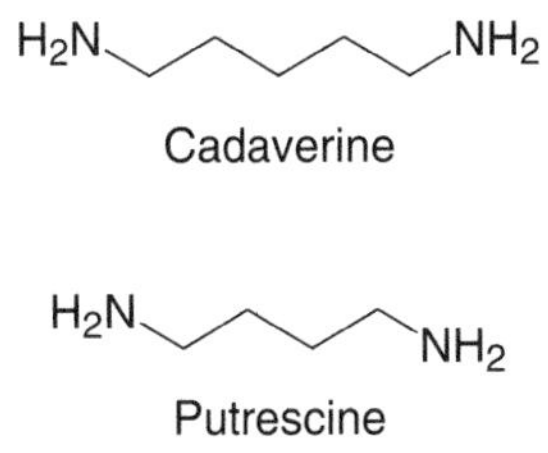

Figure 4.11. *Foul-smelling chemicals.*

Medicinal Chemicals

The Role of Chemical Synthesis in Medicine

Chemistry has revolutionized medicine in the twentieth century. In 1900 the average male life expectancy in the United States was only 49 years. This short life expectancy stemmed to a great degree from a variety of fatal bacterial infections that can now be controlled by drugs. Anesthetics and antiseptics also make possible surgical procedures that were unthinkable 40 years ago. About 8 million surgeries are performed in the United States each year to repair defective tissues and organs. Most require the administration of chemical anesthetics, antiseptics, antibiotics, and pain killers. Complex molecules with specific therapeutic properties can now be built, structurally characterized, and isolated in high purity. Even rare medicinal drugs, found in small quantities in nature, are made in bulk by pharmaceutical laboratories. Rational drug design, the molecular engineering of drugs to fit specific biological receptor sites, promises even more rapid advances in the future.

Some drugs can exist as left- and right-handed mirror image isomers. A particular biological receptor for a drug can be imagined as a glove with a certain handedness. Only a left-handed isomer will fit a left-handed receptor. The wrong isomer sometimes causes serious problems. This became clear in the late 1950s, when the drug thalidomide was prescribed widely in Europe as a sedative for pregnant women. The drug was administered as a 50/50 mixture of the mirror image isomers, which are denoted D and L. The D-isomer was an effective drug, but the L-isomer unexpectedly proved to be a potent mutagen for the developing fetus. It caused thousands of birth defects before the problem became apparent. The United States escaped this tragedy because the Food and Drug Administration questioned thalidomide's safety and denied immediate approval. Since then, government regulations require safety studies for both mirror image isomers, if a drug can exist in such forms. The thalidomide disaster stimulated chemists around the world to develop new procedures for the selective synthesis of mirror image isomers. Chemical synthesis has advanced enough so that the U.S. FDA will soon require rigorous justification when a drug is sold as a mixture of mirror image isomers. (Thalidomide returned to the news in 1993 with a report that it blocks activation of the HIV-1 virus in certain cells of the immune system. It is also being used to treat leprosy and bone-marrow transplant patients.)

The most common examples of compounds whose molecules exist in two different mirror image forms involve a carbon atom bound to four different groups. Such carbon centers are said to be chiral, and the pair of molecules are called mirror image isomers (Figure 4.12). The word *isomer* more generally refers to a compound of identical atomic composition but of different three-dimensional structure. The French chemist Louis Pasteur discovered mirror image isomers in 1848. He showed that a compound obtained from wine, tartaric acid, forms right-

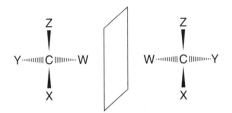

Figure 4.12. *Illustration of the two possible mirror image isomers for a carbon atom bonded to four different groups. The two molecular structures cannot be superimposed and are therefore different.*

and left-handed crystals. It took chemists many years to learn how to synthesize such isomers selectively. Only after 100 years of research have their efforts come to fruition.

One payoff to the seemingly esoteric study of making specific mirror image isomers is technology developed by chemists led by William Knowles at the Monsanto Company. In 1974 the Knowles team provided the first commercially viable synthesis of the chiral chemical L-dopa, or levodopa (Figure 4.13). This drug, an ingredient in Sinemet®, is effective in the treatment of Parkinson's disease. It was also featured as the miracle drug in the movie "Awakenings." In the brain, L-dopa undergoes a biochemical reaction to increase the amount of dopamine neurotransmitter. D-dopa, on the other hand, has no medicinal effect.

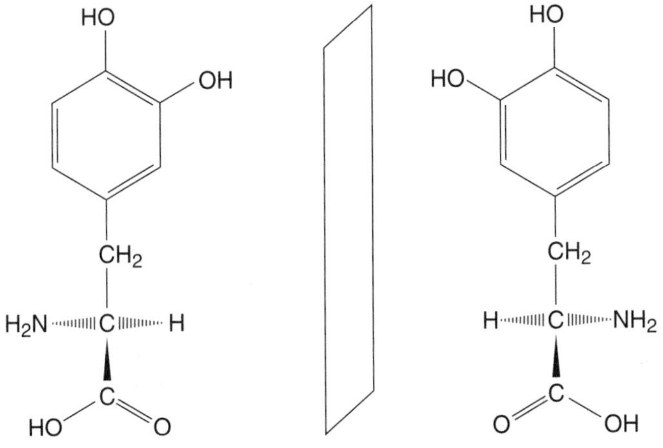

L-dopa, a drug that can cross the blood/brain barrier and increase levels of dopamine, a neurotransmitter precursor deficient in the brains of patients with Parkinson's disease.

D-dopa, the mirror image of L-dopa. This compound has no activity as a drug for the treatment of Parkinson's disease.

Figure 4.13. *The two mirror image isomers of a drug used to treat Parkinson's disease. The prefixes D and L are derived from the Latin prefixes* dexter *("right")* and laevus *("left"). They specify the direction a beam of plane-polarized light rotates when it goes through a solution of the optically active substance.*

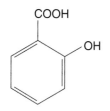

Figure 4.14.
Salicylic acid, an analgesic in willow bark extract and a topical medication used to remove warts.

Aspirin and Other Over-the-Counter Medications

Like that of many medicinal compounds, the discovery of aspirin originated in folk medicine and the healing power of natural plants. Since biblical times, willow bark extract has been used to alleviate pain and reduce inflammation and fever. In the mid 1800s, chemists were able to identify the active ingredient as salicylic acid (Figure 4.14). With the ongoing developments in organic synthetic methods, they soon worked out a synthesis that made the acid readily available. But salicylic acid had some drawbacks as a drug, because it burns the mouth and esophagus (it is now used to remove warts!). Chemists found that the medicinal properties of salicylic acid were retained and the unpleasant side effects diminished by conversion of the hydroxyl (OH) group to an ester group (acetate). The body converts aspirin back to salicylic acid after it is ingested. This new medicine, acetylsalicylic acid (Figure 4.15), didn't receive much attention until 1899, when it was introduced under the trade name aspirin by Friedrich Bayer & Co. This result showed how chemistry can be used to isolate a natural medicine and, by chemical modification, improve its therapeutic properties. Considerable research goes into modifying natural drugs to make them less toxic, more effective, and more easily absorbed. Not only must an oral drug be nontoxic, but it also must tolerate the harsh acidic conditions of the stomach without being destroyed.

Many aspirin substitutes have been synthesized that show improved properties or diminished side effects. Ibuprofen (see Figure 4.15) is an example of an analgesic compound with enhanced anti-inflammatory properties. Naproxen sodium (Naprosyn®) is a long-acting analgesic and anti-inflammatory agent for the treatment of arthritis, bursitis, tendinitis, and gout, as well as for headaches and fever. It was recently released as the over-the-counter medication Aleve®. Acetaminophen represents an alternative analgesic and antipyretic, but it does not possess anti-inflammatory properties. Acetaminophen is useful in the treatment of children with flu symptoms. In such children, aspirin may lead to a condition termed Reye's syndrome, a rare but serious degenerative disease of the brain and liver. This problem does not occur with acetaminophen.

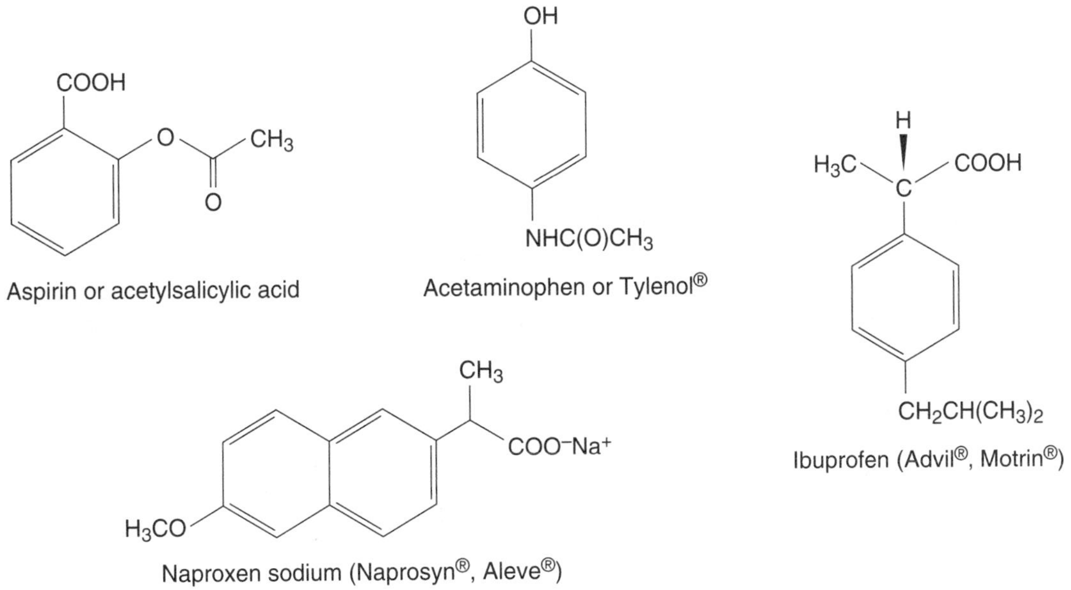

Figure 4.15. *Some widely used analgesics.*

The development of many drugs follows the pattern seen in the development of aspirin. Ephedrine, a component of the Asian Ma Huang plants belonging to the genus *Ephedra*, was found to provide cold relief. Initially it was administered as a medicinal tea made from the plant. The closely related chemical pseudoephedrine hydrochloride is better known today as the nasal decongestant medication Sudafed® (Figure 4.16). After the discovery of a molecule with desirable medical effects, chemists, biochemists, toxicologists, medical doctors, and computer modeling specialists try to improve on the design of a drug for a particular ailment.

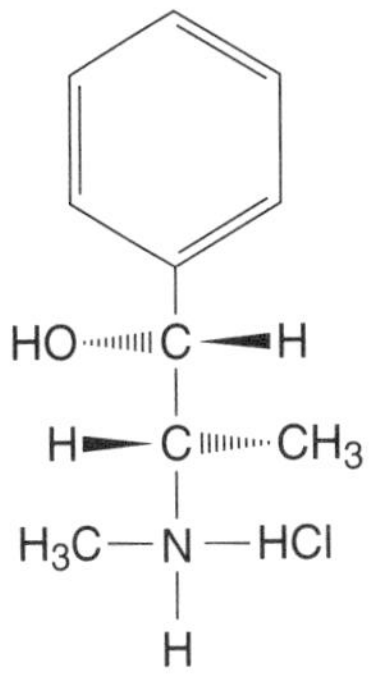

Other common medications appear in Figure 4.17. One of them, para-aminobenzoic acid (PABA), is a component of the vitamin B complex, which also absorbs ultraviolet light. It is used in many sunscreen products and cosmetics. A close chemical relative of PABA, benzocaine, both absorbs ultraviolet light and behaves as a topical anesthetic. It is an active ingredient in Bactine®. Some individuals develop allergic skin rashes from products that contain PABA, and they are sensitive to benzocaine as well.

Figure 4.16. *D-pseudoephedrine hydrochloride, the decongestant in Sudafed® and in many other cold medicines.*

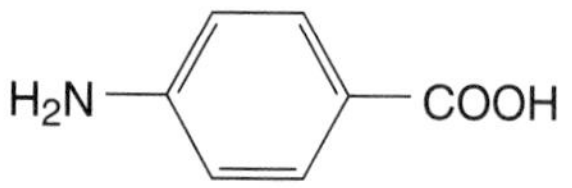

Para-aminobenzoic acid, PABA, or vitamin B_x. Used as a sunscreen.

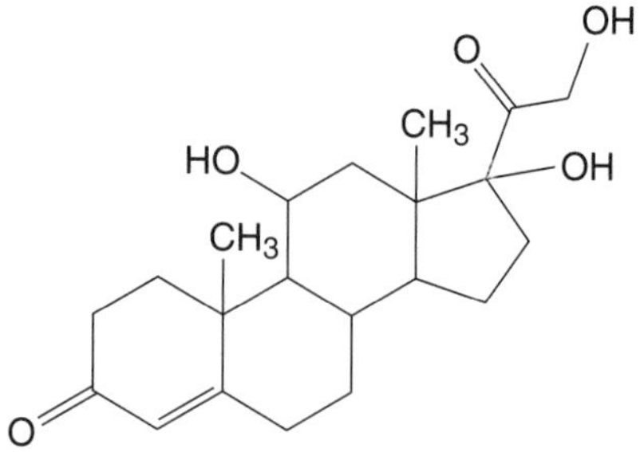

Hydrocortisone, an anti-inflammatory steroid used in skin creams for rashes.

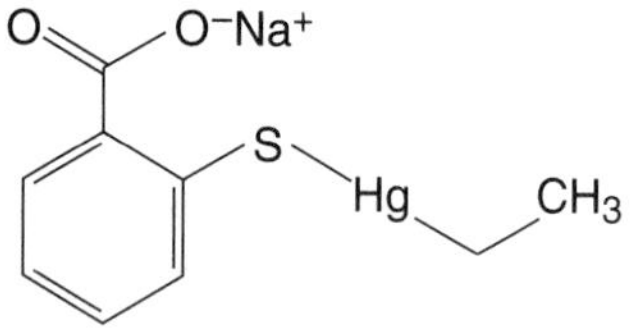

Thimerosal, a topical antiseptic and perservative in eye drops.

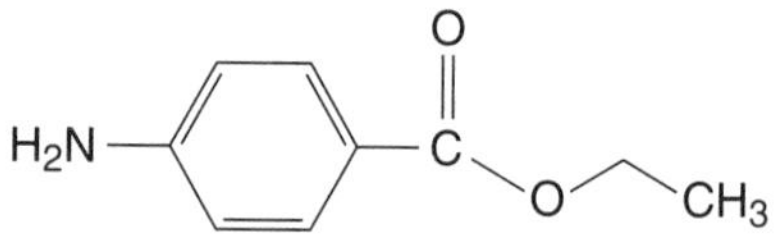

Benzocaine, a PABA derivative used as a topical anesthetic in many over-the-counter first-aid creams.

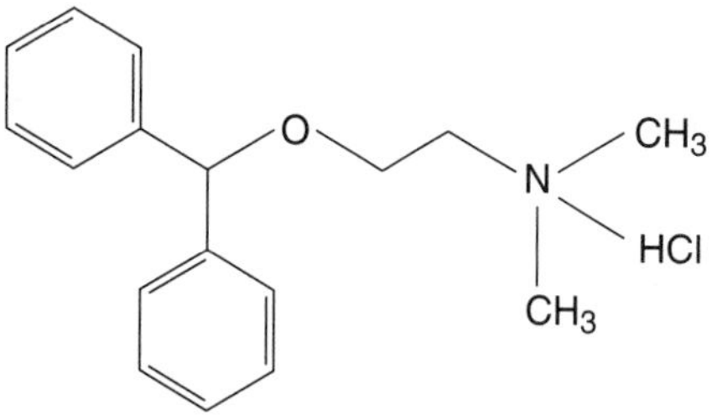

Diphenylhydramine hydrochloride, an antihistamine in allergy pills such as Benadryl®.

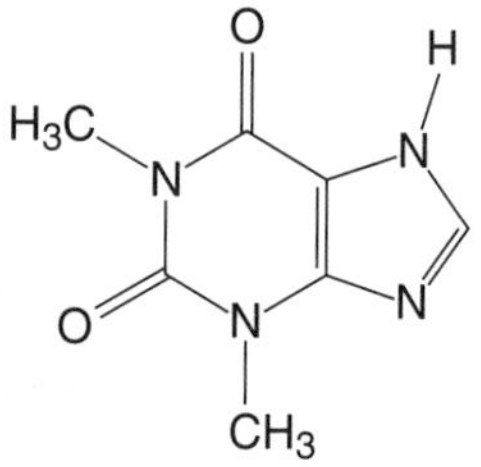

Theophylline, a diuretic cardiac stimulant, vaso-dilator, and medication for asthma and bronchitis.

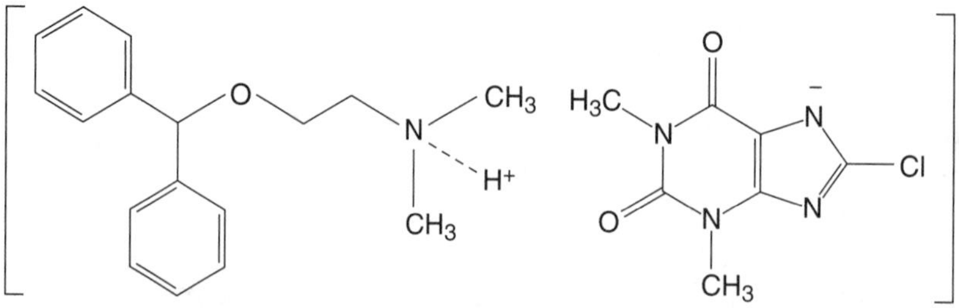

Dimenhydrinate or Dramamine®, a drug used for motion sickness.

Figure 4.17. *Chemicals used in some common nonprescription drugs.*

Hydrocortisone, the active component of "cortisone skin creams" for the treatment of itching and skin rashes, is chemically similar to the steroid hormones (see Figures 4.5 and 4.17). Hydrocortisone exhibits anti-inflammatory properties and lessens the severity of allergic reactions responsible for skin rashes. Related steroids, such as methylprednisolone, exhibit anti-inflammatory properties and are used in injections for arthritis and joint trauma. A steroid derivative known as Proscar® has been introduced recently as a treatment for enlarged prostate gland. All steroid medications cause significant side effects; they should not be used casually. Serious withdrawal symptoms can develop in persons on long-term cortisone therapies, so patients are usually cautioned against abruptly stopping their use of oral steroid medications.

Poisonous metals, such as mercury (Hg), play a key role in some antifungal and antibacterial compounds, such as thimerosol—a common preservative in eye medications. Mercurochrome is another antiseptic mercury compound used to treat cuts of the skin. Compounds that contain heavy metals are usually toxic, so it is important not to ingest them. Selenium sulfide (SeS) contains another poisonous element (selenium): It makes a useful ingredient in topical ointments and shampoos for eczema and dandruff. Small amounts of selenium are essential as a nutritional factor to prevent degenerative muscle diseases. On the other hand, too much of it causes nervousness, liver damage, and G.I. disturbances.

Antihistamines, such as diphenylhydramine or Benadryl® (see Figure 4.17), counteract the release of histamine by inflamed cells. Histamine is a molecule (Figure 4.18) that dilates blood vessels in the surrounding tissues and contributes to the swelling associated with an allergic response. Antihistamines are chemicals that block the cell's receptors for histamine. They alleviate some of the discomfort associated with allergies, such as hay fever and poison ivy. Many antihistamines are sedatives as well as antiallergy agents.

Combinations of drugs may have different uses than the individual medicines. Dramamine®, a drug prescribed for motion sickness, is composed of the antihistamine diphenylhydramine and a derivative of the antiasthmatic theophylline.

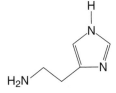

Figure 4.18. *Histamine, a potent vasodilator involved in the inflammatory response by injured cells. Increased blood flow to dilated capillaries in the affected area causes the red appearance around a skin wound or rash. Histamine is also produced in an allergic response to environmental antigens (pollen, dust, and certain foods).*

FÖR UPPTÄCKTEN AV PRONTOSILETS ANTIBAKTERIELLA VERKAN (For the Discovery of the Antibacterial Action of Prontosil)

The balding scientist could scarcely believe it! December had always been an eventful month for Gerhard Domagk, and December, 1947 was no exception. When Prof. Svartz of the Caroline Institute in Stockholm began reading the award citation, Gerhard's mind wandered back eight years ago when he had originally accepted the 1939 Nobel Prize in Physiology and Medicine. He remembered the thrill of the Nobel was all too brief because, after a week in prison with visits from the Gestapo, he had signed a second letter declining the award to comply with Nazi law. Now he was finally about to receive the medal and diploma in recognition of his work. And, although the prize money would never be his, he had something better: it was the result of an unsanctioned human test that he had carried out in December of 1933.

A pin prick injury taught Gerhard Domagk by what a fragile thread human existence hangs. His patient was but four years old and didn't realize the danger of infection. Streptococci bacteria invaded the minor puncture wound on her arm and began a hideous feast. Inflammation traveled up the child's arm, seemingly aimed at the heart. Surgeons had made incision after futile incision to try and stem the tide of infection. In spite of this, the tiny arm, scarred by fourteen cuts in all, still oozed pus and blood through the dressing. Chills shook the life from the little girl's fevered body. Her condition reminded him of the helplessness he felt as a medical student when serving in hospitals during World War I. Gerhard

decided to try his experimental drug after the surgeons recommended amputating the arm. Saw off a child's arm! The survival rate for severe blood poisoning was only 25%, at best. That wasn't much of a chance, thought Domagk. As director of I. G. Farbenindustrie's Laboratory for Experimental Pathology and Bacteriology, he well new the odds stacked against the child.

Gerhard Domagk (1953 photo by J. F. Lehman, courtesy of National Library of Medicine.)

For five years Gerhard had searched unsuccessfully for a chemical that would selectively kill bacteria. Though many were skeptical, his favorite axiom was, "I consider it my first duty in the development of chemotherapy to cure those diseases which have hitherto been incurable, so that in the first place those patients are helped, who can be helped in no other way." One day in late 1932 two chemists, Fritz Mietzsch and Joseph Klarer, gave him a red-brown dye, azosulfamide, for testing. The drug showed no initial activity against bacteria in cell cultures, but he persisted with a test in animals. All the mice infected with a streptococcal culture died within 48 hours, their internal organs ravaged by the bacteria. Gerhard was surprised, and elated, when those given azosulfamide

Prontosil or azosulfamide   Sulfanilamide

recovered completely. Domagk's expertise in pathology made him certain that he had found an antibacterial agent. The internal organs of the treated mice showed no signs of infection! Later he learned that the drug becomes active only when metabolized by the animal to a different chemical, sulfanilamide. That explained why azosulfamide didn't work in simple cell cultures. This remarkable discovery was Gerhard's scientific Christmas present on December 24, 1932.

One December later, the tests in rabbits and mice were going well. The new drug, which was called Prontosil, appeared to exhibit low toxicity. There lay the little girl, whose life was being drained by microscopic marauders. Although he was not sure that the drug had low human toxicity, Gerhard felt that Prontosil was the only hope for his stricken daughter. He was right! This was the first of many successes for Prontosil. Worldwide acclaim awaited a 1936 clinical trial of Prontosil that saved 35 out of 38 mothers infected with child-bed fever (puerperal sepsis). Sulfanilamide (Prontosil album) and many other related sulfa drugs were soon synthesized and marketed. Untreatable fatal diseases such as pneumococcic meningitis and streptococcic meningitis responded to sulfa drug therapy. Gonorrhea, septicemia, and mastoiditis all were now manageable with the medica-

tion. Bacterial dysentery, which had once had an 80–90% fatal outcome, now had a 90–95% survivability. Topical sulfa drugs also revolutionized the treatment of wounds and of deadly gas gangrene during World War II. The need for the miracle drugs rapidly outstripped their supply. A sulfa drug elixir (sulfanilamide dissolved in a toxic solvent, diethylene glycol) was hastily marketed in the United States and led to over 100 deaths. This event spurred passage of the U.S. Food, Drug, and Cosmetic Act of 1938.

By the time Domagk received the Nobel Prize in 1947, penicillin and other antibiotics had replaced sulfa drugs as the preferred antibacterial medications. Antibiotics exhibit a broader range of activity and minimal toxic side effects. Today the use of sulfa drugs is limited to bacterial infections resistant to antibiotics, to topical antibacterial agents, and to patients who might be allergic to antibiotics. Even so, the importance of sulfa drugs to medicine can't be overestimated. They were miracle drugs for a brief period of history. They showed that a chemical war waged against bacteria could be won, thereby providing the impetus for the ensuing commercial development of antibiotics. Sir Alexander Fleming, the discoverer of penicillin, said in testimony, "Without Domagk, no sulfa drugs. Without sulfa drugs, no penicillin. Without penicillin, no antibiotics."

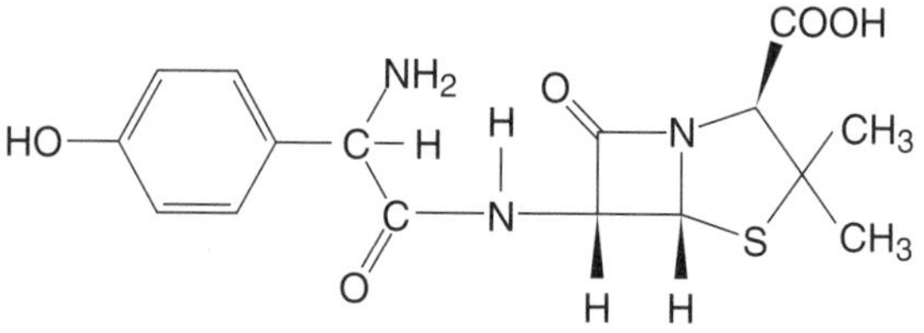

Figure 4.19.
Amoxicillin, a synthetic derivative of penicillin.

Antibiotics

Penicillins are one class of potent antibiotic chemicals produced by molds. Chemists in drug companies make small changes in the molecular structure of penicillin in order to improve its effectiveness and help circumvent drug resistance in bacteria. Amoxicillin, shown in Figure 4.19, is such a penicillin derivative used to treat ear, sinus, and urinary tract infections, as well as gonorrhea.

Antibiotic chemicals work by interfering with essential biochemical processes in bacteria. Amoxicillin and penicillin block the biochemical synthesis of bacterial cell walls. Tetracycline, streptomycin, and erythromycin interfere with essential protein synthesis inside bacterial cells. The antibacterial sulfa drugs block the synthesis of essential nucleic acids (DNA) in bacteria. Antibiotic chemicals control bacterial diseases, such as pneumonia, meningitis, and gonorrhea, as well as streptococcus and salmonella infections. They do not cure viral diseases, such as the common cold, flu, herpes, and AIDS.

Unfortunately, bacteria gradually develop a resistance to antibiotics. The incidence of gonorrhea illustrates the problem. In 1978 the Centers for Disease Control (CDC) recorded 1 million cases in the United States. Some 1000 of these were resistant to treatment with penicillin. In 1990 the total number of cases shrunk to about 700,000, but 59,000 of these did not respond to penicillin. Over the 12-year period, the percentage of penicillin-resistant cases of gonorrhea increased dramatically. Effective treatment required new drugs. The CDC issued a recommendation to use nonpenicillin antibiotics, such as ceftriaxone (Figure 4.20) to treat penicillin-resistant gonorrhea. Chemists in pharmaceutical companies expend great effort to make modified antibiotics that bacteria have not evolved to deal with.

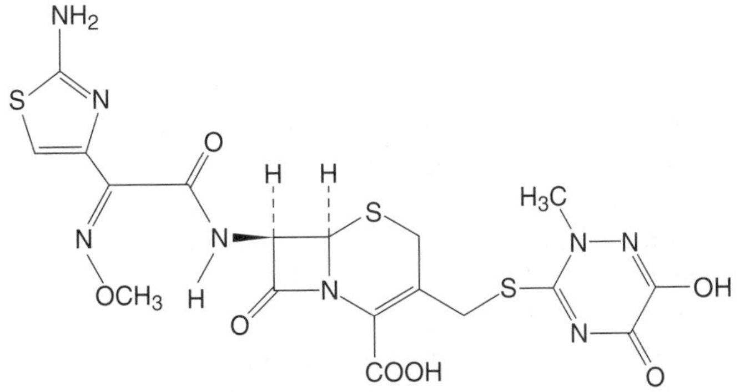

Figure 4.20. *Ceftriaxone, a cephalosporin antibiotic effective in the treatment of penicillin-resistant gonorrhea.*

Tuberculosis, or TB, a bacterial disease once regarded as eliminated in developed countries, returned recently in a "killer" form that resists the drugs (Rifampin, Isoniazid, Pyrazinamide, and Ethambutol) once used to cure it. Worldwide TB infections kill about 3 million people each year! TB-related research diminished in the developed nations, because it was thought that the disease was under control. Scientists must now rush to develop new drugs before the problem becomes epidemic. Drug-resistant TB developed because patients were unwilling to take their medication for the full 6−9 months required to eliminate the bacterium completely. Patients begin to feel better after a month of treatment. This led many to believe that they were cured. When a patient does not complete the full prescription of antibiotics, a few remaining, hardy bacteria can multiply and reinfect the patient in an evolved drug-resistant form. This is why doctors stress the importance of taking the full course of medication for all antibiotic prescriptions. Individuals with compromised immune systems (organ transplant and AIDS patients) serve as another source for the development of drug-resistant bacteria. Drug-resistant strains of pneumonia are becoming prevalent in hospitals. Staphylococcal and pneumococcus infections, which are also endemic to hospitals, have developed resistance to all conventional antibiotics except vancomycin.

The casual use of antibiotics in domestic livestock (as growth enhancers and to prevent infections) has caused concern. Such animals serve as breeding grounds for antibiotic-resistant bacteria. This can cause drug-resistant organisms that also infect humans, such as *Mycobacterium tuberculosis* and *salmonella*, to breed in poultry or livestock and enter the human food chain.

Bacteria develop resistance to antibiotics by at least four mechanisms. Some develop general pumps in their cell walls, which remove antibiotics from within the cell. This has rendered the tetracycline antibiotics ineffective. Others change their cell walls to block drugs from entering the cell. Bacteria have also learned to produce enzymes that destroy specific antibiotics, such as penicillin and cephalosporins. And sometimes the target protein an antibiotic binds to is mutated, in which case the antibiotic can no longer bind and be effective. The fight against bacterial infections requires constant, active vigilance by pharmaceutical chemists.

Anesthetics

From earliest times, humans sought ways to ease pain from serious injuries and medical procedures. Figure 4.21 shows several molecules that have anesthetic properties. Alcohol is perhaps the oldest anesthetic, but more effective and safer alternatives emerged as chemistry developed in the nineteenth century. Ether was an anesthetic used in early operating rooms, but its tremendous flammability and the formation of explosive mixtures with air introduced other hazards. Simple compounds like chloral hydrate and chloroform can also induce unconsciousness, but they damage the liver. The margin for error with anesthetics is small. It is easy to administer too much anesthetic and go beyond unconsciousness to death. It also possible to administer too little and have the patient scream in pain before the operation is over. Improved technology and more sophisticated chemicals have made the use of anesthetics much safer in modern times.

Local anesthetics, such as Lidocaine® and Novocain®, dull pain during minor surgical and dental procedures with little risk of toxic side effects. They are administered as topical anesthetics or as injections (dental surgery and spinals). General anesthetics are fascinating molecules that take a person beyond the

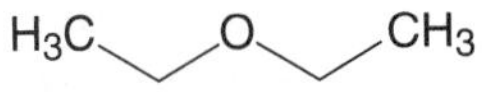

Ethyl ether, a primitive anesthetic that is highly flammable.

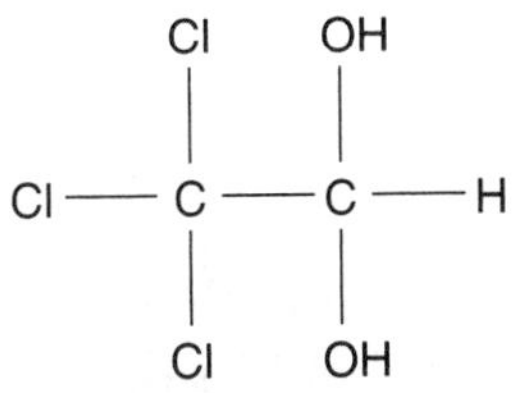

Chloral hydrate, a sedative known in detective novels as the ingredient in a Mickey Finn.

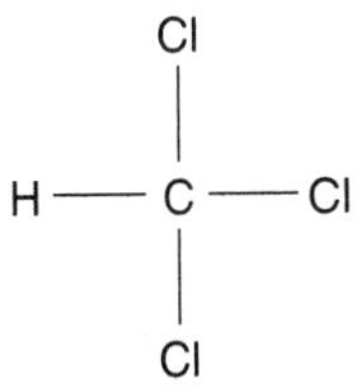

Chloroform, whose vapor can cause unconsciousness, as well as liver cancer.

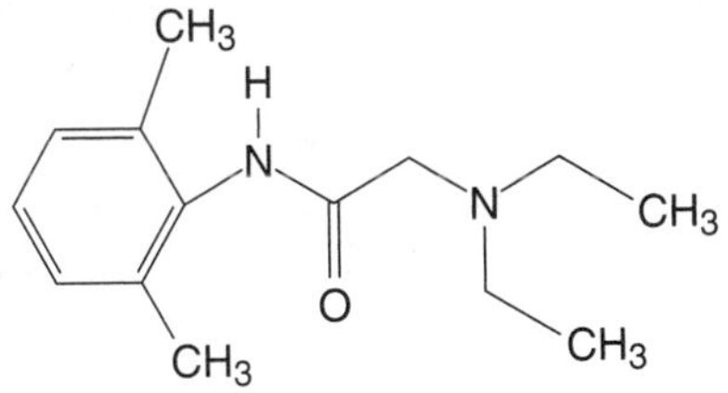

Lidocaine or xylocaine, a local anesthetic.

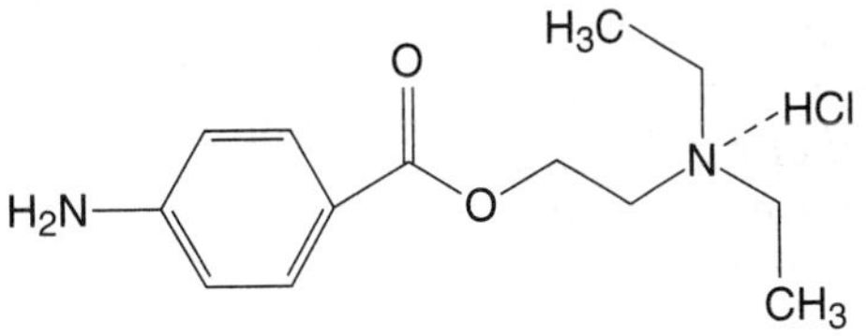

Novocain, a common local anesthetic.

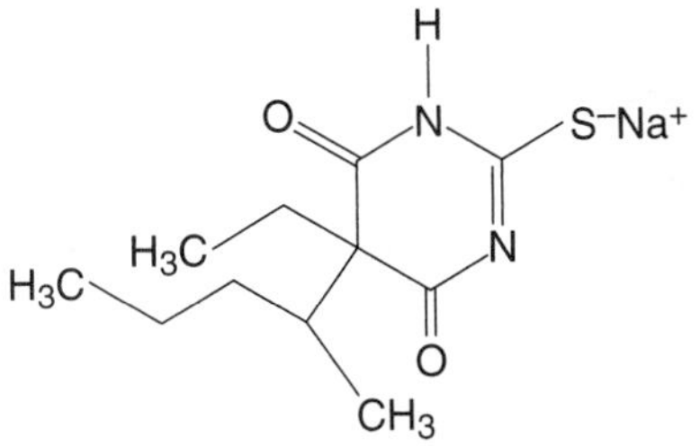

Sodium pentothal, a sulfur barbiturate, which can act as a general anesthetic.

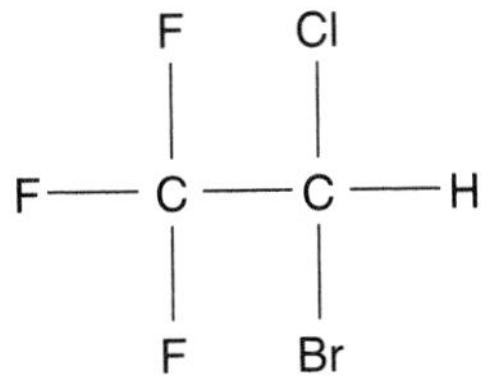

Halothane, an inhalation general anesthetic.

Figure 4.21. *Anesthetic molecules.*

brink of unconsciousness. Combinations of anesthetics are often used for long-term results. For example, intravenous sodium pentothal (see Figure 4.21) or a thiamylal sodium injection can render a patient unconscious within half a minute. The anesthetized state can be maintained for hours with halothane or isoflurane gas. Today major surgery would be unthinkable without the aid of these and other chemicals.

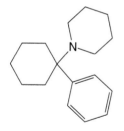

Figure 4.22.
Phencyclidine (PCP or "angel dust") is an anesthetic. It is not used clinically because it can cause permanent psychological disorders.

Phencyclidine (Figure 21), another molecule with intoxicating anesthetic properties, has serious side effects (recurring psychotic episodes and other chronic mental health problems). It is the active ingredient in the street drug PCP, or "angel dust."

Anticancer Drugs

Medical treatment for cancer relies heavily on early detection, surgery, and radiation treatments. This disease involves unchecked growth of the body's own cells, and it has been extremely difficult to find drugs that kill cancer cells without causing extensive damage to healthy tissue in the patient. Cisplatin (Platinol®), a drug containing platinum, ammonia, and chlorine, was the most widely used drug for the treatment of cancer in the United States in 1990. It is generally used in combination with other drug or radiation therapies.

A complex molecule containing platinum (Pt), two chlorine atoms (Cl), and ammonia (NH_3) can form in either of the two ways shown in Figure 4.23. The three-dimensional structure is one in which the platinum, nitrogen, and chlorine atoms all lie in the same plane. The two ammonia molecules (and the two chlorine atoms) can occupy either adjacent positions (abbreviated as *cis*) or opposite positions (abbreviated as *trans*) around the platinum. Molecules with the same chemical composition but different three-dimensional structures are called geometrical

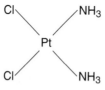

cis-Diamminedichloroplatinum(II), also known as cisplatin or Platinol®, is an effective anticancer drug.

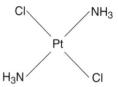

trans-Diamminedichloroplatinum(II), is a toxic compound with negligible anticancer properties.

Figure 4.23.   *The cancer drug Platinol® (cisplatin) and its inactive geometric isomer.*

isomers. For this particular platinum complex, only the *cis* isomer is effective as an anticancer drug. Cisplatin is especially valuable in the treatment of testicular, ovarian, and head and neck cancers. For some forms of cancer, cure rates approach 90%.

Like all cancer control strategies, chemotherapy represents a compromise. Cisplatin kills noncancerous cells as well as cancerous cells, although it exhibits the largest toxic effect on the rapidly dividing cancerous cells. Serious nausea, kidney failure, and neurotoxicity pose problems in Cisplatin therapy. A new drug, Carboplatin (Paraplatin®), that reduces the toxic side effects has been introduced recently (Figure 4.24). Many chemists are working on the mechanism by which the platinum drugs block cell division, since this knowledge will aid in the development of improved drugs. Although progress in the development of useful drugs for "curing" cancer is slow, significant inroads are being made.

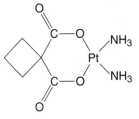

Figure 4.24. *Paraplatin® (carboplatin), a platinum cancer drug with reduced toxicity.*

The chemical taxol, isolated from the bark of the Pacific yew tree, has attracted attention recently because of its promise as a new drug for treating ovarian and breast cancer. This compound occurs only in small amounts in the yew tree. The bark contains 70–400 parts per million (ppm) of taxol, and the needles only about 50 ppm. Three 100-yr-old trees are needed to treat each patient. About 50,000–60,000 patients in the United States alone might benefit from this drug each year. The forest service estimates that about 130 million yew trees grow in national forests in the Pacific Northwest. Furthermore, these trees occur in the politically charged "old-growth" forests, the home of the endangered spotted owl, which is protected by federal law. Chemists have been working on methods to synthesize taxol (or effective analogs) from more readily available starting materials. From the molecular structure shown in Figure 4.25, it is clear why this is a difficult task. Starting with inexpensive and readily available molecules, a complete synthesis of taxol is estimated to require a sequence of 25 different chemical reactions. The overall yield in a multiple-step synthesis can be very low. The net yield from 25 reactions, each with an 80% yield, would be 0.8^{25}, which equals an 0.4% overall yield. Bristol-Myers Squibb will make

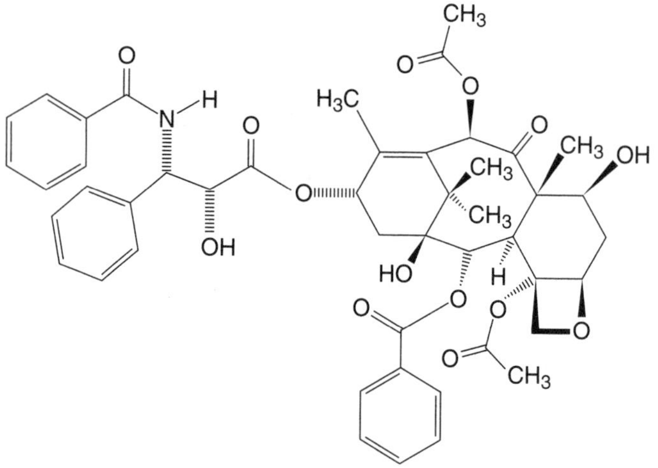

Figure 4.25. *The molecular structure of taxol, a new anti-cancer drug.*

large quantities of taxol (paclitaxel) from a chemical obtained from yew needles and twigs. This semisynthetic process requires fewer trees than the direct extraction of taxol from tree bark and is more economical than a total synthesis. The FDA recently approved the drug for the treatment of ovarian cancer.

There is also considerable interest in chemicals for cancer prevention. The anticancer benefits of a diet rich in green and yellow vegetables recently attracted public attention. Besides their role as dietary fiber sources, vegetables may contain chemicals that prevent cancer. Broccoli, cauliflower, green onions, and Brussels sprouts all contain a relatively simple molecule, sulforaphane, that has been identified in animal tests as an anti-carcinogenic component. The molecular structure of sulforaphane is shown in Figure 4.26. This chemical is thought to boost the level of protective enzymes that help prevent cancer. If future studies confirm the experiments and establish their generality, humans may one day take synthetic sulforaphane supplements to ward off cancer.

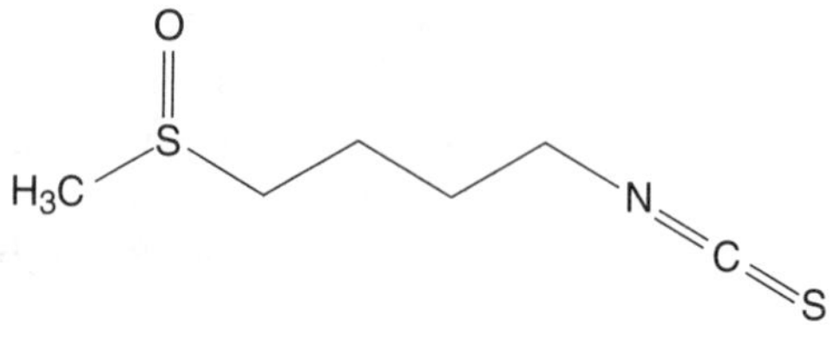

Figure 4.26. *Sulforaphane, an anticarcinogenic molecule that occurs naturally in many green and yellow vegetables.*

Antiviral Drugs

Bacterial diseases, such as scarlet fever, pneumonia, and strep throat, can be controlled with broad-spectrum antibiotics (penicillin, erythromycin, tetracycline), but the situation is grim for viral diseases, such as herpes, encephalomyocarditis, and HIV infection. Vaccines are available for a few viral diseases like polio, but this approach has been unsuccessful for most viral diseases. In general, the immune system must rise to the challenge and kill the virus (the common cold). Otherwise humans live with the virus (herpes), or succumb to it (HIV). Chemists have not yet developed broad-spectrum antiviral agents, which would be immensely valuable drugs. When HIV was discovered in 1983, only three antiviral drugs were licensed in the United States. One example of an antiviral agent discovered since then is the molecule AZT (Retrovir®). This chemical, a failed anticancer drug, has been applied in AIDS therapy, not because it's an ideal drug, but simply because not much else is available in the way of an antiviral drug.

The AZT molecule, shown in Figure 4.27, closely resembles one of the chemical building blocks for DNA. AZT fools the virus and is incorporated into its DNA. This hinders DNA syn-

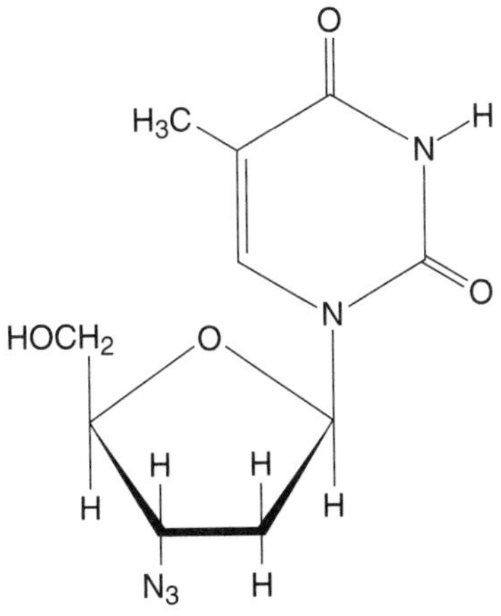

3′-Azido-3′-deoxythymidine, also known as AZT or zidovudine, is an antiviral drug used in AIDS therapy.

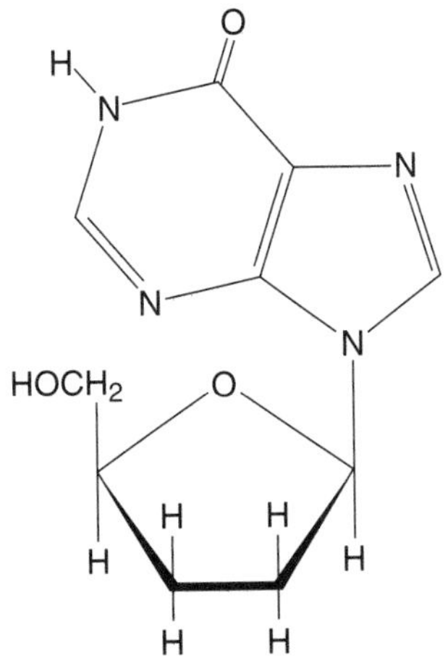

DDI, or dideoxyinosine, is a new antiviral drug approved for AIDS.

Figure 4.27. *The antiviral drugs AZT and DDI used in the war against AIDS.*

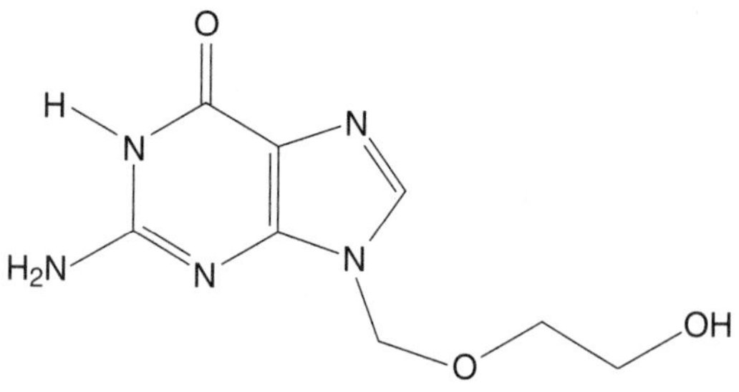

122

Figure 4.28.
Acyclovir, an antiviral drug for treating herpes.

thesis and slows down the viral infection but unfortunately does not eradicate it. The related antiviral agent DDI, also shown in Figure 4.27, and dideoxycytidine (DDC) recently won approval for the treatment of HIV infections. They show efficacy similar to or better than that of AZT, and they are less expensive. These drugs can be used in a combination therapy, which can slow down the virus's development of drug resistance. The discovery of potent antiviral agents remains one of the great challenges in pharmacology.

Acyclovir or Zovirax®, shown in Figure 4.28, has revolutionized the treatment of herpes simplex and herpes zoster infections. The drug inhibits viral DNA replication selectively with minimal side effects to the patient. Both topical and intravenous acyclovir medications help control these infections, but they do not cure them. Herpes, AIDS, and the resurgence of tuberculosis illustrate how tenuous human existence is in the face of microscopic foes. Chemists will continue to manufacture the weapons needed to fight this war.

Antiparasitic and Antifungal Drugs

Tropical parasitic diseases extract a heavy toll worldwide. For example, malaria infects 300 million people each year and kills nearly 2 million. Drugs can control the disease but not cure it. The *plasmodium* parasite responsible for malaria continues to develop improved resistance to such drugs as quinine and its derivative chloroquine. In 1972 Chinese medicinal chemists discovered that a 2000-year-old folk cure, quinghao, brewed from the *Artemisia annua* herb, contains a potent antimalarial agent, called artemisinin. Although the natural compound occurs only in trace amounts in the plant and cannot be extracted economi-

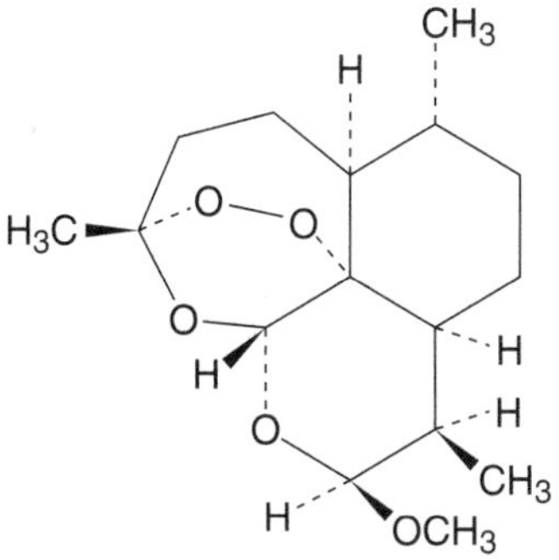

Figure 4.29.
Artemisinin, an extract from the Artemisia annua herb, which shows promise as an anti-malarial drug.

cally, it can now be prepared in the laboratory. The molecular structure of artemisinin is shown in Figure 4.29. This molecule is being studied as a possible second-generation malarial drug so urgently needed.

Immune-compromised AIDS and organ transplant patients are especially susceptible to systemic fungal infections, such as *candidiasis.* Amphotericin B, discovered in the 1950s, is still one of the most effective antifungal drugs in spite of renewed interest in the problem. The emergence of drug-resistant fungal infections makes this problem likely to receive increased attention in the future.

Drugs for Preventing Strokes and Heart Attacks

Hypertension (high blood pressure) increases the probability that a person will develop a stroke or heart attack. High blood cholesterol also increases the probability that a person will have a heart attack. These factors become increasingly problematic as a person ages. Sometimes these conditions can be alleviated with control of diet, but they may also be hereditary. Several drugs have emerged to counteract these undesirable physiological conditions and have helped preserve the quality of life for the aging.

Procardia® is a prescription drug that lowers blood pressure. It inhibits the influx of calcium ions (Ca^{2+}) in cardiac and vascular smooth muscle and is often referred to as a slow channel blocker. It decreases vascular resistance to blood flow and increases cardiac output. Cardizem® (Figure 4.30) is another calcium channel blocker. Other drugs, called beta-blockers (Blocadren®, Inderal®, Lopressor®, Toprol®, and Visken®), are also

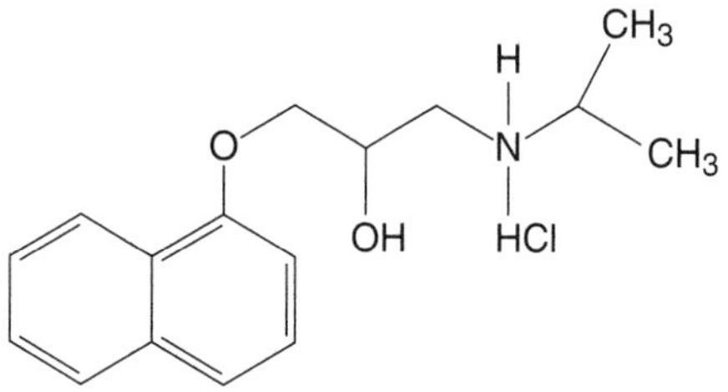

Inderal® (propranolol hydrochloride), a beta-blocker used in the treatment of hypertension and cardiac arrhythmias.

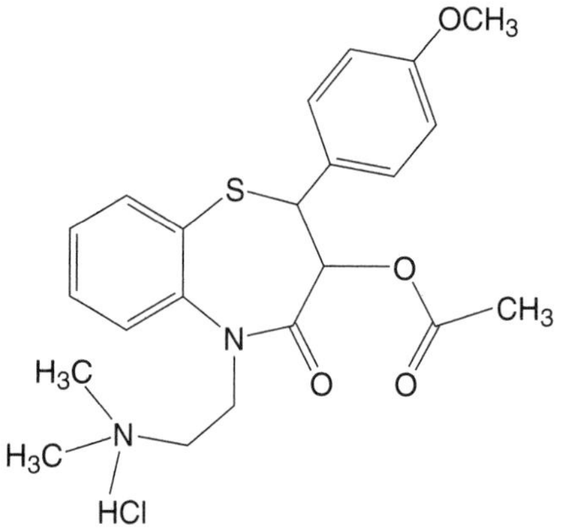

Cardizem®, a calcium channel blocker used to treat hypertension and angina.

Figure 4.30. *Molecular structures of two drugs used to treat hypertension related to heart disease.*

prescribed for hypertension. Inderal® (propranolol hydrochloride: see Figure 4.30) reduced the mortality of patients with severe heart conditions by as much as 39% in an NIH study.

Propranolol binds and blocks two of the four cellular receptors for adrenaline (called β_1- and β_2-adrenergic receptors). The exact mechanism of action is unclear, but the overall effect is to reduce cardiac output and cardiac arrhythmias. A derivative, metoprolol tartrate, used in Lopressor®, binds only the β_1 receptor. Besides being an antihypertensive drug, it also can be used to reduce by 36% the mortality risk after a heart attack.

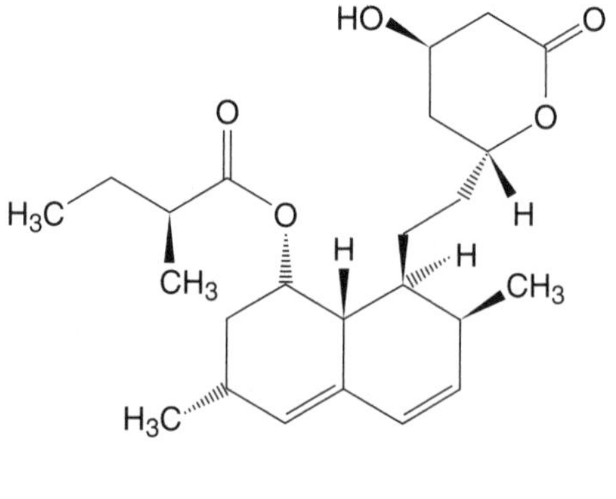

Figure 4.31.
Mevacor®, a drug that lowers serum cholesterol.

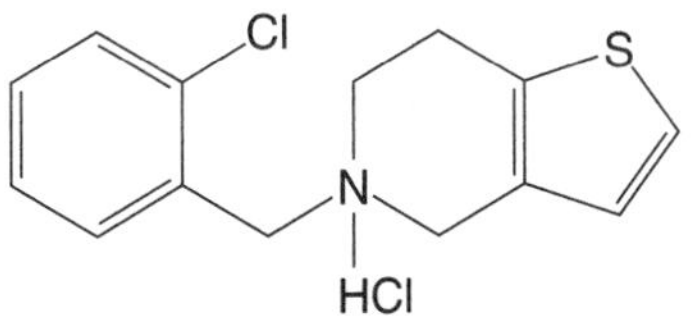

Figure 4.32. *Ticlid®*
(ticlopidine hydro-
chloride), an anti-
clotting drug used to
prevent strokes.

Elevated levels of LDL (low-density lipoprotein) cholesterol in the blood can lead to heart artery blockage. Mevacor® or lovastatin (Figure 4.31) has proved effective at reducing serum cholesterol levels in individuals with high blood cholesterol. Reductions of up to 30% can be obtained. Before the introduction of this drug, there was little help available for individuals genetically disposed toward artherosclerotic vascular diseases.

One of the more debilitating medical problems of the aged is stroke. It can lead to physical, mental, and cognitive deficits. Mild strokes and other symptoms (monocular blindness) often warn of a major stroke. Strong anticlotting agents, such as Coumadin® (warfarin) and heparin (derived from porcine intestinal mucosa!), are used to treat venous thrombosis and pulmonary embolism, but the risk of hemorrhage makes them undesirable for chronic treatment. Aspirin has been used as a milder anticlotting agent to prevent strokes. Recently, Ticlid® (ticlopidine hydrochloride, an inhibitor of blood platelet aggregation; Figure 4.32) has been introduced to help prevent stroke. It effects a reduction in the risk of stroke 24–48% greater than aspirin.

Chemical Dependencies

From Craving to Addiction: Caffeine, Nicotine, Alcohol, THC, Mescaline, and Barbiturates

Nearly everyone has seen ads for chemical dependency rehabilitation programs. Advances in chemical synthesis and purification, which make possible the development of useful drugs for treating sickness, also allow the synthesis of illegal drugs. The molecular structures of many addictive chemicals are not terribly complex, which makes them cheap to produce. Addictive chemicals span a wide range of potencies. Some, such as caffeine, may only be habit-forming, but even being deprived of a morning cup of coffee can disturb the mood of certain individuals.

Many physiologically active compounds occur naturally in plants—caffeine in coffee, nicotine in tobacco, tetrahydrocannabinol (THC) in marijuana—and are thought to be part of the natural defense of plants against animal and insect predators. They taste bad or make the predator sick. The molecular structures of these three compounds appear in Figure 4.33. Caffeine exhibits relatively mild physiological effects as a diuretic and central nervous system stimulant in the human body. There is no conclusive evidence that it has marked harmful effects. Nicotine, a by-product of cigarette manufacture, is such an effective pesticide that it is sold to farmers for that application. Cigarette smokers inhale a small amount of nicotine, which acts as a central nervous system stimulant. In larger doses it acts as a human poison. Occasionally infants suffer nicotine poisoning by chewing on cigarettes.

In contrast to caffeine and nicotine, alcohol depresses the nervous system; however, it can exacerbate hypertension. The continuous strain that high ethanol consumption places on the liver can cause alcoholic hepatitis, cirrhosis of the liver, and death. Alcohol is a factor in 10% of all deaths in the United States. Alcoholic males have a death rate about five times normal. Economic losses from diminished productivity on the job and from medical treatment costs for alcoholism amount to about $100 billion annually. This compares with about $65 billion for estimated losses from smoking and $180 billion for illegal drug use.

Tetrahydrocannabinols (or THCs) are a collection of similar molecules found in the marijuana or hashish plant. Ancient Chinese physicians (around 2800 B.C.) used *cannabis* plants for their analgesic properties. Various medicinal preparations were found in U.S. pharmacies until 1941, when its use was banned. Recently, marijuana has been shown to prevent the violent nausea associated with Cisplatin chemotherapy. Many patients were using the drug illegally, but the FDA has recently approved D-9-THC, the drug Marinol®, for chemotherapy patients. THC also causes hallucinations, which led to marijuana's popular use as a

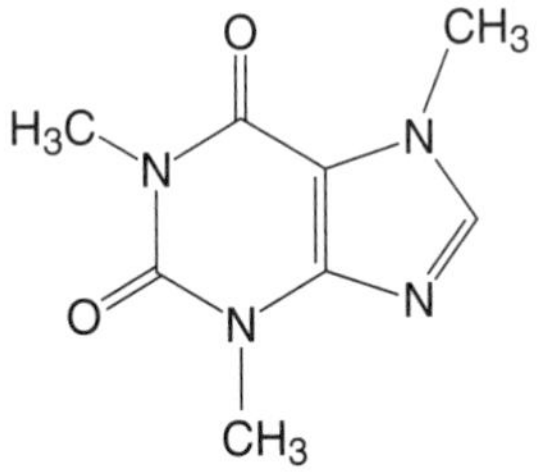

Caffeine, a stimulant that occurs in tea, coffee, and cola nuts. People can develop a craving for caffeine, but not a true addiction.

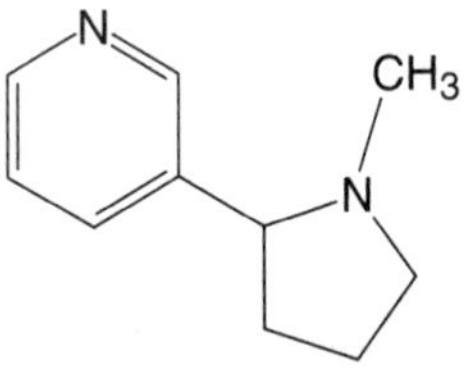

Nicotine constitutes 2–8% of dried tobacco leaves. It has a high human toxicity, and the major commercial use of nicotine is as an agricultural insecticide. It causes the physical addiction experienced by cigarette smokers.

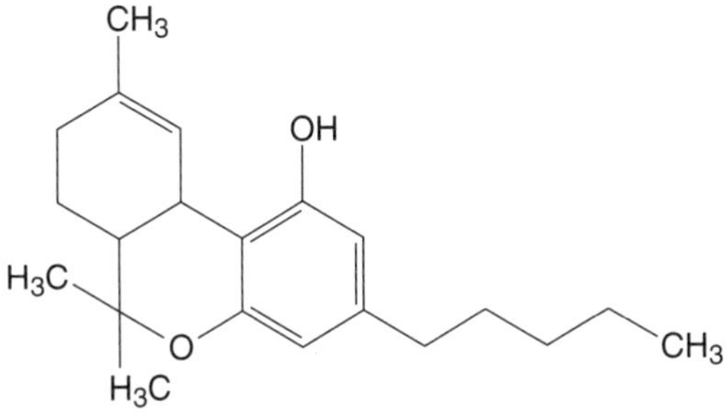

D-9-tetrahydrocannabinol is found in the marijuana plant. When the molecule is eaten or smoked, it may cause euphoria, weakness, or hallucinations. Its psychological addictiveness is a matter of debate.

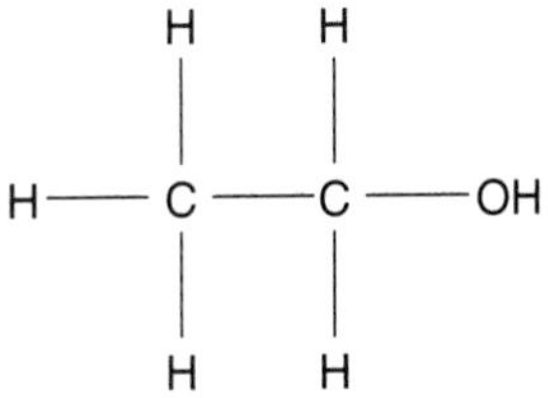

Ethanol, or ethyl alcohol, found in alcoholic beverages. Ethanol is used as a solvent and an antiseptic. It acts as a depressant in the body and can be addictive.

Figure 4.33. *Habit-forming and addictive drugs. Ethanol, caffeine, and nicotine are all legally obtained. Marijuana, the source of THC, is outlawed in many nations.*

recreational drug. In comparison to tobacco cigarettes, marijuana cigarettes contain significantly more tar and present health risks similar to those of tobacco smoking. Perhaps more serious are the dangers of hallucinations and impaired cognitive functions when operating machinery. Several serious rail disasters have been linked to the operator's use of marijuana. Like opiates, such as morphine, THC binds to a receptor in the neuronal membranes of the brain. The THC receptors are localized primarily in regions of the brain associated with short- and long-

128

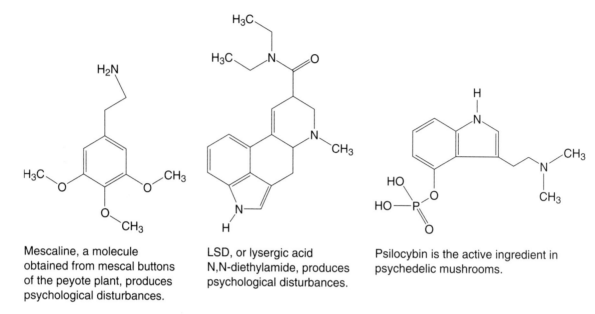

Mescaline, a molecule obtained from mescal buttons of the peyote plant, produces psychological disturbances.

LSD, or lysergic acid N,N-diethylamide, produces psychological disturbances.

Psilocybin is the active ingredient in psychedelic mushrooms.

Figure 4.34.   *Hallucinogenic drugs.*

term memory (the hippocamus and cerebral cortex) and other cognitive functions. But unlike the case of opiates, there are few THC receptors in the brain stem, a region that controls life-support functions (pulse and respiration). This is why opiates present a risk of a fatal overdose but THC does not.

There are many hallucinogens more potent than THC. They include mescaline (from the peyote plant), LSD (a modified alkaloid from the ergot fungus), and psilocybin (from psychedelic mushrooms). The structures of these three molecules are shown in Figure 4.34. Long-term effects on brain chemistry are uncertain. For LSD, where clinical information is available because of its early use in the treatment of psychological disorders, evidence suggests that it causes increased psychotic disorders (leading in the worst case to suicide or long-term hospitalization). Chromosome damage has been observed in one study of patients who formerly used LSD.

Another class of consciousness-altering drugs are sedative hypnotics, which relieve anxiety and decrease inhibitions. These include the tranquilizer diazepam, or Valium®, as well as more powerful barbiturates, such as phenobarbital and its sodi-

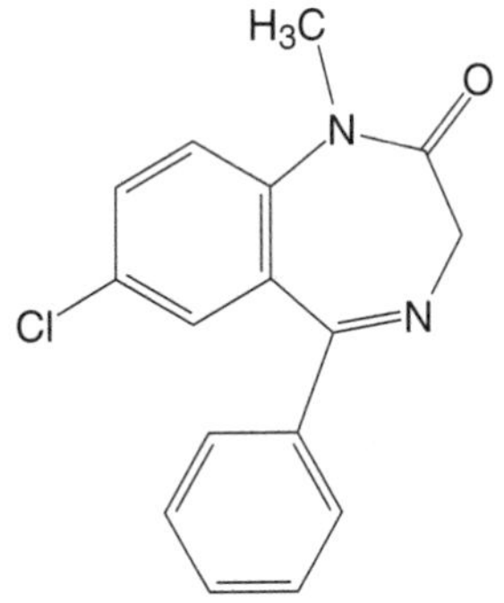

Diazepam, or Valium®, a widely abused, addictive tranquilizer.

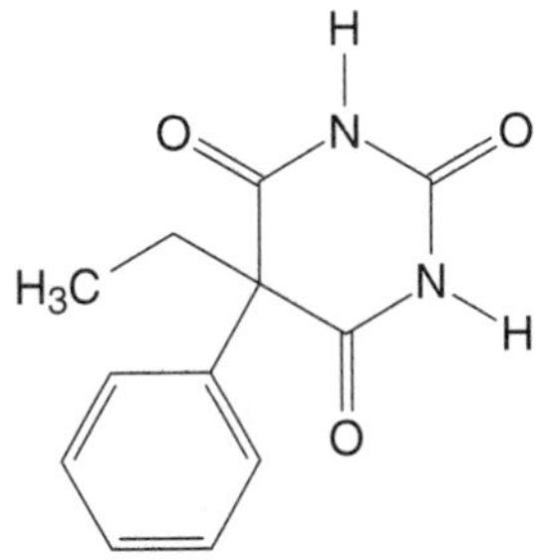

Phenobarbitol, an addictive barbiturate found in many sedative hypnotics and anticonvulsants.

Figure 4.35.
Common sedatives.

um salt. The molecular structures of these drugs are shown in Figure 4.35. Like other drugs, these chemicals appear under brand names (phenobarbital occurs in Donnatal® and a different chemical derivative, secobarbital sodium, in Seconal®). Another close derivative, sodium pentothal, is the "truth serum" of spy novels. Barbiturates are nonselective depressants of the central nervous system, that cause sedation, sleep, coma, and death as the dose increases. The lethal dose of barbiturates (about 1–2 grams of phenobarbital) is much lower if alcohol is ingested simultaneously. Barbiturates become addictive when used frequently. If they are withdrawn suddenly, convulsions and even death may occur.

Chemical analogs of Valium®, the widely prescribed sleeping pill Halcion® and the antianxiety drug Xanax® attracted attention in the news because of controversial evidence of psychological side effects. The molecular structures of these drugs are shown in Figure 4.36. In 1991 the British Department of Health banned the use of Halcion®, but a 1992 U.S. FDA advisory panel recommended its continued use. These drugs were touted for their ability to clear the body rapidly and prevent next-morning drowsiness. Adverse changes in the mood and behavior of users, such as increased aggression, depression, and short-term memory loss, have been claimed. In addition, all the drugs belonging to the benzodiazepine category (Valium®, Halcion®, Xanax®, Librium®, Dalmane®, and Serax®) can cause fetal defects if used

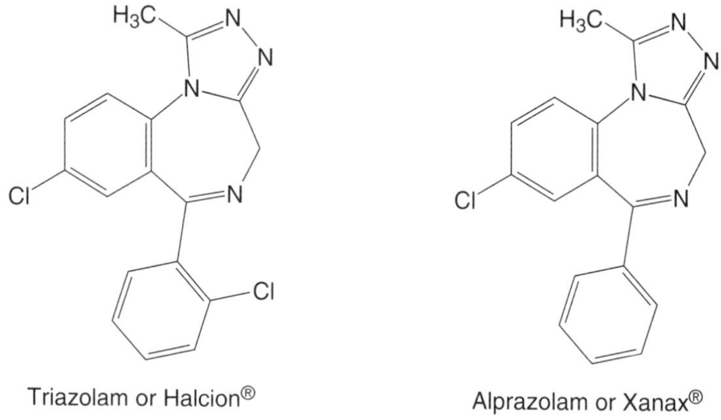

Figure 4.36. *The molecular structures of Halcion® and Xanax®.*

Triazolam or Halcion®

Alprazolam or Xanax®

in the first trimester of pregnancy. Like barbiturates, these drugs become physically addictive after only a few weeks of use. Abrupt withdrawal from benzodiazepines, after addiction, can be fatal. (Prozac® is a widely used antidepressant, chemically unrelated to the benzodiazepines.)

Addictive Sedative Narcotics

Narcotics act as even more powerful drugs that affect brain chemistry and depress the central nervous system. The molecular structures of several narcotics are shown in Figure 4.37. In medical use, narcotics find wide application to relieve the intense pain following surgery (morphine) or during birth (Darvocet®). They also cause euphoria and sedate the patient. Habitual use of these drugs causes addiction, so their availability is regulated. Heavy sedation with morphine is performed only under medical supervision, because shock, cardiac arrest, and cessation of breathing may occur. Morphine and other opiates exert their effect by binding to the endorphin receptors in the limbic system, brainstem, and spinal cord. They repress central nervous system functions and cause hallucinations.

Patients who take narcotics, barbiturates, and certain other prescription drugs are cautioned against the simultaneous consumption of alcoholic beverages. Because heavy alcohol consumption swamps the liver's detoxifying system, the amount of a drug necessary for a fatal overdose is reduced. For this reason,

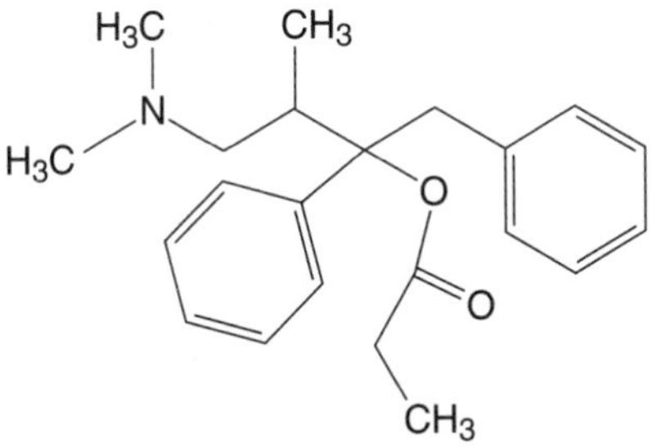

Propoxyphene, a potent narcotic pain reliever in Darvon®, Propox®, and other pain relievers. Excessive doses in combination with alcohol have been a major cause of drug-related deaths.

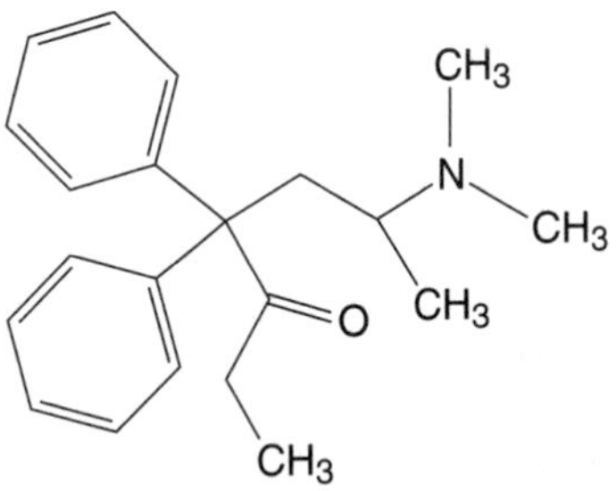

Methadone is a synthetic narcotic used to wean addicts from more dangerous narcotics, such as heroin. Methadone is also addictive.

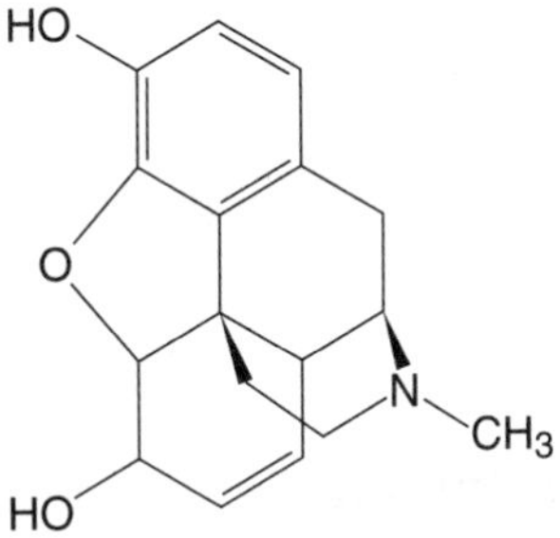

Morphine, the principal component of opium, is a narcotic used to relieve acute postoperative and chronic (terminal cancer) pain. It is highly addictive.

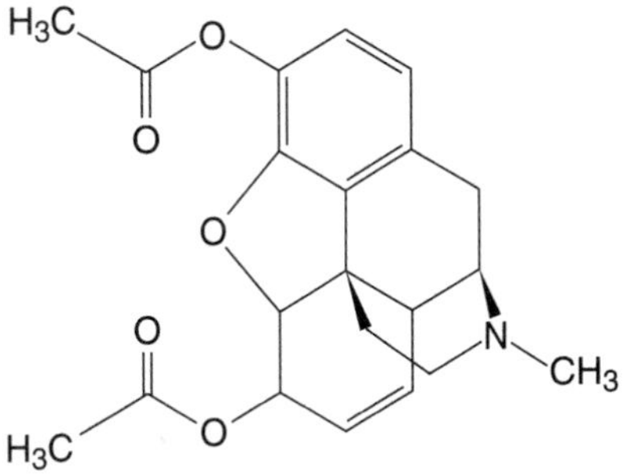

Heroin, or diacetylmorphine, is made by adding acetic acid to morphine. It is an especially addictive, illegal narcotic.

Figure 4.37. *Structures of some addictive narcotic molecules.*

accidental drug overdoses often involve the consumption of a few too many sleeping pills in combination with alcohol.

Note the similarity between the molecular structures of heroin and morphine (see Figure 4.37). The human body converts heroin to morphine, and the latter molecule exerts its narcotic effect. The enhanced addictiveness of heroin (the "high" it inflicts is shorter and more intense than that of morphine) makes it unsuitable for medical use. Drug pushers prefer selling heroin because these characteristics guarantee repeat customers. Withdrawal from heroin is so difficult that one drug rehabilitation method uses another addictive narcotic, methadone, to make the transition easier.

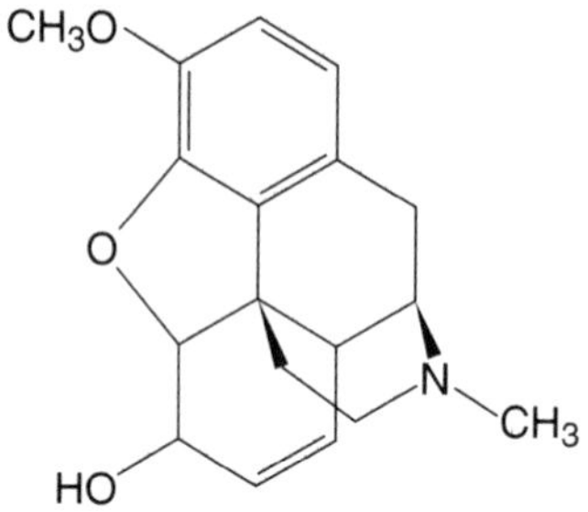

Figure 4.38.
Structures of codeine, a prescription cough medication, and dextromethorphan hydrobromide, an over-the-counter cough medication.

Codeine, a narcotic used in cough medications. It is addictive.

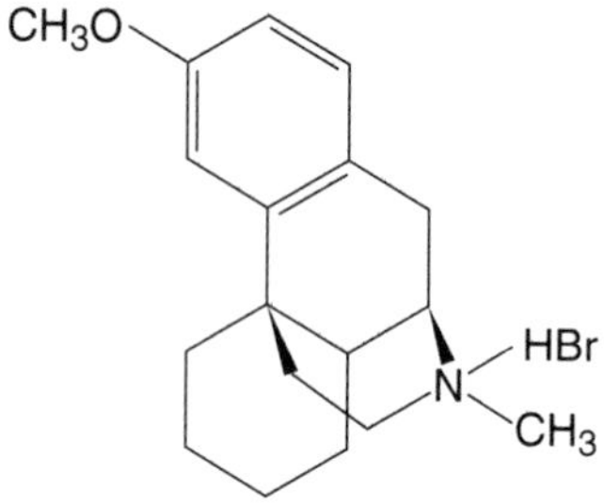

Dextromethorphan hydrobromide, an anticough medication related to codeine but not so addictive.

If one of the H atoms bound to an O in morphine is replaced by a CH_3 group, then the drug codeine is obtained. Codeine is a common ingredient in prescription cough medications. Note the chemical similarity of codeine to dextromethorphan (Figure 4.38), an over-the-counter narcotic cough medication.

Addictive Stimulants: Amphetamines and Cocaine

Other common street drugs are shown in Figure 4.39. Amphetamines and cocaine derive their effect in a different way from other abused drugs. They excite the nervous system, elevate the blood pressure, stimulate the heart, and create a dangerous high. Occasionally a blood vessel breaks in the brain to cause a stroke, resulting in permanent paralysis or death. Heart failure is also common.

Cocaine, which is largely responsible for the escalation in drug abuse during the 1980s and 1990s, is a highly addictive drug. It occurs naturally in the coca plant, as part of its internal chemical defense against insects and animals. Cocaine causes excitement, produces euphoria, and may cause insomnia. The use of cocaine by pregnant women has produced so-called Crack babies with permanent neurological damage. About a third of such infants develop brain lesions that cause behavior, learning, and coordination problems. Unlike heroin, cocaine causes relatively mild physical withdrawal symptoms; however, using it results in an overwhelming psychological addiction.

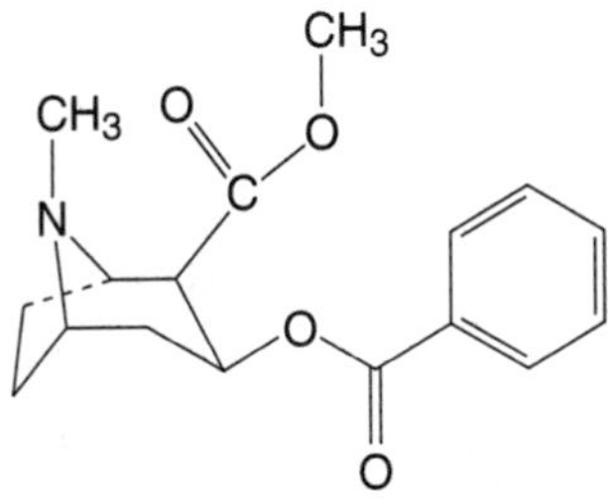

Cocaine, free base, or Crack occurs in the leaves of the coca plant and is used as a local anesthetic and vasoconstrictor. This street drug is highly addictive.

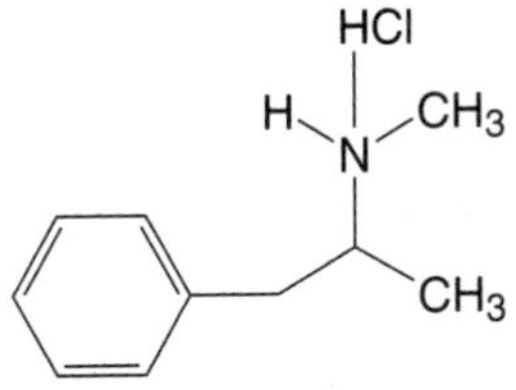

Methamphetamine hydrochloride, Speed, or Meth is a central nervous system stimulant. It has been used in diet regimes for the very obese. It can dangerously elevate the blood pressure and heart rate to cause a stroke or heart failure.

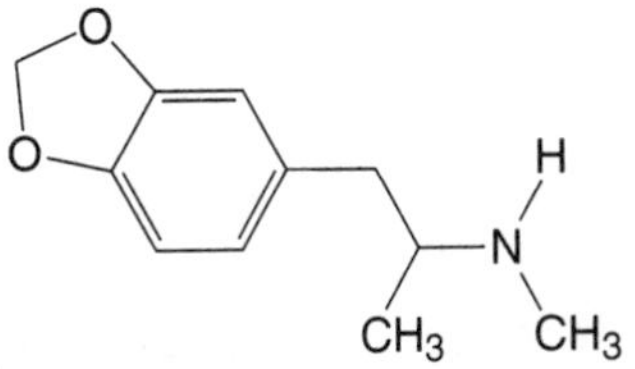

3,4-Methylenedioxymethamphetamine or MDMA, MDM, Adam, Ecstasy, or XTC. A controlled substance with amphetamine side effects, this drug is used in psychotherapy.

Figure 4.39. *Illegal addictive stimulants.*

Designer Drugs

"Designer drugs" represent an attempt to skirt the laws regulating illegal drugs. We have noted how minor chemical modifications to a given structure often produce drugs with similar properties. In an attempt to stay one step ahead of the law, street drug makers have prepared simple derivatives of controlled drugs. These new substances were not controlled by existing drug laws, because they never had been made before. It was technically legal to manufacture them. Several designer drugs and their related narcotics are shown in Figure 4.40.

Fentanyl is a common narcotic analgesic used in surgical procedures. Its effects resemble those of morphine, and it is a controlled substance by law. Designer drug makers prepared 3-methyl fentanyl and *p*-fluoro fentanyl for the street drug market. The designer drug 3-methyl fentanyl is 3000 times as potent as morphine. One tenth of a pound of the designer drug 3-methyl fentanyl is enough to prepare 200 million doses! It often sells on the street as heroin because the effects are indistinguishable.

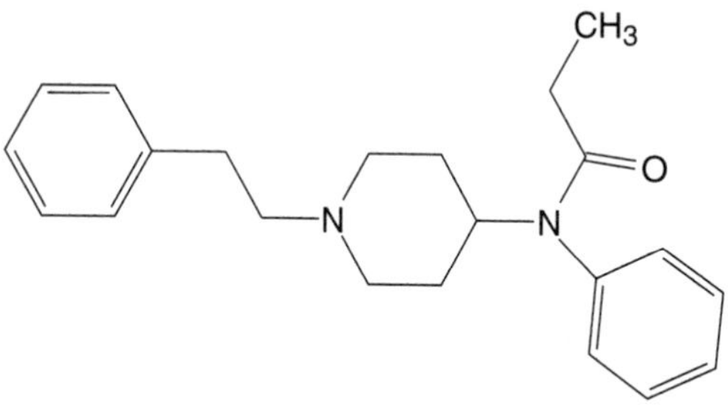

Fentanyl, a narcotic pain reliever.

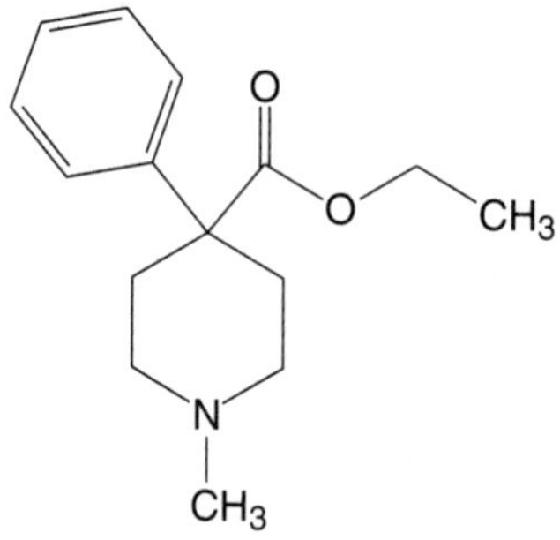

Meperidine, a narcotic pain reliever in Demerol®.

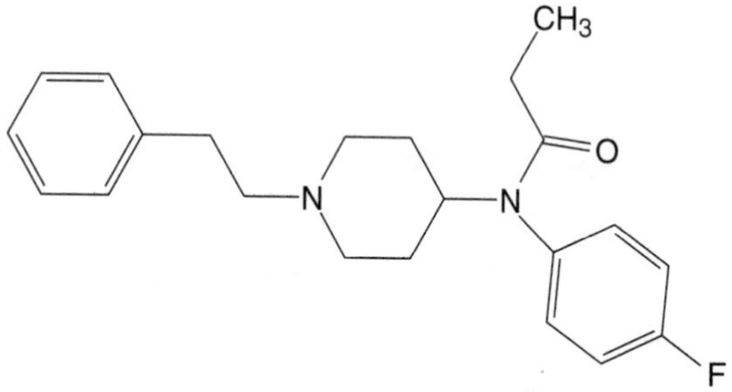

p-Fluorofentanyl, a designer drug.

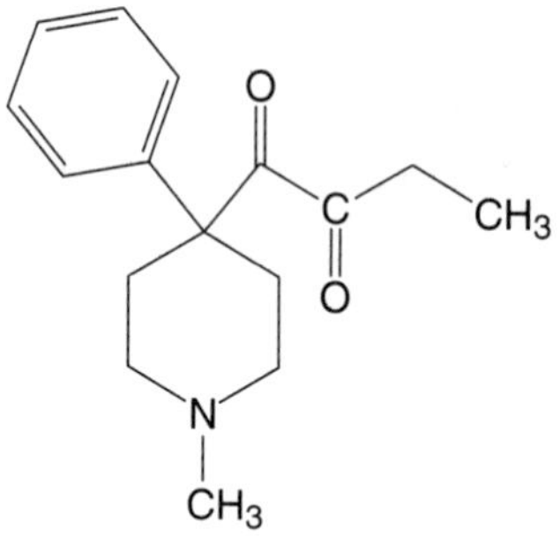

MPPP, a designer drug.

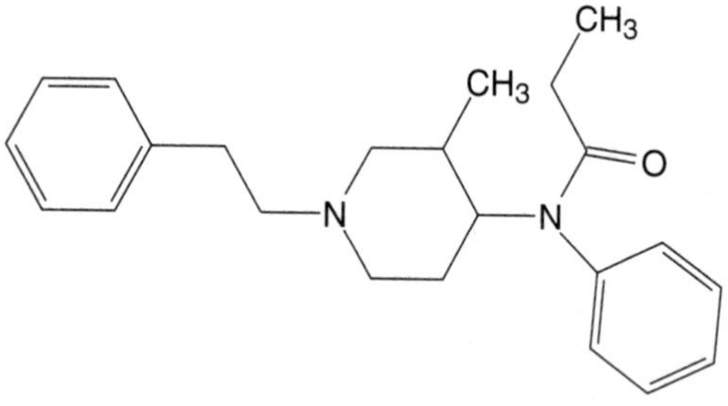

3-Methylfentanyl, a designer drug.

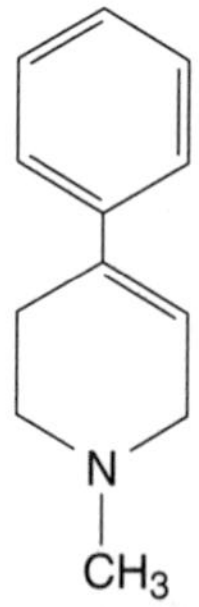

MPTP, an impurity in MPPP that
causes permanent nerve damage.

Figure 4.40.   *Narcotics, their designer drug analogs, and a toxic impurity.*

New laws that permit the classification of a substance within 30 days as an illegal drug have helped to curtail the problem of designer drugs.

The high potency of 3-methyl fentanyl spells many hazards for the supplier. Makers of such powerful drugs, working in poorly equipped labs, can overdose from breathing a small amount of the powder. If they miscalculate and prepare too concentrated a dose for distribution, then the customer dies. And there is another danger, which can arise when the maker of the drug is sloppy. The case of meperidine illustrates the risks involved.

Meperidine is a useful narcotic in medicine sold as Demerol®. A designer drug analog called MPPP appeared on the streets in the 1980s. The potency of MPPP is about 3 times that of morphine and 25 times that of meperidine itself. It was often sold as heroin, the effects being similar. Although MPPP was first made in 1947 by scientists at one of the pharmaceutical companies, it was never commercialized or scheduled as a controlled substance. If a manufacturer doesn't control carefully the conditions used to synthesize MPPP, then the product becomes contaminated with the chemical MPTP. A widely publicized incident occurred during 1982 in San Jose, California, where a few drug addicts were found paralyzed—literally as stiff as a board. These addicts purchased MPPP "synthetic heroin" contaminated with MPTP. Doctors soon learned that the MPTP contaminant causes permanent damage to the nervous system, in a manner analogous to Parkinson's disease. The lack of quality control in the manufacture of street drugs represents a danger often overlooked in their use. Street drugs also escape the extensive toxicity, mutagenicity, and cancer screening tests that must be performed before pharmaceuticals are marketed.

Molecules of War

Nearly every year an armed conflict erupts somewhere. Most of today's high-tech weapons rely on sophisticated electronics and computer guidance systems, but they must also use chemicals.

Even atomic weapons need a chemical explosion to initiate nuclear fission. Chemical technology can save lives in medicine or destroy them in war, just as radar can make air travel safer or guide a missile to its target. Decisions about applications of technology lie in the hands of national leaders around the world.

Explosives

Guncotton, or smokeless powder, was invented in 1845 as a cheap substitute for black gunpowder. Gunpowder also leaves a black cloud that reveals the origin of the projectile, an obvious disadvantage in war. Guncotton is made by the action of nitric acid on cotton (cellulose fiber) in the presence of some sulfuric acid. It has the formula $C_6H_7O_2(NO_3)$. Like nitroglycerin and many other chemical explosives, it contains nitrate (NO_3) or nitro (NO_2) groups bound to carbon. Bullets contain guncotton as the explosive to propel them.

Nitroglycerine, which is made by treating glycerine with nitric and sulfuric acids, was one of the first high explosives developed. The pure liquid is extremely dangerous because of its instability. A mixture of 75% nitroglycerin and 25% diatomaceous earth (an adsorbent sand) is the explosive called dynamite, which is safer to store and use. In 1878 Alfred Nobel developed the high explosive cordite, a mixture of guncotton, nitroglycerine, and petroleum jelly. (The millions earned by his explosives empire now fund the Nobel prizes.) On the other hand, when nitroglycerin is diluted in a capsule, where it no longer can explode, it is a useful vasodilator (vascular smooth muscle relaxant). It decreases a person's blood pressure and reduces the load on the heart when used in the treatment of angina.

Trinitrotoluene, or TNT (its molecular structure is shown in Figure 4.41), is a high-energy explosive that requires a secondary explosive to detonate it. It burns without exploding if ignited. TNT melts at 81°C and can be poured into high-explosive shells, bombs, mines, and torpedoes. The difficulty in starting a TNT explosion is important, because an artillery shell must not detonate prematurely from the shock of being shot out of a cannon.

Plastique, C4 explosive, cyclonite, and RDX (whose molecular structure is shown in Figure 4.41) are all names for sub-

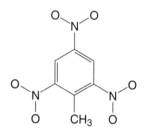

The high explosive TNT (trinitrotoluene).

RDX (cyclonite) is used in plastique explosives. It is also a rat poison.

Figure 4.41. *The molecular structures of the explosives TNT and RDX.*

stances that contain an even more energetic explosive. Cyclonite is used in shaped-charge weapons and in small hand-held rockets. Plastique, or military C4 explosive, contains about 6% of a polymer mixed with the active explosive, so the material has the consistency of putty. It is used for small-scale demolitions, as well as by terrorists. Composition B, a military explosive consisting of TNT and RDX mixtures with a binding agent, finds application in large-scale demolition efforts (buildings and bridges). All the chemical high explosives consist of unstable molecules that react internally. Heat or shock sets their decomposition in motion. Explosive molecules evolve a tremendous amount of energy and rearrange their atoms to form gaseous products, which include nitrogen, carbon monoxide, water vapor, and carbon dioxide. The hot, expanding gases that are produced on detonation move faster than the speed of sound and create the shock wave characteristic of explosions.

Lead azide, $Pb(N_3)_2$, can be used as a shock-sensitive primer to detonate a secondary explosive like TNT. Mercuric fulminate, $Hg(CNO)_2$, also finds use as a detonator for high explosives and in percussion caps for firearm rounds, but lead azide is preferred because of its greater chemical stability.

Chemical Weapons

Chemicals are also used as toxic-warfare agents. The lacrimator 2-chloroacetophenone (also known as MACE, military CN, and tear gas) is the primary chemical agent for riot control. It incapacitates individuals, rather than kills. Tear gas really isn't a gas but a volatile liquid, applied as a fine mist spray or as a stream of liquid from hand-held MACE weapons. The molecular structure of MACE is shown in Figure 4.42.

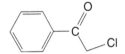

Figure 4.42.   2-Chloroacetophenone, commonly called MACE or tear gas.

Chlorine, phosgene, and mustard gas (Figure 4.43) made their debut as chemical-warfare agents during World War I. These highly reactive molecules destroy lung tissue and may even burn the skin. Death occurs when a victim's damaged lungs fill with fluid (pulmonary edema). Of these three chemical-warfare agents, only chlorine and phosgene are completely gaseous under normal conditions. Chlorine was first used by Germany in World War I in 1915 at Ypres, France. This event initiated an

Cl — Cl

Chlorine, a green gas
that is toxic.

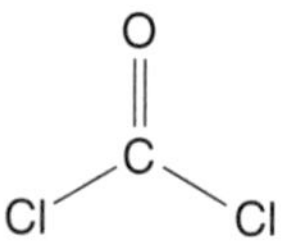

$$O$$
$$\|$$
$$C$$
$$Cl \quad Cl$$

Phosgene, a toxic,
colorless, volatile liquid
that smells like hay.

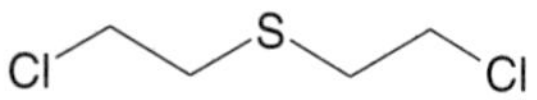

Cl ⌐⌐⌐ S ⌐⌐⌐ Cl

Mustard gas burns the
eyes, skin, and lungs.

Figure 4.43.
*Principal chemical-
warfare agents used
in World War I.*

escalation in technology. Immediately the gas mask, a defensive measure against chlorine, was developed. Unfortunately, other poisons were employed to circumvent this defense.

Phosgene, a liquid at 0°C (32°F), boils at 8.2°C (46.8°F) to form gas readily. Unlike chlorine, which exhibits a characteristic pungent odor, phosgene's hay-like smell makes it difficult to detect. Fatal concentrations of phosgene are not noticeably irritating, and death by pulmonary edema may occur days later. Soldiers at the front during World War I were equipped with gas masks that could be donned at the first whiff of chlorine. But with the advent of phosgene, a soldier could inhale a fatal dose of the substance before realizing that a chemical attack was under way. Phosgene and chlorine are both highly toxic and lethal gases. Breathing air that contains as little as 0.5 milligram of phosgene per liter for 10 minutes is fatal. In addition to their military use, chlorine and phosgene are important industrial chemicals. For example, chlorine is used to purify drinking water, and phosgene is essential in the manufacture of urethane foams, plastics, and coatings.

Mustard gas, so named because of its faint mustard odor, causes debilitating chemical burns to the skin and eyes rather than death. At harmful concentrations this agent exhibits no perceptible odor. It penetrates protective rubber boots and gloves, and ill effects appear only after several hours. Of all the chemical-warfare agents used in World War I, mustard caused the most casualties. The effectiveness of mustard gas as a warfare agent lies in its persistence. Mustard gas is in fact a liquid that doesn't boil until above the temperature of boiling water. It lingers on skin, on clothing, and on the ground, evaporating very slowly. During World War I, soldiers' uniforms had to be cleaned every night to remove traces of mustard agent. Otherwise the slowly releasing vapor could poison them while they slept or blister their skin from the heat of physical exertion (armpit burns were commonplace). Germany's use of mustard gas in World War I prompted the United States to initiate a large-scale chemical-weapons program at Edgewood Arsenal, Maryland. Saddam Hussein of Iraq used mustard agent against the Kurds to suppress political demonstrations, and in the Iran–Iraq war during the 1980s.

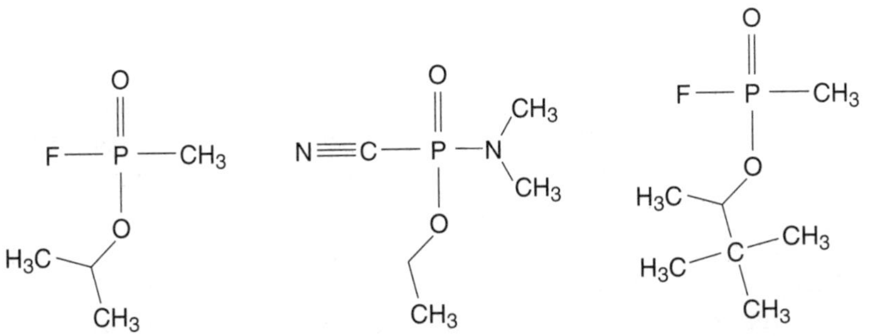

Figure 4.44. *Nerve agents.*

Sarin or Agent GB   Tabun   Soman or Agent GA

Nerve gases are a collection of high-boiling liquids (not gases) that contain the element phosphorus (P). Many phosphorus compounds are essential for living systems, but nerve agents contain phosphorus in a deadly combination with other atoms. The molecules in a nerve agent react with a specific protein in the body that controls the relay of nerve impulses. Initially it overexcites the nerves, which then become overstimulated and fatigue as paralysis sets in. Many insecticides, such as Malathion, are nerve gases tailored to react with the proteins in insects but are much less toxic to humans. Military nerve gases, however, were designed for human mortality. They are among the most poisonous of all chemicals. As little as half a milligram (enough to wet the tip of a needle) can be fatal. A pound contains fatal doses for about 5 million people! The molecular structure of three of the simpler nerve agents, Sarin (GB), Tabun, and Soman (GA) appear in Figure 4.44.

Even more effective nerve gases (VX) have been prepared that have an estimated toxicity (classified) several hundred-fold greater than that of phosgene, the most toxic chemical weapon used in World War I. The extreme toxicity of nerve agents makes them dangerous for the user as well as for the enemy. So-called binary chemical weapons attempt to avoid this problem. Here the spinning action of an artillery shell serves to mix two chemicals that react to form a toxic nerve agent while the shell flies to its target.

Nerve agents were discovered in Germany shortly before World War II. By 1942 the Germans had manufactured large stockpiles of Tabun, but these weapons were never used. Much military research effort has been devoted to the development of nerve agents; however, they have not been employed in any large-scale conflict. The strategic military value of chemical weapons remains questionable. Uncertain effects of weather complicate their deployment, and residues pose a threat to any advancing army that might use them. Military analyses of World War I assess the role played by chemical weapons as minor. Protective chemical measures can be devised to nullify chemical threats, but bullets and bombs can never be completely neutralized.

Given the indiscriminate killing caused by chemical weapons (like bombs, they kill civilians and soldiers alike), they were banned, though ineffectively, by the Hague Declaration of 1899 and the Geneva Protocols of 1925. The 1993 Chemical Weapons Convention, signed in Paris by 130 nations, promises a ban on chemical weapons and the destruction of existing stockpiles. The success or failure of its implementation should be evident over the next decade.

CHAPTER 5

Chemical Reactivity

How did life originate? This question has captivated the minds of scientists, philosophers, and religious leaders. Scientists who study the origin of life believe that life on Earth began about 4 billion years ago, in the late Hadean or early Archean eon. The planet at that time was young and violent. Erupting volcanoes, torrential rains, and thunderstorms raged over a surface covered by shallow, warm salt water. The atmosphere contained virtually no ozone, so the surface of the planet was unprotected from the intense ultraviolet radiation of the Sun. And without oxygen, life as we know it could not exist. The exact ingredients of the atmosphere at this point in the Earth's history are still controversial. Most cosmochemists agree that the Earth had a reducing atmosphere. Nitrogen, carbon dioxide, carbon monoxide, ammonia, methane, hydrogen, and water vapor were probably present. How could complex, fragile organic molecules, which life depends on, be made under these harsh conditions?

An interesting possibility was suggested in 1953 by Stanley Miller, a research fellow at the University of Chicago. Miller wanted to find out what molecules came into being when lightning discharged in the atmosphere of the primordial Earth. In a sealed glass container, he mixed ammonia, methane, hydrogen, and water vapor, the gases he assumed were present in the primeval atmosphere. Solar ultraviolet radiation was simulated

141

with a small ultraviolet lamp. Lightning was simulated by using an electric spark. In a matter of days, Miller's sample chamber contained a plethora of organic molecules. Careful analysis revealed that several of the building blocks of biological molecules were present. Perhaps these simple starting materials were used to make living systems.

Miller's experiments suggest that the fundamental building blocks for biological molecules were constantly being synthesized by natural events on the primordial Earth. But how these molecules assembled into living beings is unknown and remains an active area of research. On an even more fundamental level, it is interesting to ponder the question of whether life can be defined in a purely molecular context. One possible definition for a living molecular system is the ability to undergo self-replication. Several chemical systems have already been discovered that exhibit this behavior.

The chemical reactions that may have created life, and that are responsible for its continuation, fall into two broad categories: oxidation–reduction reactions and acid–base reactions. In this chapter we will examine both.

Redox Reactions

Oxidation and reduction reactions involve the transfer of electrons from one molecule to another. Molecules that lose electrons are called *reductants.* Those that gain electrons are called *oxidants.* Reactions between oxidants and reductants are termed **red**uction–**ox**idation, or redox, reactions. Combustion processes are redox reactions that involve oxygen as the electron acceptor (oxidant) and a fuel as the electron donor (reductant).

Acid–base reactions do not involve the transfer of electrons from one species to another. An acid–base reaction involves the transfer of a proton (an H^+ ion). The molecule that transfers the H^+ ion is called the acid. The molecule that receives the H^+ ion is called the base. Acid–base reactions are of immense importance in industrial chemistry, in environmental chemistry, and in the processes of life.

Oxidants

Oxidation reactions derive their name from early observations of substances reacting with the element oxygen, O_2. At the present time, oxygen makes up 21% of the atmosphere and gives the Earth an oxidizing environment. Photosynthesis by plants and blue-green algae produces most of the atmospheric oxygen. Before the emergence of these organisms some 4 billion years ago, the atmosphere had an oxygen composition of only 0.002%. A consequence of the oxidizing atmosphere today is that most metals, and even nonmetals, occur in nature as oxides. Humans take advantage of the abundance of oxygen and use it as an oxidant for energy-producing reactions in cars, airplanes, and furnaces— and in the body.

When heated, most metals react with oxygen to yield a solid metal oxide. Iron burns in oxygen (equation 1) at an elevated temperature to form Fe_3O_4, a magnetic oxide found in nature as the mineral magnetite. It is a black, water-insoluble material that conducts electricity.

$$3Fe_{(s)} + 2O_{2(g)} \rightarrow Fe_3O_{4(s)} \tag{1}$$

Iron (reductant) reacts with oxygen (oxidant) to form the mineral magnetite.

When iron combines slowly with air in the presence of moisture, a different reddish oxide called rust forms, as shown in equation (2). Rust does not have a unique chemical formula because it can absorb varying amounts of water. The symbols $\cdot n$ are placed in front of the symbol for water, H_2O, to indicate that the solid contains a variable amount, n, of water. By heating rust, it is possible to drive all the water away.

$$4Fe_{(s)} + 3O_{2(g)} + nH_2O_{(l)} \rightarrow 2Fe_2O_3 \cdot nH_2O_{(s)} \tag{2}$$

Iron metal reacts with oxygen and water to form rust.

Steel is used widely in the construction of buildings. Steel is largely iron, which needs to be protected from oxidation to maintain its strength. A layer of paint can be used to protect the metal surface from attack by water and oxygen. Alternatively, steel can be coated with a corrosion-resistant metal, such as chromium. This process is called chromium plating. Steel can also be protected by creation of a stainless steel alloy with chromium. The relatively high cost of chromium limits the use of stainless steel.

Oxygen also reacts with nonmetallic elements. Elemental sulfur and carbon burn in air to form the oxides shown in equations (3) through (5).

$$S_{8(s)} + 8O_{2(g)} \rightarrow 8SO_{2(g)} \tag{3}$$

Sulfur (reductant) reacts with oxygen (oxidant) to form sulfur dioxide.

$$2C_{(s)} + O_{2(g)} \rightarrow 2CO_{(g)} \tag{4}$$

Carbon (reductant) reacts with oxygen (oxidant) to form carbon monoxide.

$$C_{(s)} + O_{2(g)} \rightarrow CO_{2(g)} \tag{5}$$

Carbon (reductant) reacts with oxygen (oxidant) to form carbon dioxide.

The oxide that forms may depend on the oxygen supply. In combustion processes, carbon monoxide, an odorless gas, forms when a limited amount of oxygen is available (equation 4). Carbon monoxide is toxic to humans. Carbon dioxide forms when excess oxygen is present (equation 5). In contrast to carbon monoxide, carbon dioxide is nontoxic at moderate concentrations. It is widely used as a refrigerant in its solid form (dry ice), as the gas in carbonated beverages, and in many other ways.

In the combustion of hydrocarbons, such as oil, natural gas, and gasoline, all the carbon atoms in the molecules undergo oxidation (if there is an abundance of O_2) to form CO_2. Furthermore, all the hydrogen atoms in hydrocarbon fuels undergo oxidation to water, as illustrated by the combustion of methane (CH_4) shown in equation (6).

$$\underset{H}{\overset{H}{\diagdown}}\hspace{-0.3em}C\hspace{-0.3em}-\hspace{-0.3em}H \; + \; 2\,O\!\!=\!\!O \; \rightarrow \; O\!\!=\!\!C\!\!=\!\!O \; + \; 2\;\underset{H}{\diagup}\overset{O}{\diagdown}{}_{H} \tag{6}$$

Methane (natural gas) reacts with oxygen to form carbon dioxide and water.

Once again, when the oxygen supply is limited, combustion reactions may generate carbon monoxide as a by-product. For this reason, it is important to clean the air intake filter to a furnace regularly and check to make sure the burner burns with a blue flame. With an insufficient oxygen supply, natural gas burns with an orange, sooty flame and may produce enough carbon monoxide to poison anyone who is nearby.

Complex hydrocarbons such as isooctane (a gasoline component) burn similarly and produce mainly water and carbon dioxide, as shown in equation (7). The water produced in the combustion of organic fuels may not be apparent, because water vapor is colorless.

Gasoline (isooctane) reacts with oxygen to form carbon dioxide and water.

Look for water dripping from the tailpipe the next time you start your car outdoors on a cool day. Each gallon of gasoline produces more than a gallon of water when it burns. For this reason, it is not wise to run an automobile briefly and then not use it for an extended period. Water will condense in a cold engine and exhaust system and initiate internal corrosion. The white clouds of steam that rise from chimneys on a cold day nicely illustrate the water produced in combustion reactions.

Elements other than oxygen can act as oxidants. An oxidant is *any* species that takes electrons away from other compounds. The transfer of electrons may be complete to form an ionic com-

146

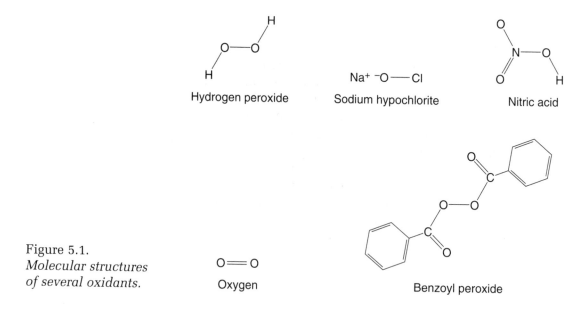

Figure 5.1.
*Molecular structures
of several oxidants.*

pound such as a metal oxide, or it may be partial to yield polar
covalent bonds, as in water or carbon dioxide. Fluorine, chlorine, bromine, iodine, and sulfur generally behave as oxidants.
In the reaction of sodium and chlorine to form Na^+Cl^- salt, the
element chlorine behaves as the oxidant and accepts electrons
from metallic sodium, the reductant. Many chemical compounds are oxidants. Hydrogen peroxide (available in drug
stores as a 3% solution in water), sodium hypochlorite (available
as a water solution in Clorox® bleach), benzoyl peroxide (used
to bleach flour and in acne cream medications), and nitric acid
(an industrial acid) are commonly used as oxidants. The molecular structures of several oxidants are shown in Figure 5.1.

Reductants

In addition to an oxidant, every oxidation–reduction reaction
requires a reductant. The reductant transfers electrons to the oxidant. In the reactions of oxygen in equations (1) and (2), the
reductant was iron, whereas in equation (3) it was sulfur.
Common elemental reductants are carbon, hydrogen, and phosphorus—and all the metallic elements are reductants. Fuels such
as hydrocarbons and many other chemical compounds also are

reductants. Certain reductants are sometimes referred to as antioxidants.

One cancer prevention strategy involves the administration of antioxidants, because oxidation reactions in the body can lead to the production of highly reactive molecules known as free radicals. Free radicals are unstable compounds that contain an extra electron not involved in a chemical bond. In the methyl radical, formed by removing a hydrogen atom from methane, the carbon atom contains one electron (signified by the dot in Figure 5.2) that is not involved in a chemical bond. Free radicals often behave as oxidants. They are known to attack DNA in cells and may cause gene mutations that lead to cancer. Reductants can provide electrons to free radicals, which renders them harmless. Analysis of the urine from cigarette smokers has shown greater amounts of by-products derived from free-radical damage to DNA than occur in the urine of nonsmokers. Unfortunately, a study of Finnish smokers showed that an increased intake of the antioxidants vitamin E and β-carotene (Figure 5.3) did not diminish the lung cancer rate (or the incidence of heart problems) in smokers. In fact, a high intake of β-carotene appeared to increase slightly the rate of lung cancer in smokers. The verdict is not yet in on whether the ingestion of antioxidants diminishes free-radical concentrations inside the body and helps prevent other forms of cancer.

It has also been suggested that free radicals are a factor in aging. One theory postulates that aging is due in part to cumulative damage to the body by free radicals. Thus vitamin C, vitamin E, and β-carotene antioxidants could be antiaging chemi-

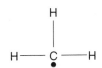

Figure 5.2. *The structure of the methyl radical, a free radical formed by the removal of a hydrogen atom from the methane molecule.*

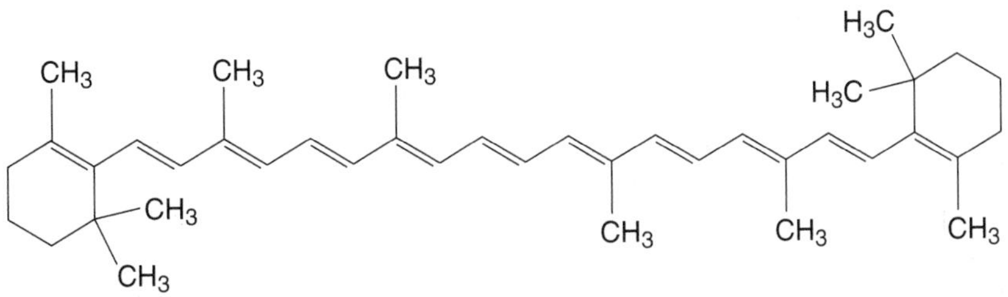

Figure 5.3. *β-Carotene, a component in carrots that the human body converts to vitamin A. It is also used as a yellow coloring agent in food and as an antioxidant.*

cals, but no one is sure because the fundamental chemical origin of aging is unknown. In taking vitamins, it is important to remember that too large an excess of vitamin A (β-carotene is a safe source) causes serious liver damage. And excess vitamin C may cause mild stomach distress and diarrhea.

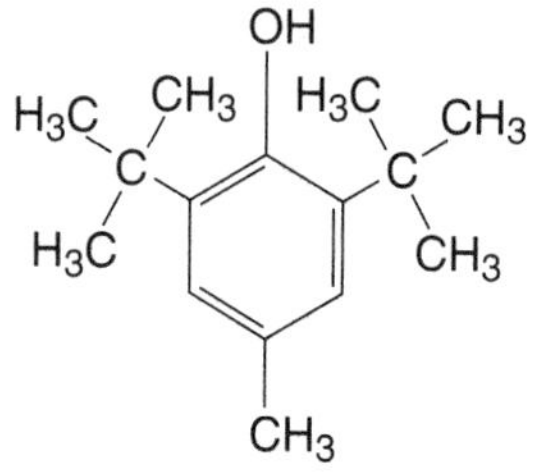

Antioxidants are also useful as food preservatives. The spoilage of foods, such as the onset of staleness and rancidity, results from air oxidation of fats and oils in food. Air oxidation may create carbon-centered free radicals (such as the methyl radical) in food. These react with oxygen in air to form bad-tasting chemicals. Antioxidants intercept free radicals and slow down this process. Vitamin C (ascorbic acid) or its salt, sodium ascorbate, can be added to bread to keep it fresh longer. The antioxidant food additive, known as butylated hydroxytoluene (BHT, Figure 5.4), also is used in paints, inks, vegetable oils, soap, rubber, and plastics; it retards the slow air oxidation of these substances. Antioxidant food additives do not prevent biological spoilage by molds and bacteria, so different chemicals are added for that purpose.

Figure 5.4. *BHT (butylated hydroxytoluene), an antioxidant used to retard the air oxidation of food, oils, rubber, and plastics.*

Chemical Energy and Redox Reactions

The redox reaction between oxygen, inhaled from the air, and carbohydrate or fat molecules provides energy for the body. A different redox reaction in the firefly generates light. Another produces electricity in the electric eel. It is important to understand how energy is linked to chemical reactions.

The branch of chemistry concerned with the energy changes that accompany a reaction is called *thermodynamics*. One of the laws of thermodynamics states that a chemical reaction can occur spontaneously only when it proceeds from less-stable reactants to more-stable products. In terms of chemical bonding, this means that a reaction generally proceeds in the direction that corresponds to the breaking of weak bonds and the forming of stronger bonds. Consider a bicycle rider at the top of a hill. He or she can proceed spontaneously downhill (coasting) and at the bottom will have acquired the gravitational potential energy that has been liberated in the form of increased speed or kinetic energy. This situation resembles a spontaneous chemical reaction that gives off heat. Conversely, the cyclist at the bottom of a hill cannot coast to the top. Energy must be provided (stored as gravitational potential energy) for the cyclist to return to the high-energy state at the hilltop. Like the bicycle rider, chemical reactions can be forced "uphill," to yield less stable chemicals, via an input of energy into the reaction. This type of energy storage occurs when a battery is charged with the aid of electrical energy.

Chemists measure energy in calories (abbreviated cal) or in kilocalories (abbreviated kcal); 1 kcal = 1000 cal. One calorie corresponds to the energy required to raise the temperature of 1 gram of liquid water by 1 degree Celsius (1°C). A more common engineering unit, used to rate air conditioners and heating systems, is the British thermal unit, or Btu. One Btu is the energy needed to raise the temperature of 1 pound of water by 1°F. The Btu equals 252 cal, or 0.252 kcal. The Calorie (with a capital C) is used by nutritionists to indicate the energy content of foods. The Calorie is the same as a kilocalorie; that is, 1 Cal = 1000 cal. We will use small calories (cal) or kilocalories (kcal) for energy units. The amount of energy liberated in a reaction, such as combustion, depends on the amount of reactant consumed. Chemists therefore often specify energies on a per-mole basis (kcal/mole), just as nutritionists rate the energy content of foods on a per-unit-of-weight or a per-serving basis.

Chemical reactions that liberate heat are termed *exothermic*, and those that absorb heat are called *endothermic*. Most—but not all—spontaneous reactions are exothermic. Energy changes during chemical reactions may be included in the equation. A plus sign for the energy on the product side of the equation signals the production of energy (usually heat), as in the combustion of methane shown in equation (8). When we write the equation in reverse, as in equation (9), we signal that energy needs to be provided to enable the reaction to take place by placing the energy term on the left side of the equation.

$$CH_{4(g)} + 2O_{2(g)} \rightarrow CO_{2(g)} + 2H_2O_{(l)} + 213 \text{ kcal} \qquad (8)$$

*Methane reacts with oxygen to form carbon
dioxide and water, and energy is liberated.*

$$213 \text{ kcal} + CO_{2(g)} + 2H_2O_{(l)} \rightarrow CH_{4(g)} + 2O_{2(g)} \qquad (9)$$

*Carbon dioxide and water react to form
methane and oxygen, and energy is absorbed.*

Reversing the direction of a chemical reaction simply reverses the flow of energy involved. In terms of chemical bonds, equation (8) is a favorable reaction because the two C=O and the four O–H bonds in the products are stronger than the two O=O and the four C–H bonds in the reactants. Though all reactions can be written in two different directions, only the one that "goes downhill" energetically can occur spontaneously.

In comparing different foods and fuels in terms of practical energy content, it is more useful to make comparisons on a per-

Table 5.1.　*Heat of Combustion Per Gram of Foods and Fuels*

Food or Fuel	Heat of Combustion in calories for 1 Gram	Food or Fuel	Heat of Combustion in calories for 1 Gram
Hydrogen	29,000	Nuts	6,000
Methane	12,000	Sunflower seeds	5,700
Dry wood (cellulose)	6,000	Ice cream	2,100
Coal	8,000	Bacon	5,600
Gasoline	10,700	Eggs	1,600
Heating oil	10,600	Chicken	1,500
Methanol	5,400	Steak	3,300
Ethanol (200 proof)	6,400	Tuna (in water)	1,400
Ethanol (100 proof)	3,200	Glucose	3,700
Beer	400	Sucrose (table sugar)	4,000
Table wine	900	White bread	2,600
Lard	9,200	Cereal (corn flakes)	3,700
Salad oil	9,200	Potatoes	700
Butter	7,000	Green vegetables	300
Margarine	7,000	Fruits	600
Peanut butter	6,300	Soda pop	400
American cheese	3,700	Vinegar	130
Chocolate	5,000	Water	0
Potato chips	5,700		

unit-of-mass basis. A big molecule like isooctane, C_8H_{18}, has an unfair advantage over a small one like methane, CH_4, because a mole of the larger molecule contains more mass. Table 5.1 compares the energy liberated during the combustion of some common foods and fuels on a per-gram basis. The net energy change is about the same whether the reaction proceeds directly by burning in air or through a complex biological process. One gram of butter liberates about the same amount of energy when metabolized in the body as when melted and burned in a frying pan. This table shows why natural gas (chiefly methane) makes an excellent energy source and it shows the exceptional energy stored in the H_2 molecule in hydrogen gas, thereby explaining why the *Hindenburg* burned so violently. Table 5.1 also reveals why fats in the diet, such as butter, can easily provide more mass than can be metabolized. When that occurs, body fat accumulates. A 175-pound male needs about 2500 kcal, and a 120-pound female about 2000 kcal, to meet the body's daily energy requirement.

Rocket Fuels

A rocket needs a tremendous amount of energy to escape the Earth's gravitational field (Figure 5.5). This constitutes the major barrier to space exploration: Most of a rocket's weight consists of fuel. Payloads must be small, and expensive, lightweight materials must be used throughout. The possible hazards associated with an errant rocket eliminate the option of nuclear–powered rocket engines. All modern rockets rely on solid or liquid chemicals for propulsion. Gases do not provide a dense enough energy source to be useful and are always liquefied first. Solid rocket fuels offer the advantage of simplicity in design but suffer the drawback that they burn completely, with little control possible, after ignition. Solid rocket fuels also operate just below the explosion threshold and severely stress the motors designed to contain them. A blown O-ring seal in one of the two solid rocket motors on the *Challenger* space shuttle led to the rupture of an adjacent liquid hydrogen fuel tank and a catastrophic explosion. Astronauts ride atop an awesome pile of chemical energy to reach outer space.

Figure 5.5. *The redox reactions that power the engines of the space shuttle release tremendous chemical energy. (Courtesy NASA, National Aeronautics and Space Administration.)*

All rocket fuels rely on redox reactions to provide energy for thrust. The goal in the design of rocket fuels is to obtain a reaction with maximal energy generation that is just short of an explosion. Because rockets must also operate outside the Earth's atmosphere, where oxygen is not available, their fuel must include both a reductant and an oxidant. Solid rocket fuels consist of an intimate mixture of oxidant and reductant in a single chemical. The U.S. space shuttle uses such a fuel that consists mainly of the ionic salt ammonium perchlorate (Figure 5.6). The ammonium cation is the reductant, and the perchlorate anion is the oxidant. Ammonium perchlorate is widely used as a rocket propellant. Solid propellants contain a polymer binder mixed with ammonium perchlorate to hold the powder in a rigid form and dilute it below the explosion threshold. The mixture is about 70% ammonium perchlorate, 16% aluminum powder, and 14% an organic polymer binder. A trace of iron catalyst also is added to speed up reactions in the ignited propellant. The Sidewinder air-to-air missile, the Minuteman and MX ICBMs, and the Trident submarine-launched missiles (including the Tomahawk Cruise missile) all use an ammonium-perchlorate–based propellant. The former Soviet Union's SS-20 series ICBMs and LRBMs used a similar solid propellant. Other solid-fuel weapons include the hand-held Stinger antiaircraft missile; the Phoenix, Harpoon, HARM, Hawk, Patriot, Sparrow, Maverick, and AMRAAM missiles; and the Delta II and Titan IV rockets. All rely on ammonium perchlorate for energy.

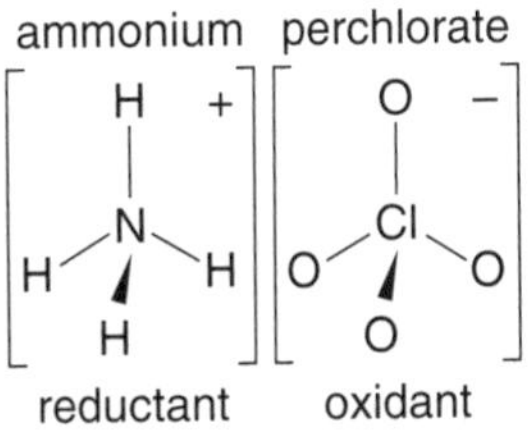

ammonium perchlorate

reductant   oxidant

Figure 5.6. *Molecular structure of a solid rocket fuel, ammonium perchlorate.*

Ammonium perchlorate exists as a white crystalline salt at room temperature, but it may explode on heating. Eight million pounds of ammonium perchlorate accidentally exploded at a plant that made fuel for the space shuttle in Henderson, Nevada, on May 4, 1988. A mushroom cloud visible for miles formed, and even a jetliner cruising near 10,000 feet felt the shock wave from the explosion. The event registered 3.5 on the Richter scale on seismographs in California. Only two people were killed, but over 350 residents of the small town nearby suffered injuries. Damage from the blast extended as far as 20 miles away!

Rapid decomposition of ammonium perchlorate occurs according to equation (10).

$$2NH_4ClO_{4(s)} \rightarrow N_{2(g)} + Cl_{2(g)} + 2O_{2(g)} + 4H_2O_{(g)} \qquad (10)$$

Ammonium perchlorate gives off hot gases (nitrogen, chlorine, oxygen, and water) when it decomposes.

The $Cl_{2(g)}$ and $O_{2(g)}$ products of equation (10) are also reactive oxidants. The organic binder may combine with the excess oxygen to form additional gases (CO_2, CO, and water). To take full advantage of the oxidants, finely divided aluminum metal powder (a reductant) is added to improve the propellant's efficiency. The aluminum metal reacts with the chlorine and oxygen gases as described in equations (11) and (12), and the heat released by these reactions provides additional thrust for the rocket. The more energetic (fast-moving) the escaping gas molecules, the more thrust the rocket motor develops.

$$4Al_{(s)} + 3O_{2(g)} \rightarrow 2Al_2O_{3(s)} + heat \qquad (11)$$

Aluminum (reductant) reacts with oxygen (oxidant)
to form aluminum oxide and give off heat.

$$2Al_{(s)} + 3Cl_{2(g)} \rightarrow 2AlCl_{3(g)} + heat \qquad (12)$$

Aluminum (reductant) reacts with chlorine (oxidant)
to form aluminum chloride and give off heat.

In reality, the chemistry isn't as clean as these reactions suggest. Hot, expanding gaseous products (mainly H_2O and N_2, as well as some NO and HCl) escape from the rocket motor. Each space shuttle launch uses about 1.5 million pounds of solid rocket propellant. Together the U.S. space and military programs require over 60 million pounds of ammonium perchlorate per year.

Liquid rocket fuels consist of separate liquid oxidants and reductants (Table 5.2). Both liquid components mix in the rocket nozzle, where they vaporize and react. Early ICBMs, such as the U.S. Titan II and the Soviet SS-18 and SS-19, used liquid

Table 5.2.   *Some Liquid Oxidant/Liquid Reductant Combination Rocket Fuels.*

Oxidant	*Reductant**	*Use*
Oxygen, O_2	Hydrogen, H_2	Space shuttle, Saturn V (stage II and III)
Oxygen, O_2	Ammonia, NH_3	X-15 rocket plane
Oxygen, O_2	Kerosene (RP-1)	Titan I ICBM, Atlas, Saturn V (stage I)
Dinitrogen tetraoxide, N_2O_4	50/50 Hydrazine/UDMH	Titan II ICBM and Lunar Module
N_2O_4	MMH	Apollo Main Module
85% HNO_3, 15% N_2O_4	80% Diethylenetriamine, 20% UDMH	Air-to-air missiles

*UDMH stands for unsymmetrical dimethyl hydrazine, $H_2N=N(CH_3)_2$. MMH stands for monomethyl hydrazine, $H_2N=NH(CH_3)$.

rocket fuels. Reductant fuels tend to be chemicals rich in hydrogen. Oxidants tend to be chemicals rich in oxygen atoms. The reductants used in liquid rocket fuels are hydrogen, hydrocarbons, or molecules rich in N–H bonds. Liquid oxygen, an energetic and inexpensive oxidant, boils at –218.9°C (–362°F). Liquid hydrogen, the best reductant, boils at an even lower temperature, –252.8°C (–423°F). Liquid hydrogen and oxygen must be loaded from refrigerated containers, which limits their application to rockets that are fueled just before launch. The N_2O_4 liquid oxidant boils at 21.2°C (70°F) and can be stored for long periods in closed containers. Along with nitric acid mixtures, N_2O_4 is used as a liquid oxidant in applications where long-term storage is necessary.

Explosions

An explosion occurs when a highly exothermic redox reaction proceeds rapidly and generates hot gaseous products. These superheated gases expand against the surrounding atmosphere, and the accompanying shock wave sounds like a plane breaking the sound barrier. Several kinds of explosions can occur. If a volatile organic solvent like gasoline or an organic gas such as methane mixes with air, then the mixture may undergo explosive combustion upon ignition. That is why organic vapors should not be allowed to build up in an enclosed area and why gas pipeline leaks are so dangerous. A series of explosions devastated Guadalajara, Mexico, in 1992 when natural gas and hydrocarbon vapors formed an explosive mixture with air in the city's sewer system. Explosions in illicit drug laboratories usually originate with fumes from a volatile cocaine-processing solvent, ethyl ether. The structure of ethyl ether is shown in Figure 5.7.

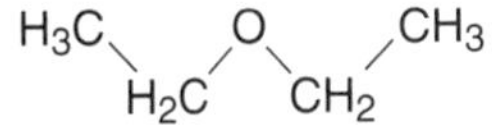

Figure 5.7. *Ethyl ether, a volatile solvent whose vapors form explosive mixtures with the oxygen in air.*

The oxidizing ability of the oxygen in air creates other explosion hazards. Finely divided dust with a high carbon and hydrogen atom content behaves as a reducing agent much like liquid hydrocarbons. Metal powders also are reducing agents that combine readily with oxygen. Powdered coal, flour, metals, starch, and grain all burn vigorously in the presence of oxygen. Finely divided reductants of this sort suspended in air are extremely dangerous, as evidenced by the tremendous explosions that still occur from time to time in grain elevators.

Designed explosives, on the other hand, place oxidant and reductant in intimate contact so that once reaction begins, it continues rapidly enough to detonate. Gunpowder (or black powder), the earliest solid explosive discovered (about A.D. 1200), consists of a 75:15:10 mixture of white saltpeter (potassium nitrate, KNO_3), black powdered charcoal, and yellow powdered sulfur. Saltpeter acts as the oxidant. The reductants charcoal (essentially elemental carbon) and sulfur become oxidized to form gaseous CO, CO_2, and SO_2. The sulfur smell of firecrackers and other fireworks comes from the acrid SO_2 vapor (it is a toxic lung irritant). Modern fireworks use black powder as both the propellant and explosive charge. Commercial fireworks displays often begin with bright flashes of white light and deep booming explosions from aerial bombs. Their explosive charge differs and consists of finely divided magnesium or aluminum metal as the reductant and potassium perchlorate ($KClO_4$) as the oxidant. Temperatures in the explosive flash may exceed 3000°C. The extremely high energy acquired by the vaporized metal atoms in the explosion causes them to emit intense light.

The common household safety match is in one sense a miniature explosive. It consists of potassium chlorate ($KClO_3$) powder as the oxidant mixed with a sulfur–glue–dye reductant. The redox reaction follows equation (13) to emit foul-smelling sulfur dioxide.

$$2KClO_{3(s)} + 3S_{(s)} \rightarrow 2KCl_{(s)} + 3SO_{2(g)} \qquad (13)$$

Potassium chlorate reacts with sulfur to form
potassium chloride and gaseous sulfur dioxide.

Ignition procedures vary with match construction. The tips of the "strike anywhere" wooden matches contain P_4S_3, a compound that ignites when heated by the friction produced by rapidly drawing it across a rough surface. In safety matches, the striking surface contains some red phosphorus mixed with glue and ground glass. When the match is dragged along this surface, it ignites.

Texas City

At the end of World War II, the U.S. government began shipments to Europe of fertilizer-grade ammonium nitrate (FGAN). The great war had devastated European agriculture, and the hungry survivors desperately needed chemical fertilizers to increase food production. Although ammonium nitrate (NH_4NO_3) was the active ingredient in a number of blasting agents, FGAN was thought to be nearly impossible to detonate. It consisted of ammonium nitrate granules mixed with 3.5% inert clay and coated with wax to prevent caking. It was ironic that a blasting agent that had spread death in World War I would help nourish the continent after World War II. Such might have been the thoughts of a chemist watching the loading of the French freighter *Grandcamp.*

For five days the ship had been moored at slip 2 at the expansive docks of Texas City, Texas, just across the bay from Galveston. She contained cotton, peanuts, and oil-drilling equipment. Early on the morning of April 16, 1947, the first officer walked among the drill pipe secured topside. Soon they would be ready to set to sea. At 8 A.M. a working detail of longshoremen arrived to finish loading 2280 tons of seemingly innocuous bags of FGAN into the ship's hold. The crew oversaw the effort, with an occasional pause to light a cigarette.

Not only did the boom city of 15,000 boast a thriving port, but Monsanto had built a large chemical plant there during the war to produce styrene for artificial rubber. With the loss of access to rubber tree plantations in the Pacific, the synthetic rubber kept Allied vehicles rolling throughout the war. Texas City was also an important terminal for the state's oil industry. The Atlantic, Humble, and Petrotex refineries and Stone Oil, Republic Oil, and Sid Richardson plants surrounded the docks, as did the tank farms of the Carbide and Carbon division of Union Carbide. Texas City Terminal Railways serviced a convenient link among industry, the docks, and the Santa Fe Railroad line. Cotton, sulfur, grain, and tin products poured through its rail yards. As Fred Brumley and John Norris flew above the city in their small plane, they couldn't help but marvel at the raw economic power laid out below them.

Around 8:30 A.M. a cry of "Fire!" rang out from the No. 4 hold of the *Grandcamp.* With portable fire extinguishers, the crew was able to snuff the flames momentarily, but the fire rekindled itself. The first officer, fearing damage to the paper-wrapped cargo in the ship's hold, ordered no water when the smoke returned. Instead, the hatch and vents were shut to deny the fire oxygen, and steam was released into the cargo area to help dampen the fire. Several hundred onlookers gathered across the docks to watch the excitement. At 9:12 A.M., just as the Texas City Fire Department seemed to have the fire under control, an explosion erupted. Within a second the *Grandcamp* disappeared. Lethal pieces of metal rained down on the dock, and some fragments landed 4 miles from the blast site. Two miles away, falling debris killed an unlucky couple in their car. Fred Brumley

Texas City after the Grandcamp *explodes. (AP/Wide World Photos.)*

and John Norris's plane was literally blown from the sky and crashed near the blast site. Sea birds rained down on Texas City—similar victims of the airborne shock wave. All over town, roofs were punched in by the blast. Windows blew out of buildings 11 miles across the bay at Galveston.

Industries surrounding the docks suffered severely. Giant oil storage tanks crumpled from the force of the blast, their contents ignited by white-hot metal fragments. Tanks further away, some 50 in all, were shot full of holes. Drill pipe fragments launched from the *Grandcamp's* deck became red-hot missiles. Hot, expanding gases in the fireball from the explosion even swept a steel oil barge docked near the *Grandcamp* up onto the street. The charred remains of a fire truck came to rest precariously atop it. The Monsanto chemical plant, a mere 500 yards away, was flattened. Seventy percent of the chemists and chemical engineers in the plant perished. Deadly chlorine gas leaked from storage containers at the plant, and explosions of volatile solvent vapors filled the air with waves of fire.

Early rescuers arrived to find the streets littered with dozens of bodies. Dazed survivors of the concussion stumbled out of buildings, blood streaming from their noses and ears. They were fortunate compared to those left screaming, beyond help, in the fiery Monsanto plant. Bloody tracks trailed through the streets near the docks, a graphic map of injuries from flying metal, glass, and concrete. Black smoke from the 2-mile-wide scorched area rose as high as 4000 feet in the air. But the carnage had not ended. A freighter two docks away from the *Grandcamp*, loaded with 960 tons of ammonium nitrate, caught fire. Aptly named, the *SS High Flyer* blew up 13 hours after the initial explosion to extract an additional toll on would-be rescuers. The Pan American Refining Corporation joined the conflagration. A once-prosperous industrial city lay in ruin.

President Truman immediately pledged federal aid and offered the nation's condolences. The Red Cross dispatched 20 disaster workers and appropriated $250,000 for emergency aid. Two hundred and forty nurses from the Galveston chapter of the Red Cross were already at work under battlefield conditions. Striking Southwestern Bell Telephone employees returned to their jobs to provide emergency communication and repair service. Blood plasma was flown in from stockpiles throughout the nation. General Jonathan M. Wainwright, a veteran of the war in the Pacific, arrived to coordinate the Army's relief effort. He was quoted as saying, "I have never seen a greater tragedy in all my experiences." Hal Boyle, war correspondent for the Associated Press, began his news account, "In four years of war coverage I have seen no concentrated devastation so utter, except Nagasaki, Japan, victim of the second atom bomb, as presented today by flaming Texas City." *The New York Times* quoted Senator W. Lee O'Daniel of Texas, who introduced a resolution calling for a Senate investigation: "I realize that this series of explosions could have been accidental, and I do not want to charge or insinuate that anybody could have been so mean or debased as to have planned and executed this disaster. But in view of the high tension on international affairs, and due to a rather large number of fires, explosions, railroad wrecks, and other disastrous occurrences in this nation lately, all so nearly resembling disastrous occurrences which preceded our entry into the last war, due largely to communistic underground activity, I believe it is the duty of this Senate to conduct a full and complete investigation into this Texas City disaster and to start the investigation immediately."

By Saturday, April 19, the fires were at last contained, but the grim hunt for casualties continued. The exact cause of the explosion was never fully understood, but thereafter FGAN was handled with far greater caution. When the smoke cleared, there were over 600 fatalities and 300 seriously injured. The faith of the nation in its industrial economy had been shaken to its foundation.

The frame of the burned Monsanto plant. (Courtesy of the Texas City Sun.*)*

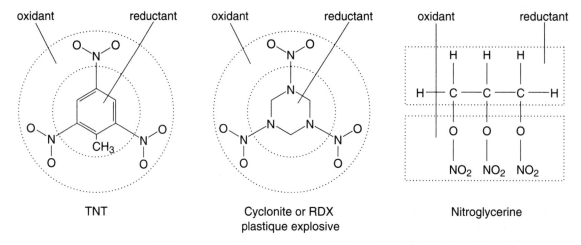

Figure 5.8. *Three common high explosives. Each molecule contains separate regions that act as oxidant and reductant.*

Military high explosives contain nitro (NO_2) or nitrate (NO_3) oxidant groups directly bound to a hydrocarbon fragment that acts as the reductant. Because the explosive reaction occurs within each molecule, the speed of reaction is remarkably fast, which causes a very powerful explosion. The structures of several explosives are shown in Figure 5.8. The parts of the molecule corresponding to the oxidant and reductant are highlighted.

We cannot overemphasize the dangers associated with explosives, including legal fireworks. All these materials contain an unstable mixture of an oxidant and a reductant, just waiting for the opportunity to explode. Sometimes a spark of static electricity sets them off, with tragic consequences to those nearby. Amateur pyromaniacs are often severely injured while mixing up homemade gunpowder or packing it into a container. The du Pont black powder mills of the 1800s consisted of concrete block houses with a weak roof facing the Brandywine River, a layout designed to direct the force of the inevitable accidental blasts upward over an uninhabited area. Special hazards exist in the manufacturing of many explosives. Just the friction and heat of mixing, or a spark from a metal tool, can detonate the mixture.

Why Some Reactions Are Fast and Others Slow

How long does it take for a chemical reaction to occur? Why do some energetically favorable reactions proceed with explosive violence and others remain quiescent for ages? The speed of a reaction is called the *reaction rate*. Just as the speed of a car is measured in miles per hour, chemical reactions have a speed or rate that reflects the amount of substance reacted in a particular length of time.

It is important to understand that the speed of a chemical reaction does not necessarily reflect the overall energy change that occurs in the reaction. For example, the oil in salad dressing can burn in air and emit a large amount of thermal energy. At normal temperatures, this favorable process does not occur. Heated salad oil will, however, begin to smoke and then catch fire. After ignition it continues to burn, even if the heat source is removed. Why? Most chemical reactions, even favorable ones, have an energy barrier they must overcome on their way to products. Chemists call this energy barrier the reaction's *activation energy*. It arises because molecules may have to break or stretch a bond before the reaction can proceed, which calls for an initial input of "start-up" energy.

Figure 5.9 depicts the energy changes for three reactions: (A) an energetically downhill reaction with no activation energy barrier, (B) an energetically downhill reaction with a small activation energy barrier, and (C) an energetically downhill reaction with a large activation energy barrier. The vertical axis represents the energy content of all the atoms in the reactants that occur along the path to products. The horizontal axis represents the path that atoms follow to make products. In case A, the reaction proceeds instantly. In case B, the reaction occurs at some measurable rate. In case C, the reaction proceeds so slowly that it would not be perceptible in a human lifetime.

The chemistry of everyday life depends on reaction rates. Heat energy usually provides the "kick" that gets molecules over activation energy barriers. Once a molecule crosses the activation barrier, it rides downhill and liberates the energy it needed to get over the barrier, in addition to the energy difference

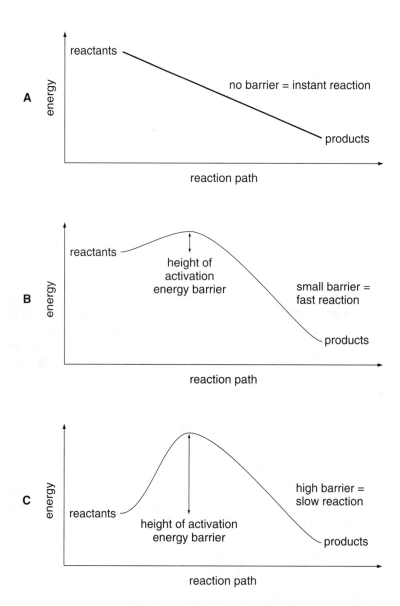

Figure 5.9.   *Diagrams showing reactions that exhibit the same change in energy between reactants and products but that have different activation energy barriers. In reaction A there is no barrier, in reaction B there is a small barrier, and in reaction C there is a large barrier.*

between reactants and products. For exothermic reactions (such as the combustion of gasoline), the emitted energy pushes several other molecules over the activation energy barrier. These molecules in turn release more energy, which sustains the reaction once it is started.

Increased temperatures provide greater thermal energy, which causes an increased rate of reaction. Chefs rely on heat to speed up chemical reactions in food. Baking a cake or heating a roast initiates reactions in the food that form more tasty chemicals. These reactions occur faster at higher temperatures. A rough rule is that a reaction rate doubles for every 10°C (about 20°F) increase in temperature. This rule doesn't always apply directly to cooking, because heat conduction to the center of a cake or roast plays a role as well. An area of cooking where it does apply is in steaming or boiling foods with water. This is why we have to cook food longer at higher altitudes. At higher elevations, atmospheric pressure is less than at sea level. The temperature at which the pressure of the water vapor above the heated water equals atmospheric pressure (the temperature at which the heated water boils) lies lower. At an elevation of 10,000 feet, the atmospheric pressure amounts to only two-thirds that at sea level. Water boils at 89°C (192°F) at this altitude, instead of the 100°C (218°F) at sea level. According to our rule, it should take twice as long to cook hard-boiled eggs or to steam carrots at an elevation of 10,000 feet. Check it out the next time you're in the mountains.

Converting Chemical Energy to Electrical Energy

A battery uses a redox reaction to provide electrical energy. All batteries work by physically separating an oxidant from a reductant so that they cannot combine directly. The device to be powered is connected by wires to the separated oxidant and reductant. The wires define the path by which the electrons travel from the reductant to the oxidant. To complete the electrical circuit, there needs to be another conducting path, inside the battery, between the separated oxidant and reductant. A conducting ionic solution or paste usually serves this purpose. These key

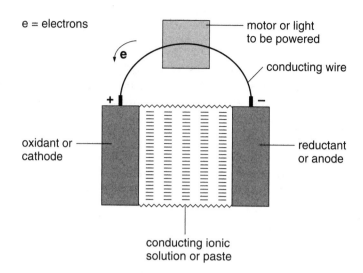

e = electrons

motor or light
to be powered

conducting wire

+

−

oxidant or
cathode

reductant
or anode

conducting ionic
solution or paste

Figure 5.10.
Schematic of a bat-
tery-powered electri-
cal circuit.

features of a battery-powered electrical circuit appear in Figure 5.10. The oxidant or positive electrode of a battery is called the cathode, and the reductant or negative electrode the anode. Most batteries rely on a reactive metal as the anode, which oxidizes to form an ionic compound as current flows.

The anode of the lead–acid storage battery used in automobiles consists of metallic lead alloyed with about 3% arsenic and antimony. The cathode is solid lead dioxide, PbO_2. Both are immersed in sulfuric acid, H_2SO_4, which provides two H^+ and one SO_4^{2-} to conduct electricity inside the battery. The net reaction follows equation (14).

$$Pb_{(s)} + 2H_2SO_{4(l)} + PbO_{2(s)} \rightarrow$$
$$2PbSO_{4(s)} + 2H_2O_{(l)} + 2.0 \text{ volts} \qquad (14)$$

Lead (reductant), sulfuric acid (electrolyte), and lead oxide (oxidant) react to produce voltage and the discharged battery products lead sulfate and water.

The voltage defines the cell's electrical energy output. Battery voltages can be increased by connecting several anodes and cathodes in series. The net voltage equals the sum of the anode–cathode pairs connected. A 12-volt car battery contains six 2-volt cells connected in series. Another important consideration is how much current (measured in amps) the battery provides. This depends on the size of the battery. The maximum available battery current increases as the electrode surface area increases. The power (energy output per second) obtainable from a battery is given by the product of the voltage and the current, measured in watts.

Equation (14) shows that as the reaction proceeds, the sulfuric acid (H_2SO_4) liquid is used up and liquid water forms. This provides a method for determining a battery's state of charge. The density of pure sulfuric acid is 1.84 grams for each cubic centimeter of liquid (g/cm^3). The corresponding density for water is only 1.00 g/cm^3. Mechanics determine the state of charge of a car battery by measuring the density of liquid in the battery with a device called a hygrometer. The density of battery liquid decreases as the battery discharges. Because sulfuric acid is corrosive, it is essential to wear protective goggles when working around car batteries or when checking the fluid level in the six cells.

Batteries eventually exhaust all their reactants during use and run down. Some batteries can be recharged. Recharging reverses the direction of chemical change by applying electricity from another source. The energy (voltage) applied always exceeds what the battery puts out when it is used. In a car battery, the chemical reaction for charging is the reverse of equation (14), as shown in equation (15).

$$2PbSO_{4(s)} + 2H_2O_{(l)} + >2.0 \text{ volts} \rightarrow$$
$$Pb_{(s)} + PbO_{2(s)} + 2H_2SO_{4(l)} \qquad (15)$$

Lead sulfate reacts with water when electrical energy greater than 2 volts is applied to produce lead, lead dioxide, and sulfuric acid.

The source of electricity to charge a car battery is the electrical generator (or alternator) run by a fan belt from the engine. This maintains the battery in a highly charged state. Car batteries eventually die when solid Pb, PbO_2, and $PbSO_4$ reactants flake off the electrodes after repeated recharge cycles. When the sludge at the bottom of the battery becomes deep enough to touch both electrodes, there is a short circuit. The only solution is to buy a new battery.

A dangerous situation may develop if a damaged battery becomes overcharged. The electrical energy converts sulfuric acid to hydrogen by the process shown in equation (16). In this equation, $2e^-$ stands for the two electrons that are pumped into the cathode by the car's alternator.

$$H_2SO_{4(l)} + 2e^- \rightarrow H_{2(g)} + SO_4{}^{2-}{}_{(aq)} \qquad (16)$$

Sulfuric acid reacts with two electrons to form hydrogen gas and solvated sulfate anions.

Hydrogen gas can explode when mixed with oxygen from the air. If this occurs in the battery, it sprays acid on anyone nearby. That is why instructions for jump-starting a car that has a dead battery should be followed carefully. If the electrical cables spark near the battery in either car, and if the battery near the spark contains hydrogen, then it may explode with tragic consequences.

The inexpensive, nonrechargeable flashlight battery, or dry cell, uses a zinc metal anode and a manganese dioxide (MnO_2) cathode. The chemistry of this battery involves the reaction shown in equation (17).

$$Zn_{(s)} + 2H^+ + 2MnO_{2(s)} \rightarrow$$
$$Zn^{2+} + Mn_2O_{3(s)} + H_2O + 1.5 \text{ volts} \qquad (17)$$

Zinc (reductant), protons (electrolyte), and manganese dioxide (oxidant) react to generate voltage and the discharged battery products (zinc cations, manganese oxide, and water).

Constructing a practical device based on this reaction requires taking into account some special considerations. As shown in Figure 5.11, the battery is composed of an outer can of zinc metal separated by a moist cardboard liner from a paste that contains the MnO_2 oxidant. The center rod of the battery consists of electrically conducting carbon, which makes electrical contact with the MnO_2 paste cathode. These batteries should never be recharged. Zinc metal crystals develop upon recharging and pierce the cardboard separator to cause a short circuit. Also, the generation of hydrogen from overcharging a sealed battery could build up enough pressure to cause it to explode.

The alkaline cell involves a design modification of the dry cell that increases the battery's shelf life. The anode and cathode

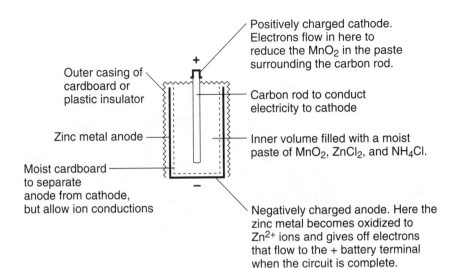

Positively charged cathode. Electrons flow in here to reduce the MnO_2 in the paste surrounding the carbon rod.

Outer casing of cardboard or plastic insulator

Carbon rod to conduct electricity to cathode

Zinc metal anode

Inner volume filled with a moist paste of MnO_2, $ZnCl_2$, and NH_4Cl.

Moist cardboard to separate anode from cathode, but allow ion conductions

Negatively charged anode. Here the zinc metal becomes oxidized to Zn^{2+} ions and gives off electrons that flow to the + battery terminal when the circuit is complete.

Figure 5.11. *Schematic of a dry-cell battery.*

materials are identical with those in the dry cell. The difference between the two batteries is that the alkaline cell contains potassium hydroxide, KOH (a basic, or *alkaline*, substance), which provides ions for conduction in the paste near the zinc electrode. The alkaline cell also puts out 1.5 volts. Like the dry cell, the alkaline cell is not a rechargeable battery.

The rechargeable NiCad, or nickel–cadmium, battery provides nearly the same voltage output as the preceding two cells. It relies on cadmium metal, Cd, as the electron-donating anode and on basic nickel oxide, NiO(OH), as the electron-accepting cathode. These metals cost more than Zn and MnO_2, so such batteries are expensive. The net reaction of the NiCad battery is given in equation (18).

$$Cd_{(s)} + 2NiO(OH)_{(s)} + 2H_2O_{(l)} \rightarrow$$
$$Cd(OH)_{2(s)} + 2Ni(OH)_{2(s)} + 1.4 \text{ volts} \qquad (18)$$

Cadmium (reductant), basic nickel oxide (oxidant), and water react to generate voltage and the discharged battery products cadmium hydroxide and nickel hydroxide.

The solid reactants and products remain attached to the electrodes, and the reaction can be readily reversed upon recharging. Cadmium is a toxic heavy metal. Nickel oxide may be carcinogenic.

Figure 5.12. *The 1995 Solectria* Force *Sedan has rechargeable nickel cadmium batteries. The left photo shows the battery pack under the front hood, and the right, the exterior. The car has a top speed of 75 mph and a 100-mile range. The Solectria* Sunrise, *planned for production in 1997, will employ Ovonic nickel metal hydride batteries and achieve a 200-mile range. (Left, David L. Ryan/The Boston Globe; right, courtesy of Solectria Corporation, Wilmington, MA.)*

Mercury button cells find use in specialty applications, such as watches and hearing aids. Mercury cells have a Zn/Hg alloy anode and a HgO/C paste cathode separated by a KOH-moistened separator. They exploit the reaction shown in equation (19).

$$HgO_{(s)} + Zn_{(s)} \rightarrow ZnO_{(s)} + Hg_{(l)} + 1.35 \text{ volts} \qquad (19)$$

Mercuric oxide (oxidant) reacts with zinc (reductant) to generate voltage and the discharged battery products zinc oxide and mercury.

Considering the toxicity of the components, it would make sense to develop recycling programs for both the NiCad battery and the mercury battery.

In addition to these common batteries, the search continues for a battery with a high ratio of energy to density for electric automobiles (Figure 5.12). California has adopted legislation requiring that 10% of all vehicles sold have zero pollution emission by the year 2003. Conventional lead–acid car batteries store too little energy for their weight and bulk to be practical for powering long-range vehicles. One possible lightweight battery uses liquid sodium (reductant) and liquid sulfur (oxidant) separated by a solid alumina (β-Al_2O_3) electrolyte. Alumina has the unique property of allowing Na^+ ions to move through it and conduct electricity at elevated temperatures. Before this battery will be practical, however, the problems of keeping it at a high temperature (300°C, or 572°F) and making a "fail-safe" case for the highly reactive chemicals must be solved with inexpensive technology. A nickel hydride battery has also been proposed for this application; it is currently used as a rechargeable power source in laptop computers. Without an advance in battery technology, the first generation of electric cars will use improved lead–acid batteries for vehicles of limited range.

Electroplating: Using Electrical Energy to Cause a Redox Reaction

The use of electricity to carry out chemical changes is termed *electrolysis.* Recharging a battery uses electrical energy to make a chemical reaction occur in such a way that energy is stored. It is one example of electrolysis. Another example occurs in the electroplating of an iron bar. Electroplating involves immersing the iron bar in a solution that contains cations of a metal to be plated on the surface of the iron. Common ions used for this purpose include $Cr^{3+}_{(aq)}$ (chromium plating), $Ag^+_{(aq)}$ (silver plating), and $Zn^{2+}_{(aq)}$ (galvanizing).

In electroplating, the anode of an electrical power supply is connected to the iron bar to charge it negative. The cathode of the power supply is connected to a second electrode in the solution, which becomes positively charged. This completes the circuit. The power supply forces electrons into the iron bar, and these electrons are transferred to the positive ions ($Cr^{3+}_{(aq)}$ or $Zn^{2+}_{(aq)}$) in solution, causing them to plate out on the iron surface as neutral Cr or Zn metal. Usually, only a very thin coating of an expensive metal is necessary to protect the surface of an inexpensive metal. An atomic layer of gold coated onto the surface of a brick costs less than a penny. The price of a solid gold brick, on the other hand, is about $150,000!

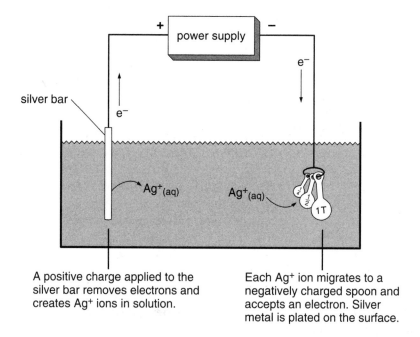

Figure 5.13.
*Schematic of the
electrolysis cell used
for electroplating sil-
ver on spoons made
of a less expensive
metal.*

A positive charge applied to the
silver bar removes electrons and
creates Ag⁺ ions in solution.

Each Ag⁺ ion migrates to a
negatively charged spoon and
accepts an electron. Silver
metal is plated on the surface.

Another application of electrolysis is silver plating. The procedure for electroplating a silver spoon is sketched in Figure
5.13. Here electrical work strips electrons away from a bar of silver metal, which causes it to dissolve as cations in solution.
These silver ions move through the solution to the other electrode (the object to be plated with silver), where they accept
electrons and plate out. In industrial silver plating, cyanide ion
is added as the salt sodium cyanide (NaCN). The NaCN dissolves
in solution to form $Na^+_{(aq)}$ and $CN^-_{(aq)}$. These ions increase the
conductivity of the solution and also help stabilize the dissolved
silver ions.

Acid–Base Reactions

Acids

Common acids behave as hydrogen ion donors in chemical reactions. In water solution, the hydrogen ion, $H^+_{(aq)}$, consists of a
bare proton solvated by polar water molecules. The (aq) subscript signifies a water, or aqueous, solution. Many acidic com-

pounds contain an O–H group in their structures. Figure 5.14 shows the molecular structures of some common acids. When these acids are added to water, the O–H group ionizes to produce $H^+_{(aq)}$. The electron that belongs to the hydrogen atom remains behind on the oxygen, giving it a negative charge. For example, when dissolved in water, nitric acid (HNO_3) dissociates to produce $H^+_{(aq)}$ and the nitrate ion, $NO_3^-_{(aq)}$. The acidic functions of the acids shown in Figure 5.14 are enclosed in boxes. Citric acid and phosphoric acid contain several acidic hydrogens. These molecules can provide more than one $H^+_{(aq)}$. Phosphoric acid can ionize in water to provide three $H^+_{(aq)}$, leaving behind the phosphate anion, PO_4^{3-}. Phosphoric acid is an acidic flavor component in many soft drinks.

Uric acid occurs in the urine of all carnivores, as well as in the excrement of birds and reptiles. When a proton is lost from uric acid, the low solubility of urate salts (with Na^+ as the other ion) causes it to precipitate. If excessive levels of urate build up in the blood of humans, it crystallizes out in the joints to cause a painful condition known as gout.

The acids shown in Figure 5.14 exhibit a wide range of acid strength. Sulfuric, nitric, and hydrochloric acids are strong industrial acids that can seriously burn the skin. Acids have a characteristic sour taste, and many foods and drinks owe their tart flavor to traces of mild acids. Acetic acid is a weak acid in water: A 5% solution of acetic acid in water is called vinegar, which is used as a salad dressing. And the sour taste of spoiled wine is caused by acetic acid. Lactic acid, another weak acid, builds up in the blood and muscles as a by-product of metabolism during vigorous exercise, as oxygen levels in the cells decrease. The acid buildup leads to muscle fatigue, which prompts the organism to rest until oxygen levels return to normal. Lactic acid also occurs in sour milk, beer, molasses, apples, and wine. The malolactic fermentation used in the production of "buttery" chardonnay wines converts malic acid into lactic acid (Figure 5.15). Citric acid occurs naturally in lemons, oranges, and tomatoes; it gives them their sour taste.

Acids are classified as strong or weak by their ionization behavior in water. Strong acids dissociate (ionize) completely,

172

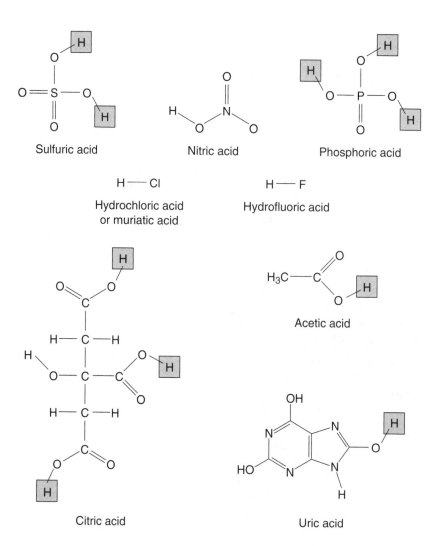

Figure 5.14. *The molecular structures of several common acids. For molecules that contain more than one hydrogen atom, the boxes highlight the hydrogen atoms responsible for their acidity.*

similar to ionic salts dissolving in water. Equation (20) represents the change that occurs when HCl dissolves in water.

$$HCl_{(g)} \xrightarrow{H_2O} HCl_{(aq)} \rightarrow H^+_{(aq)} + Cl^-_{(aq)} \qquad (20)$$

Gaseous hydrochloric acid dissolves in water and then dissociates completely into a hydrated proton and a hydrated chloride ion.

Gaseous $HCl_{(g)}$ dissolves initially to form an HCl molecule in water $HCl_{(aq)}$. This dissolved molecule then immediately dissociates into a proton and a chloride ion solvated by water molecules.

Weak acids, such as acetic acid, do not dissociate completely in water. The dissociation reaction for acetic acid is shown in
equation (21).

$$H_3CCOOH_{(aq)} \rightleftharpoons H_3CCOO^-_{(aq)} + H^+_{(aq)} \tag{21}$$

Acetic acid partially dissociates in water
to form the acetate ion and a proton.

Note the difference between the arrows used in equations (20)
and (21). The double arrow, $\rightleftharpoons$, is a shorthand notation signify-
ing that a measurable balance between reactants and products
occurs. Such reactions usually involve small changes in energy,
so neither reactants nor products are favored strongly. The reaction proceeds only to the extent that it is energetically favorable.
When the optimal balance between reactants and products is
obtained, the reaction is said to have reached equilibrium. At
equilibrium, the rates of the forward and reverse reactions are
identical. In vinegar, only about 0.4% of the acetic acid molecules dissociate to yield acetate ion and $H^+_{(aq)}$. The other 99.6%
exist as unionized $H_3CCOOH_{(aq)}$ in the water solution.

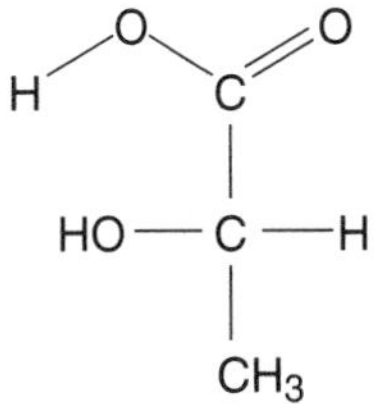

The acidity of a solution is often quantified in terms of pH.
The pH unit is a measure of the concentration of protons in solution. The pH acidity scale is logarithmic, like the Richter scale
for earthquakes. This means that each change of 1 unit in the pH
corresponds to a 10-fold change in the concentration. By convention, a lower number signifies a greater concentration of protons. Pure water itself undergoes a small amount of self-ioniza-
tion, as shown in equation (22). Only about one water molecule
in every billion dissociates in pure water.

$$H_2O_{(l)} \rightleftharpoons H^+_{(aq)} + OH^-_{(aq)} \tag{22}$$

Molecules in liquid water partially dissociate
into protons and hydroxide anions.

The natural pH of water is equal to 7 at 25°C. This value is
defined as that characteristic of a neutral solution. The addition
of acids or bases changes this value. Substances that behave as
acids in water all have a pH less than 7. This reflects the fact that
the solution contains a larger concentration of $H^+_{(aq)}$ than that

Lactic acid

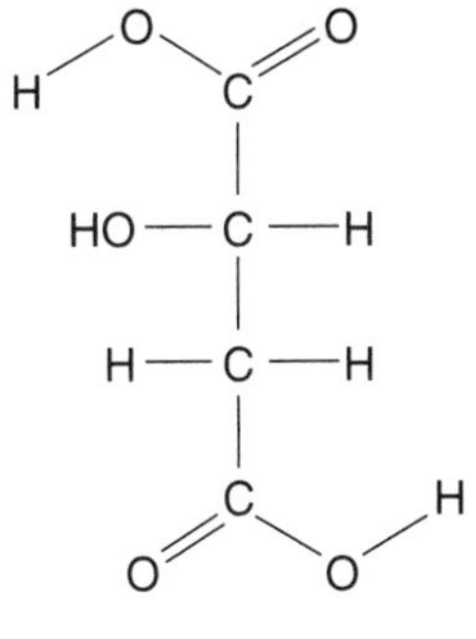

Malic acid

Figure 5.15.
Comparison of the
molecular structures
of the naturally
occurring acids lactic
acid and malic acid.

found in pure water. Solutions regarded as mildly acidic have a pH between 2 and 7. Strongly acidic solutions exhibit a pH between −1 and 2.

Hydrogen cyanide, H−C≡N$_{(g)}$, behaves as a weak acid when dissolved in water, producing only a small amount of CN$^-_{(aq)}$. On the other hand, sodium cyanide salt, NaCN$_{(s)}$, dissolves completely in water to yield Na$^+_{(aq)}$ and CN$^-_{(aq)}$. Both HCN$_{(g)}$ and sodium cyanide solutions contain CN$^-_{(aq)}$, which is deadly to humans. In the "gas chamber" used to administer the death penalty, a pellet of NaCN or KCN is dropped into a container of sulfuric acid, a strong acid. The strong acid protonates the CN$^-_{(aq)}$ and produces deadly HCN gas, which escapes from the water solution.

Bases

Basic water solutions have a bitter taste and a characteristic slippery feel. Soap forms a basic solution in water. So does the free base of cocaine. (How many times have you seen a scene in the movies or on TV in which a drug is identified by its bitter taste?) Quinine, shown in Figure 5.16, is a basic chemical found in the bark of the cinchona tree; it is responsible for the bitter taste of tonic water. In quinine, the two nitrogen atoms serve as proton acceptor (basic) sites.

Chemists define a base as a substance that reacts with H$^+_{(aq)}$, thereby increasing the pH of the solution. Basic compounds are often referred to as alkaline substances. The strongest common bases are metal hydroxides, such as sodium hydroxide (NaOH) and potassium hydroxide (KOH). Sodium hydroxide is the main ingredient in Drano® drain cleaner. Sodium hydroxide ionizes completely in water, as shown in equation (23).

$$NaOH_{(s)} \xrightarrow{H_2O} Na^+_{(aq)} + OH^-_{(aq)} \qquad (23)$$

Sodium hydroxide dissociates completely in water
to form the sodium cation and the hydroxide anion.

The OH$^-$ ion, or hydroxide ion, combines with the proton of any acid it encounters. A neutralization reaction then occurs to produce water. The reaction between sodium hydroxide and either

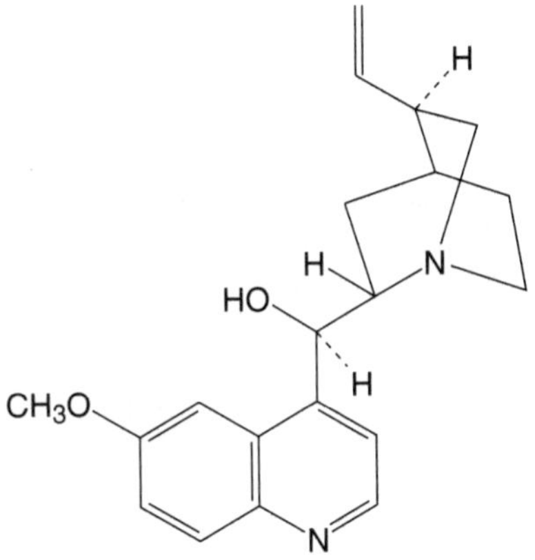

Figure 5.16.
Quinine, the bitter flavoring agent in tonic water and a medication used to treat malaria and muscle cramps.

a weak acid (acetic acid) or a strong acid (hydrochloric acid) proceeds completely as shown in equations (24) and (25).

$$Na^+_{(aq)} + OH^-_{(aq)} + H_3CCOOH_{(aq)} \rightarrow$$
$$Na^+_{(aq)} + H_3CCOO^-_{(aq)} + H_2O_{(l)} \quad (24)$$

*Dissolved sodium hydroxide (base) and acetic acid
(acid) react to form a water solution of sodium acetate.*

$$Na^+_{(aq)} + OH^-_{(aq)} + H^+_{(aq)} + Cl^-_{(aq)} \rightarrow$$
$$Na^+_{(aq)} + Cl^-_{(aq)} + H_2O_{(l)} \quad (25)$$

*Dissolved sodium hydroxide (base) and dissociated hydrogen
chloride (acid) react to form a water solution of table salt.*

In both cases, an aqueous solution of an ionic salt results. Equation (25) shows that mixing equal amounts of the strong base sodium hydroxide and the strong hydrochloric acid (either of which would be deadly) produces a neutral solution of sodium chloride (table salt). This neutralization reaction also evolves considerable heat. The sodium chloride "product solution" may even boil if concentrated solutions are used in the reaction.

Basic compounds cause the pH of water to increase above the value of 7. Weakly basic water solutions have a pH between 8 and 11, whereas strongly basic solutions lie in the range of 12 to 15. The pH values of some common acids and bases are given in

Table 5.3. *The pH of Some Common Foods and Chemicals*

Approximate pH	Substances Near This pH Range
−1	Industrial acids (nitric, sulfuric, muriatic)
0	Toilet bowl cleaners, some bathroom cleansers
1	Stomach acid
2	Lemon and lime juice
3	Cola beverages, wine, vinegar, pickles
4	Carbonated water, tomato juice, beer, acid rain
5	Coffee, bananas, beans, spinach, squash
6	Rainwater, tuna fish, salmon, oysters
7	Pure water, blood, milk, saliva, tears
8	Sea water, egg white
9	Baking soda in water
10	Soaps and detergents, milk of magnesia
11	TSP, trisodium phosphate floor cleaner
12	Ammonia cleaner, lime
13	Household bleach
14	A 1 mole/liter solution of lye or caustic soda
15	Drain cleaners

Table 5.3. With the exception of stomach acid, biological fluids exhibit a nearly neutral pH of 7. Serious damage occurs to cells when their pH changes by more than 0.5 pH unit. Strong acids or bases, substances below a pH of 3 or above a pH of 9, cause chemical burns. Tearless shampoos are made by adding a carefully measured amount of acid to the basic shampoo soaps and detergents until the pH decreases to 7. This resembles the neutralization reactions of equations (24) and (25). A pH 7 shampoo does not greatly irritate the eye, because it now has the same pH as the fluids in the eye.

Serious chemical hazards around the home arise from the acids and bases stored in garages and under kitchen sinks. Everyone has experienced the intense pain from getting soap (about pH 10) in the eyes. Bases react with oils in the skin to produce soaps, which are slimy and difficult to rinse from the eyes and skin. Floor cleaners, ammonia window cleaners, muriatic acid, bleach, and drain cleaners contain strong acids and bases that should never be splashed near the face. A chemist wouldn't think of handling these materials in the laboratory without wearing safety glasses (which are also required by federal regulations). A single drop of a highly acidic or basic home chemical could burn the eye's cornea or even cause blindness. Contact lens wearers face an especially high hazard, because the corrosive substance wets the lens and it becomes trapped between the lens and the cornea.

Applications of Acid–Base Chemistry

Baking

Many baking procedures take advantage of the equilibrium reaction of carbon dioxide and water. When carbon dioxide gas dissolves in water, most of it exists as $CO_{2(aq)}$. The amount present in solution depends on the pressure of gas above the solution. The greater the pressure of $CO_{2(g)}$, the more $CO_{2(aq)}$ forms. This simple reaction is represented by equation (26).

$$CO_{2(g)} \overset{H_2O}{\rightleftharpoons} CO_{2(aq)} \qquad (26)$$

Carbon dioxide gas partially dissolves to form carbonated water.

An increase in the pressure of $CO_{2(g)}$ causes the reaction to respond by shifting to the right. A decrease in the pressure of carbon dioxide above the solution causes the reaction to shift to the left. We observe this effect whenever we open a bottle of soda sealed under a pressure of $CO_{2(g)}$. When the bottle depressurizes, $CO_{2(g)}$ bubbles form in the liquid, as the reaction represented by equation (26) shifts to the left.

After dissolving, some of the $CO_{2(aq)}$ undergoes a reaction with water. This reaction follows equation (27) to form a proton and the bicarbonate ion. Because this reaction generates $H^+_{(aq)}$, highly carbonated beverages are acidic (pH 4).

$$CO_{2(aq)} + H_2O_{(l)} \rightleftharpoons H^+_{(aq)} + HCO_3^-_{(aq)} \qquad (27)$$

Carbon dioxide and water partially react to form a weak acid solution that contains the bicarbonate ion.

Baking soda is the ionic solid sodium bicarbonate $NaHCO_3$. When sodium bicarbonate dissolves in milk- or water-based batters, it dissociates into sodium cations and bicarbonate anions, as shown in equation (28).

$$NaHCO_{3(s)} + H_2O_{(l)} \rightleftharpoons Na^+_{(aq)} + HCO_3^-_{(aq)} \qquad (28)$$

*Sodium bicarbonate dissolves in water to
form sodium cations and bicarbonate unions.*

Baking soda is often required in recipes for baked goods. It works as follows: Once dissociated, the bicarbonate anion undergoes further reaction. As indicated by the reverse of equation (27), $HCO_3^-_{(aq)}$ reacts slowly with protons in the batter (from the lactic acid in milk, for example) to produce dissolved $CO_{2(aq)}$. This in turn must react according to the reverse of equation (26), forming $CO_{2(g)}$ bubbles in the batter. Heating the cake speeds up these reactions and lowers the solubility of CO_2 gas in water, so even more bubbles form with heating. This causes the cake to rise.

Pancakes require a fast-acting ingredient to help them rise, because they cook quickly and have less time to establish the foregoing series of reactions. In this application, baking powder is used instead of baking soda. Baking powder is made by adding a solid acid, calcium dihydrogen phosphate, $Ca(H_2PO_4)_2$, to baking soda. $Ca(H_2PO_4)_2$ ionizes in water to form $Ca^{2+}_{(aq)}$ and $2H_2PO_4^-_{(aq)}$. Note that the dihydrogen phosphate anion derives from phosphoric acid (Figure 5.14) by ionizing one of its three acidic hydrogens. In pancake batter the $H_2PO_4^-$ ion gives up one of its two protons to the bicarbonate anion, HCO_3^-, to form $CO_{2(g)}$ by the reaction sequence of equations (29) and (30).

$$H_2PO_4^-_{(aq)} + HCO_3^-_{(aq)} \rightarrow HPO_4^{2-}_{(aq)} + H_2O_{(l)} + CO_{2(aq)} \quad (29)$$

*The dihydrogen phosphate anion reacts with bicarbonate to
form hydrogen phosphate, water, and dissolved carbon dioxide.*

$$CO_{2(aq)} \rightleftharpoons CO_{2(g)} \qquad (30)$$

Dissolved carbon dioxide partially vaporizes to form carbon dioxide bubbles.

This reaction occurs rapidly because the added acid provides protons to react with the bicarbonate ion, which produces $CO_{2(g)}$

bubbles in the batter. Unfortunately, this neutralization reaction also occurs slowly in a can of baking powder as it absorbs water from the atmosphere. That is why an opened can of baking powder loses its activity after a year or so. These mechanisms for rising differ from the use of yeast, a living organism, to make bread rise. Yeast organisms consume sugar in bread dough to make ethyl alcohol and the needed $CO_{2(g)}$.

Indigestion, Ulcers, and Stomach Acid

The stomach's pH can approach 1, which is a strong enough acid to strip zinc off galvanized steel! A constantly regenerating mucous cell membrane lines the stomach's inner wall and protects it. If a break in this barrier occurs, an ulcer (an acid burn of the stomach wall) results. The acid may even eat a hole through the stomach's wall in serious cases (perforated ulcer). Often the stomach produces excess acid in response to food and drink. If acid leaks from the stomach up the esophagus, then the acid burn of the esophagus causes the sensation of "heartburn." One way to address this problem is to use basic chemicals to neutralize the excess acid. Sodium bicarbonate works for this purpose. The neutralization process also generates large amounts of $CO_{2(g)}$, which causes belching. Alka Seltzer®, another stomach acid remedy, consists of tablets that contain citric acid in addition to sodium bicarbonate. Upon dissolving in water, the citric acid reacts immediately with the sodium bicarbonate to produce gas bubbles. This concoction doesn't neutralize acid any better than straight sodium bicarbonate, but the fizz looks impressive and the carbonated water may help induce belching.

Milk of Magnesia employs a different mild base, magnesium hydroxide, $Mg(OH)_{2(s)}$. This compound doesn't dissolve very well in water, so it is sold as a fine, milky suspension (hence the name) that has to be shaken before use. Once in the stomach, it neutralizes excess acid according to equation (31).

$$Mg(OH)_{2(s)} + 2H^+{}_{(aq)} \rightarrow Mg^{2+}{}_{(aq)} + 2H_2O_{(l)} \qquad (31)$$

*Magnesium hydroxide reacts with protons
to generate magnesium cations and water.*

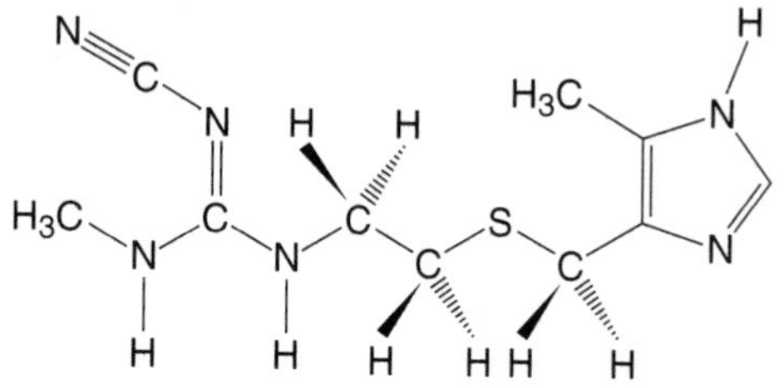

Figure 5.17. *The molecular structure of cimetidine—brand name Tagamet®—an antiulcerative drug.*

The magnesium ion, Mg^{2+}, which is a reaction product, is an essential nutrient. This reaction generates no gaseous products. Maalox® consists of both solid $Mg(OH)_2$ and $Al(OH)_3$ pressed into a pellet with a binder.

The antacid Tums® consists primarily of calcium carbonate, $Ca[CO_3]$, and magnesium carbonate, $Mg[CO_3]$, components of limestone. It uses carbonate, CO_3^{2-}, to accept two protons and thus counteract excess stomach acid, as shown in equation (32).

$$CO_3^{2-}{}_{(aq)} + 2H^+{}_{(aq)} \rightleftharpoons H_2O + CO_{2(aq)} \tag{32}$$

In water, the carbonate anion partially reacts with two protons to form water and dissolved carbon dioxide (carbonated water).

Like bicarbonate remedies, Tums® generates carbon dioxide gas. Ingesting the essential minerals calcium and magnesium, instead of sodium, may benefit those on a low-salt diet. The calcium provided in Tums® is also helpful in combatting osteoporosis; however, ionized calcium stimulates the production of stomach acid.

Prescription ulcer drugs, such as cimetidine (Tagamet®, Figure 5.17) represent a more sophisticated biochemical approach to the control of stomach acid. Tagamet® and related drugs, such as Zantac® and Pepcid®, inhibit the secretion of stomach acid. This prevents the underlying problem and permits recovery of the stomach's mucosal lining in patients with a peptic ulcer. These chemicals have proved to be miracle drugs for ulcer patients. They may even be available soon as over-the-counter drugs for the relief of excess stomach acid. A newly discovered bacterium has been identified as a cause of ulcers. Antibiotic cures for ulcers are being developed rapidly.

Retarding the Decay of Books

Both comic book and rare book collectors worry about the acidity of paper. Acids in paper accelerate the addition of water to cellulose, the natural wood fiber component in paper. This reaction breaks the long-chain cellulose molecules in the paper, which causes it to crumble. About one-third of the books in the U.S. Library of Congress cannot be circulated because they are too brittle. Many books in research libraries across the United States have been estimated to be in a similar state of decay. Oddly enough, books printed earlier than the eighteenth century, such as the Gutenberg Bible, often survive in better condition than their nineteenth-century counterparts. This correlates with a change in paper manufacturing techniques. Early paper was usually a high-quality grade made from rags. In the eighteenth century, however, the large-scale production of paper from wood pulp produced an inferior product with tiny cavities and holes in the surface. These surface defects caused the ink from a printing press to wick and spread out beyond the area where the type applied it. Paper made from wood pulp must be sealed with a sizing agent to prevent ink creep. Unfortunately, the inexpensive ionic salt known as alum, $Al_2(SO_4)_3$, the puckering agent in pickle juice, was selected along with rosin for this application. When alum dissolves in water or absorbs moisture in high humidity, it ionizes to form solvated aluminum cations and the sulfate anion, as shown in equation (33). The aluminum cations react with water to produce an acidic solution, as shown in equation (34).

$$Al_2(SO_4)_{3(s)} \overset{H_2O}{\rightleftharpoons} 2Al^{3+}_{(aq)} + 3SO_4^{2-}_{(aq)} \qquad (33)$$

Aluminum sulfate partially dissolves in water to generate aluminum cations and sulfate anions.

$$2Al^{3+}_{(aq)} + 2H_2O_{(l)} \rightleftharpoons 2Al(OH)^{2+}_{(aq)} + 2H^+_{(aq)} \qquad (34)$$

Aluminum cations partially react with water to form aluminum hydroxide and protons.

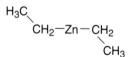

Figure 5.18. *The molecular structure of diethylzinc, a chemical used to neutralize the acidity of paper.*

This sequence of reactions occurs in the pores of paper. The acidic products break down the cellulose fibers in paper. An obvious solution is to neutralize the acid in paper, but this has not been easy to accomplish.

The neutralization of acidity in an old book requires a base to penetrate every small pore on every page. It would be too time-consuming to use a page-by-page procedure. One neutralization method that is being tested by the Library of Congress, and is already used by the Harvard University Libraries, involves diethylzinc, shown in Figure 5.18. This volatile base partially vaporizes at room temperature.

Gaseous diethylzinc initially reacts with water in a hydrolysis reaction to form the mild base zinc oxide, ZnO, as shown in equation (35). The ZnO can neutralize acid in the paper, as shown in equation (36).

$$Zn(C_2H_5)_{2(g)} + H_2O \rightleftharpoons ZnO_{(s)} + 2C_2H_{6(g)} \qquad (35)$$

Diethylzinc reacts with water to form zinc oxide and ethane gas.

$$ZnO_{(s)} + 2H^+_{(aq)} \rightleftharpoons Zn^{2+}_{(aq)} + H_2O_{(l)} \qquad (36)$$

Zinc oxide reacts with acids to form zinc cations and water.

One difficulty with this method arises from the violent exothermic reaction between diethylzinc and oxygen. Diethylzinc vapor burns on contact with air! This complicates the treatment of books, since the treatment chamber must be free of oxygen (or else the books burn up!). This method also doesn't effectively neutralize glossy, coated paper.

For new books, there has been increased use of acid-free printing paper. A nonacidic filling agent for the pores in paper, such as white titanium dioxide (TiO_2), calcium carbonate (chalk, $CaCO_3$), or white clay, makes paper that has low acidity. Books, journals, and manuscripts printed on such paper will survive in libraries much better than their alum-filled counterparts.

CHAPTER 6

Wall Street Chemistry

Chemical! The word conveys images of an industrial complex spewing noxious fumes into the environment. Just the thought of industrial accidents, such as the tragedy at Texas City, makes us long for simpler days when the Earth was "pure." This attitude, of course, overlooks the innumerable benefits derived from chemistry, which society cannot afford to give up. If all chemical plants were to disappear, the entire economy would come to a screeching halt!

The development of chemical synthesis and purification methods made the medical revolution of the twentieth century possible. For our great grandparents, the death of a child was a tragic but common experience. Pure drugs and antibiotics virtually eliminated the deaths of babies, children, and young adults from pneumonia, tuberculosis, scarlet fever, and a host of other diseases. The quality of modern life would also diminish greatly without pain relievers, anesthetics, and antibiotics. Pesticides are essential for the control of malaria and other insect-borne plagues in flood-stricken areas. Chemicals help people survive in the face of bacterial, parasitic, and viral diseases.

Also in this century, chemists met the challenge of producing nontoxic fluids for household refrigerator compressors. Refrigeration greatly reduced bacterial diseases associated with eating spoiled foods. The alternative preservation methods, such as smoking and salting foods, introduced more chemicals into the diet in "the good old days" than we commonly see today!

Improved food preservation methods have been credited with a dramatic decrease in stomach cancer rates during this century. Chemical toxicity studies of food additives now allow the rational selection of preservatives that pose minimal health risks. People are safer than in the 1800s, when confectionery products contained toxic and carcinogenic metal salts as coloring agents. Examples of the old food additives include red mercury sulfide, blue copper sulfate, white lead acetate or lead carbonate, and green copper arsenite—the same pigments used in artists' paints! Inert plastic piping has replaced lead-soldered piping, which was used in U. S. household water supplies until 1987. The use of lead pipes and copper pipes with lead-soldered joints has been linked to heavy-metal contamination of drinking water.

Chemical fertilizers and pesticides help produce an abundance of inexpensive food in many countries. The alleviation of food shortages indirectly preserves the wilderness. When faced with starvation, humans hunt animals to extinction and clear every available forest for farmland. Synthetic chemical fibers, dyes, and leather help make clothing less expensive, and the advent of synthetic fibers helped end the widespread use of animal furs and skins. Lightweight, synthetic-chemical sleeping bags, tents, Gore-Tex® parkas, and plastic-wrapped, freeze-dried food help modern backpackers enjoy the outdoors with minimal disruption of the environment. Energy-efficient houses also require efficient chemical insulation materials. Imagine the lung problems that early *Homo sapiens* endured with only a smoky wood fire to warm a poorly ventilated cave!

Chemistry pervades nearly every aspect of society. Human survival dictates that the chemical industry will always exist. People work in buildings made of concrete and drive on asphalt chemical highways. Artists' paints, writers' ink, doctors' medicines, farmers' fertilizers, teachers' papers, and the computer scientists' microchips all require chemical technology in their manufacture. Humans are complex chemical-processing plants who take molecules in from food, break them into smaller fragments, and then recombine the fragments into biochemicals. Along with the opportunities that chemistry provides to society, risks may be involved in the application of a particular technology. The best use of chemistry is made by a society that makes informed decisions about the applications to which it puts chemistry—and by a society that learns from its mistakes.

Big Business and Chemistry

Nearly every business uses chemicals, so they are produced on a grand scale in developed nations. Table 6.1 lists the top 30 chemical companies, along with their 1992 sales in billions of dollars. These numbers do not include gasoline or oil revenues per se, but only the sales of chemicals made from them. The chemical sales volume (not including oil and gasoline) at Exxon, number 7 on the chemical list, accounts for only 9% of total sales by the corporation. Chemical operating profit margins in 1992 range from deficits for Shell, ENI, and British Petroleum to a high of 15.2% for General Electric's chemical operations. The top 30 chemical producers accounted for $267 billion in sales in 1992.

It is no accident that the leading economic powers in the world host sizable chemical industries. An automobile cannot be built, nor a computer manufactured, without a wide array of chemical plastics, inks, adhesives, and semiconductors. For this reason chemical company stock exchange prices rise early during an economic recovery. In the 1950s the United States led the world in chemical production, but now Germany produces more chemicals than any other country. Of the top 50 chemical companies, the U.S. leads with 15, and Japan follows closely with 10. If Europe is considered as an economic unit, it leads the world with 25 companies. All the top 50 chemical companies are based in Europe, the United States, or Japan! Economic competitiveness in today's world market clearly requires a strong chemical industry. In 1991 the U.S. chemical industry had the greatest positive balance of trade ($18.8 billion), and chemicals accounted for 10.2% of all U.S. exports. This industry directly creates 1.1 million high-paying jobs. The U.S. chemical industry even enjoys a healthy trade surplus ($2.4 billion in 1991) with Japan. A commitment to research and development (about 5% of sales) may explain the industry's economic competitiveness. For compari-

Table 6.1.   *Leading Chemical Producers Worldwide in 1992.*
(Chem. Eng. News, p. 18, July 26, 1993)

Rank	Firm Name and Country	Chemical Sales ($ billions)	Percent of Company's Total Sales
1	BASF (Germany)	22.88	80
2	Hoechst (Germany)	22.47	76
3	ICI (U.K.)	18.41	86
4	Bayer (Germany)	16.28	62
5	Du Pont (U.S.)	15.55	41
6	Dow Chemical (U.S.)	12.92	68
7	Exxon (U.S.)	10.63	9
8	Shell (U.K., Netherlands)	10.33	11
9	Rhône–Poulenc (France)	9.98	65
10	Ciba–Geigy (Switzerland)	9.60	61
11	ENI (Italy)	9.50	23
12	Elf Aquitaine (France)	9.31	25
13	Asahi Chemical (Japan)	7.85	100
14	Akzo (Netherlands)	7.72	81
15	Solvay (Belgium)	6.79	86
16	Veba (Hüls) (Germany)	6.58	16
17	Mitsubishi Kasel (Japan)	5.60	100
18	Air Liquide (France)	5.54	100
19	Monsanto (U.S.)	5.38	69
20	Sekisui Chemical (Japan)	5.32	100
21	British Petroleum (U.K.)	5.23	8
22	Henkel (Germany)	5.12	57
23	BOC (U.K.)	5.06	100
24	DSM (Netherlands)	4.94	98
25	Sumitomo Chemical (Japan)	4.93	100
26	Union Carbide (U.S.)	4.87	100
27	General Electric (U.S.)	4.85	9
28	Norsk Hydro (Norway)	4.77	51
29	Toray Industries (Japan)	4.58	100
30	Takeda Chemical (Japan)	4.46	100

son, a successful consumer electronics company, Sony, devotes 5.7% of sales to research and development. Pharmaceutical companies, which must develop, test, and manufacture drugs, devote as much as 15–20% of sales to research and development.

Raw Materials for Chemicals

The chemical industry must begin with the minerals available on Earth, and elements do not usually exist in a pure state. The conditions present when the Earth formed, and the exposure to atmospheric gases, sunlight, and water, caused elements in the crust to react and form more stable products. Only a few unreactive metals, such as copper, gold and silver, exist naturally in their elemental state. Early civilizations valued these elements for their unusual properties, because other metals (iron and aluminum) were unknown. And no less than those early civilizations, modern society needs to consider the availability of raw materials in long-term economic planning.

A few elements and compounds found on Earth exist in reactive chemical forms. These materials are generally formed in processes associated with living systems. Oxygen gas, made by plants during photosynthesis, and the coal or oil derived from dead plants and animals provide examples of reactive terrestrial chemicals that are important raw materials. Huge deposits of elemental sulfur occur beneath salt-dome structures in and about the Gulf of Mexico. This important source of reactive sulfur originated eons ago from the action of anaerobic bacteria on sedimentary sulfate deposits.

Compounds on the planet that contain carbon illustrate the raw materials problem. Figure 6.1 shows some natural sources of carbon. The most abundant forms of carbon are the CO_2 in the atmosphere and the carbonate ion (CO_3^{2-}) in minerals such as limestone ($CaCO_3$). Pure unreacted carbon in the form of graphite or diamond is rare and results from its geologic formation underground in the absence of oxygen.

The forms of carbon that are most useful to organic chemists are the more complex hydrocarbons found in limited deposits of natural gas, oil, and coal. Chemists use oil to synthesize plastics, textiles, pharmaceuticals, and other organic compounds. Companies that make organic chemicals and polymers from oil often show reduced profits when the price of their raw material rises. Stock values in these companies reflect the fluctuations in the costs of raw materials.

Figure 6.1. *In nature, carbon occurs in diamond and graphite as pure carbon and in carbonate minerals such as the limestone found in caves. (Left, courtesy of General Electric; right, courtesy of W.T. Lee, U.S. Geological Survey.)*

Unlike carbon, such elements as platinum, gold, and rhodium are rare in all forms. Some rare elements are quite useful. Platinum/rhodium/palladium alloys play an important role in millions of catalytic converters in automobiles. The most widely used cancer drug, cisplatin, also contains platinum. The global distribution of natural deposits of rare elements can also have political significance. Almost 75% of the Western World's supply of platinum comes from the South African Bushveld Igneous Complex. During the economic embargo of South Africa in the 1980s, trade in the essential metals platinum and rhodium silently continued. Rare elements cannot be made, except by small-scale nuclear processes. Conservation and recycling of compounds that contain rare elements is of paramount importance. About 90% or more of the key industrial metals chromium, cobalt, manganese, platinum, palladium, and rhodium are imported by the United States.

Table 6.2 lists many important elements, gives their natural chemical forms, and indicates their abundance. These data make

Table 6.2   *Chemical Forms of Selected Elements on Earth*

Element	Useful Chemical Forms in Nature	Abundance
Al—*aluminum*	$Al_2O_3 \cdot nH_2O$—bauxite	High
Ar—*argon*	0.9% in air	Low
As—*arsenic*	By-product of Cu and Pb smelting	Low
B—*boron*	$Na_2[B_4O_5(OH)_4] \cdot nH_2O$—kernite ($n = 2$), borax ($n = 8$)	Medium
Br—*bromine*	NaBr in brine	Low
Cd—*cadmium*	Impurity in Zn ores	Rare
Ca—*calcium*	$CaCO_3$—limestone, marble, chalk, aragonite, $CaSO_4 \cdot 2H_2O$—gypsum	High
C—*carbon*	C—graphite, C_xH_x—coal, C_xH_{2x}—oil, CH_4—natural gas	High
Cl—*chlorine*	NaCl—rock salt	High
Cr—*chromium*	$FeCr_2O_4$—chromite	Low
Co—*cobalt*	$CoAs_2$—smaltite; CoAsS—cobaltite	Low
Cu—*copper*	$CuFeS_2$—chalcopyrite; $Cu_2CO_3(OH)_2$—malachite	Medium
F—*fluorine*	CaF_2—fluorite	Medium
Au—*gold*	Au—native gold	Very rare
He—*helium*	In natural gas	Rare
H—*hydrogen*	H_2O—water	High
I—*iodine*	NaI in brine; $Ca(IO_3)_2$—lautarite	Low
Fe—*iron*	Fe_2O_3—hematite; Fe_3O_4—magnetite, FeS_2—pyrite	High
Pb—*lead*	PbS—galena; $PbSO_4$—anglesite; $PbCO_3$—cerussite	Medium
Mg—*magnesium*	$MgCa(CO_3)_2$—dolomite; $MgCO_3$—magnesite; $MgSO_4 \cdot 7H_2O$—epsomite	High
Mn—*manganese*	MnO_2—pyrolusite; $MnCO_3$—rhodochrosite	Medium
Hg—*mercury*	HgS—cinnabar	Rare
Mo—*molybdenum*	MoS_2—molybdenite	Rare
Ne—*neon*	0.002% in air	Rare
Ni—*nickel*	$Ni_xMg_ySi_4O_{10}(OH)_8$—laterites	Medium
N—*nitrogen*	N_2—78% of atmosphere; $NaNO_3$—saltpeter	High
O—*oxygen*	O_2—21% of atmosphere; H_2O and in silicate and carbonate rocks—Earth's most abundant element	High
P—*phosphorus*	$3Ca_3(PO_4)_2 \cdot CaX_2$—apatites, where X = F, Cl, or OH	High
Pt—*platinum*	Impurity in Cu and Ni ores	Very rare
K—*potassium*	KCl—sylvite; $KCl \cdot MgCl_2 \cdot 6H_2O$—carnallite	Medium
Rh—*rhodium*	Impurity in Ni and Cu sulfide ores	Very rare
Si—*silicon*	SiO_2—quartz, sand	High
Ag—*silver*	Ag_2S—argentite	Rare
Na—*sodium*	NaCl—rock salt; $(Na_2CO_3)_2(NaHCO_3)(H_2O)_2$—trona	High
S—*sulfur*	S_8—yellow sulfur; H_2S in natural gas and oil; gypsum	High
Th—*thorium*	Monzanite sands	Very rare
Sn—*tin*	SnO_2—cassiterite	Low
Ti—*titanium*	$FeTiO_3$—ilmenite; TiO_2—rutile	High
W—*tungsten*	$CaWO_4$—scheelite, $FeWO_4$—wolframite	Rare
U—*uranium*	U_3O_8—pitchblende; $K_2(UO_2)(VO_4)_2 \cdot 3H_2O$—carnotite	Very rare
Zn—*zinc*	ZnS—sphalerite; $ZnCO_3$—smithsonite	Medium

it clear that many elements exist in combination with the elements S, O, and Cl. These minerals and the atmospheric gases provide the starting point for the chemical industry.

Commodity Chemicals

The U.S. production alone of a high-volume chemical amounts to 2–100 billion pounds per year, which equals ½–25 pounds for every person on the planet! The lifestyles that have evolved in developed countries require large amounts of chemicals to process gasoline, make clothes, grow food, and publish newspapers, and to build, paint, and carpet houses and workplaces. In developed countries, everyone indirectly uses hundreds of pounds of industrial chemicals each year, from shoe polishes and cosmetics to the rubber in automobile tires.

Putting high-volume chemistry into practice involves many considerations. Mechanical separation of minerals may be necessary. Some raw materials must be purified. Reaction temperatures and pressures must be controlled, and recycling over a catalyst may be needed. Products must be separated, corrosion of chemical reactors controlled, product quality monitored, and waste products removed. To meet these needs, the construction cost of a large-volume chemical plant runs into hundreds of millions of U.S. dollars. Such plants represent a unique combination of chemistry, metallurgy, physics, engineering, and computer science.

In the following sections, we examine several classes of chemicals produced in high volume, noting which chemicals in each category are most important, how they are synthesized or extracted, and the major uses to which they are put.

Acids

Because of its low cost of synthesis, the strong acid most widely used by the chemical industry is 100% *sulfuric acid*. Figure 6.2 shows an industrial reactor used to manufacture sulfuric acid.

Figure 6.2. *A large sulfuric acid manufacturing plant. (Courtesy of Peridot Chemicals (New Jersey), Inc.)*

Sulfuric acid is made by the sequence of reactions given in equations (1) to (3). Recall that the subscripts (s), (g), and (l) denote solid, gaseous, and liquid components.

$$S_{(s)} + O_{2(g)} \rightarrow SO_{2(g)} \qquad (1)$$

Sulfur reacts with oxygen to form sulfur dioxide.

$$2SO_{2(g)} + O_{2(g)} \rightarrow 2SO_{3(g)} \quad \text{(requires a } V_2O_{5(s)} \text{ catalyst)} \qquad (2)$$

Sulfur dioxide reacts with oxygen in the presence of the catalyst vanadium pentoxide to form sulfur trioxide.

$$SO_{3(g)} + H_2O_{(g)} \rightarrow H_2SO_{4(l)} \qquad (3)$$

Sulfur trioxide reacts with water to form sulfuric acid.

$$S_{(s)} + 1.5O_{2(g)} + H_2O_{(l)} \rightarrow H_2SO_{4(l)} \qquad (4)$$

In the overall process, sulfur reacts with oxygen and water to form sulfuric acid.

All the raw materials for the synthesis of sulfuric acid are abundant and inexpensive. The process burns sulfur with the aid of a solid vanadium pentoxide ($V_2O_{5(s)}$) catalyst, followed by a reaction with water.

Vanadium pentoxide is one of many *catalysts* with industrial uses. Chemical industries use catalysts to speed up reactions in 90% of their processes. A catalyst speeds up a reaction by providing a reaction pathway that has a lower activation energy barrier (see Figure 5.9). The overall reaction is unchanged. Industry tends to use solid heterogeneous catalysts, which stay in the reaction vessel where they are used repeatedly. The word *heterogeneous* means the state of the catalyst (a solid) differs from that of the reactants (gases or liquids) and products (gases or liquids). In equation (2), solid $V_2O_{5(s)}$ acts as a heterogeneous catalyst for the reaction between gaseous sulfur dioxide ($SO_{2(g)}$) and oxygen ($O_{2(g)}$) to produce gaseous sulfur trioxide ($SO_{3(g)}$). Many catalysts were initially discovered by accident, or by trial and error. Catalysis now constitutes an advanced science in the chemical industry. Environmental concerns demand improved reaction efficiencies (yields near 100%) to eliminate waste products. The economic competitiveness of the chemical industry and the desire to have a clean environment require first-rate catalysis technology.

The major use of sulfuric acid is to prepare *phosphoric acid* according to equation (5). Gypsum or calcium sulfate (the principal component of wallboard for the building industry) forms as a by-product in this reaction.

$$3Ca_3(PO_4)_2 \cdot Ca(OH)_{2(s)} + 12H_2SO_{4(l)} + 18H_2O \rightarrow$$
$$6H_3PO_{4(l)} + 12CaSO_{4(s)} \cdot 2H_2O_{(l)} \qquad (5)$$

Phosphate rock reacts with sulfuric acid
to form phosphoric acid and gypsum.

The 84% by weight solution (16% water) of phosphoric acid obtained in this process has a syrupy consistency. Phosphoric acid, which is used mainly to make phosphate fertilizers, also finds application in the synthesis of phosphate detergents and household TSP (trisodium polyphosphate) used in floor cleaners and in dishwasher detergents. Phosphoric acid, an additive (0.01–0.05% by weight) to cola and root beer soft drinks, imparts a tart flavor.

Nitric acid is a strong acid and a strong oxidant. Chemicals derived from it include fertilizers, explosives, and nitrated organic chemicals. Nitric acid can be prepared by air oxidation of ammonia (NH_3), which in turn is made from N_2 in the atmosphere. Ammonia gas burns readily with oxygen gas at 900°C over a platinum (93%)–rhodium (7%) metal screen catalyst to produce nitric oxide (NO), which combines with oxygen and water to form nitric acid; the net reaction is given by equation (6). The synthesis of nitric acid is referred to as the Ostwald process to honor its inventor. Nitric acid forms as a 60% aqueous solution. When the solution is heated, a fraction boils off, or distills, at a constant temperature of 120°C; the fraction consists of 68.5% HNO_3 by weight.

$$NH_{3(g)} + 2O_{2(g)} \rightarrow HNO_{3(l)} + H_2O_{(l)} \qquad (6)$$

Ammonia reacts with oxygen to form nitric acid and water.

Hydrochloric acid occurs as a by-product in the chlorination of organic chemicals by reaction with Cl_2. It also forms by protonation of sodium chloride with the strong acid H_2SO_4 at 550–600°C, as given in equation (7). Here the continuous removal of the HCl gaseous product helps drive the reaction to the right.

$$2NaCl_{(s)} + H_2SO_{4(l)} \rightleftharpoons Na_2SO_{4(s)} + 2HCl_{(g)} \qquad (7)$$

Sodium chloride reacts with sulfuric acid to form sodium sulfate and hydrochloric acid.

Aqueous hydrochloric acid, obtained by distillation at 109°C as a mixture 20.2% by weight, is used to clean the oxide scale from steel. In home supply stores, a 31% solution of hydrochloric acid in water (by weight) is sold as muriatic acid to clean and etch bricks or concrete and to adjust the acidity in swimming pools. Fumes of corrosive $HCl_{(g)}$ escape from solution when the acid is used.

Acetic acid is often prepared by fermentation, as in wine vinegar. White vinegar consists of 95% water and 5% acetic acid by weight. Pure acetic acid (called glacial acetic acid) is an important industrial solvent that dissolves both polar and nonpolar compounds. Photographers use glacial acetic acid to prepare a "stop bath" for developing photographic film and prints. Glacial acetic acid, although chemically categorized as a weak acid, causes severe skin and eye burns. Its synthesis involves two catalyzed reactions. First, carbon monoxide reacts with hydrogen over a $ZnO_{(s)}$ heterogeneous catalyst to form methanol, as shown in equation (8). The methanol is then liquefied and allowed to react with dissolved carbon monoxide gas under pressure. This reaction, shown in equation (9), requires a soluble rhodium iodide homogeneous catalyst. This catalyst is called a homogeneous catalyst because it is in the same state (dissolved in solution) as the reactants and products.

$$CO_{(g)} + 2H_{2(g)} \rightarrow CH_3OH_{(g)} \tag{8}$$

Carbon monoxide reacts with hydrogen to form methanol.

$$CH_3OH_{(l)} + CO_{(dissolved)} \rightarrow CH_3COOH_{(l)}$$
$$\text{(requires a dissolved } RhI_3 \text{ catalyst)} \tag{9}$$

Methanol reacts with carbon monoxide in the presence
of the catalyst rhodium triiodide to form acetic acid.

Bases

Calcium oxide (CaO), or *lime*, forms according to equation (10) when limestone is heated to 600°C. The thermal decomposition of metal carbonates, such as lime, to liberate $CO_{2(g)}$ is a common reaction. The reaction proceeds to the right as the gaseous carbon dioxide escapes.

$$CaCO_{3(s)} \rightleftharpoons CaO_{(s)} + CO_{2(g)} \tag{10}$$

Limestone (calcium carbonate) partially reacts
to form lime (calcium oxide) and carbon dioxide.

Addition of water to $CaO_{(s)}$ produces calcium hydroxide, $Ca(OH)_{2(s)}$, which is called hydrated lime or slaked lime. Slaked lime, available in most hardware stores, is spread on fields and lawns to neutralize soil acidity. Soil pH must be near 7 for the best availability of minerals and for optimal plant growth. The inexpensive mild base $Ca(OH)_2$ is well suited for this agricultural application.

Lime also makes up about 60% of the dry weight of Portland cement; the other major components are silica (SiO_2) and alumina (Al_2O_3). Portland cement was originally made in nineteenth-century England by heating a mixture of chalk and clay. Concrete is a mixture of Portland cement, sand, and crushed gravel. The construction of buildings, bridges, and highways requires immense quantities of concrete.

Ammonia is another alkaline substance of agricultural importance. It provides the starting point for the manufacture of nitrogen-based chemicals. The synthesis of ammonia has political origins. Before World War I, natural deposits of sodium nitrate in Chile were the major raw material for the synthesis of nitric acid and nitrogen-based explosives (gunpowder, TNT, and dynamite). A few years before World War I two German chemists devised a way to make ammonia from the nitrogen gas in the atmosphere. The process, developed by Fritz Haber and Carl Bosch, relies on the combination of atmospheric $N_{2(g)}$ with $H_{2(g)}$ at elevated temperatures and pressures over an iron–iron oxide heterogeneous catalyst to produce ammonia (equation 11).

$$N_{2(g)} + 3H_{2(g)} \rightarrow 2NH_{3(g)} \quad \text{(iron–iron oxide catalyst)} \quad (11)$$

*Nitrogen reacts with hydrogen in the presence
of an iron–iron oxide catalyst to form ammonia.*

The ammonia could then be converted to nitric acid by the Ostwald process (see equation 6). Nitric acid is used directly in the synthesis of high explosives. Figure 6.3 shows a typical chemical plant for the synthesis of ammonia.

The Haber–Bosch process came along just in time, because the natural deposits of sodium nitrate could not supply enough fertilizer to keep pace with the population increase of the twentieth century. Ammonia, an inexpensive source of nitrogen, has replaced $NaNO_3$ in the manufacture of fertilizers. The Haber–Bosch process helped prevent mass starvation as the world's population tripled.

Sodium hydroxide (NaOH), or *lye*, is an inexpensive base obtained from the electrolysis of brine solutions. When electric

Figure 6.3. *An ammonia synthesis facility. (Courtesy Monsanto, Luling, LA.)*

current passes through a NaCl solution, $Cl_{2(g)}$ evolves at the anode and $H_{2(g)}$ at the cathode to leave an aqueous solution of NaOH. The main application of NaOH is as a base in the chemical and pulp-paper industries, but its use to make "lye soap" has been known for centuries. In frontier times, sodium and potassium carbonates found in wood ash served as the alkaline component in the making of soap. Soap, a metallic salt of a fatty acid, forms when fat is heated with NaOH or KOH. Equation (12) represents the saponification reaction of tristearin, a fat abundant in beef and mutton tallow, to form sodium stearate soap and glycerin. Sodium soaps feel hard to the touch, whereas potassium soaps feel soft. Glycerin, perfume, and other additives are combined to produce the many varieties of soaps available on the market. A soap that floats can be made by whipping air into molten soap as it cools.

$$(C_{17}H_{35}COO)_3C_3H_{5(l)} + 3NaOH_{(s)} \rightarrow$$
$$3C_{17}H_{35}COONa_{(s)} + C_3H_5(OH)_{3(l)} \quad (12)$$

Tristearin fat reacts with sodium hydroxide
to form sodium stearate soap and glycerin.

Soaps and detergents gain their cleaning ability by forming micelles (Figure 6.4) in water. Recall the rule that "like dissolves like." The nonpolar C–H bonds of the sodium stearate molecule

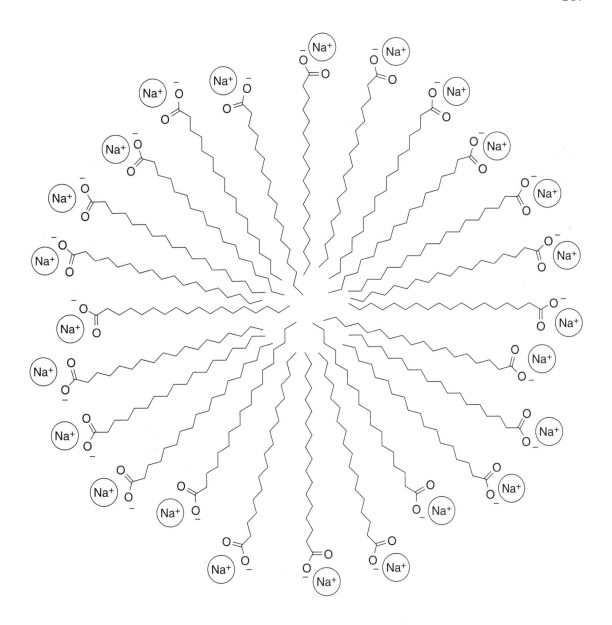

Figure 6.4. *A cross section of the micelle formed by sodium stearate. The hydrocarbon backbone of each soap molecule points to the center of the micelle, and the charged ends lie on the surface where they can interact with solvent water molecules. The cleaning action of the micelle comes from its ability to dissolve nonpolar (oily) compounds in the interior nonpolar hydrocarbon region of the micelle.*

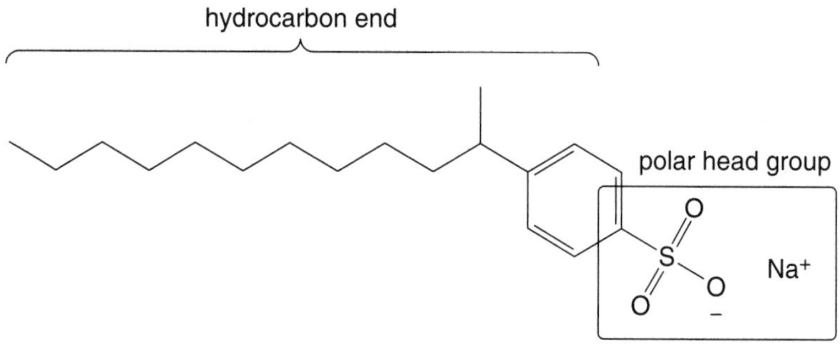

hydrocarbon end

polar head group

Na+

Figure 6.5. *A linear alkylsulfonate surfactant, or detergent.*

avoid water. But the ionized polar end of the soap molecule dissolves in water. To satisfy both tendencies, the soap molecules cluster in a spherical conglomerate called a micelle. Nonpolar hydrocarbon tails from thousands of sodium stearate molecules interact with each other in the nonpolar interior of the micelle. The polar *head groups* of the soap molecule lie on the surface of the micelle. Here they can interact with polar water molecules. Micelles remove greasy dirt from clothes by dissolving nonpolar grime in the nonpolar interior of the micelle.

Detergents, which are also called surfactants, resemble soap molecules in that they also contain a long hydrocarbon chain attached to a charged *head group*. They differ in that they are not derived from naturally occurring fatty acids. The linear alkyl-sulfonate molecule shown in Figure 6.5 is used in many detergent formulations. Detergents don't work well in hard water, and chemicals are added to soften the water. These water-softening agents are commonly called builders. A typical laundry detergent is about 40% detergent and 35% builder. About 10% is sodium sulfate, an inexpensive inert filler. The remaining 15% might consist of whitener, bleach, corrosion inhibitor, antisudsing agent, fragrance, and enzymes. Before the trend toward concentrated solid laundry detergents, as much as 40% consisted of the inert filler. This occurred because consumers mistakenly focused on the size of the box per dollar spent, rather than on the number of wash loads.

Sodium tripolyphosphate is a builder used extensively in laundry detergents and dishwashing detergents. The molecular structure of this builder is shown in Figure 6.6. Detergents that contain sodium tripolyphosphate as a builder are referred to as

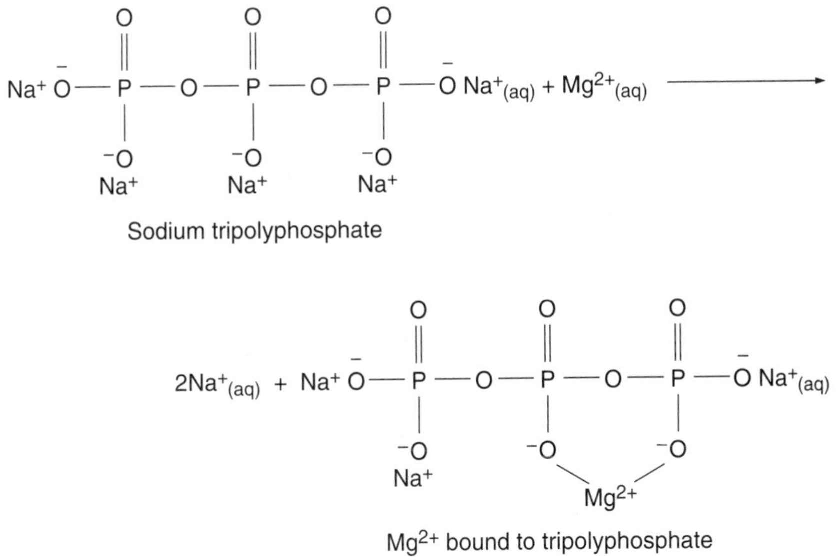

Sodium tripolyphosphate

Mg^{2+} bound to tripolyphosphate

Figure 6.6. *The chemical action of the builder molecule sodium tripolyphosphate. The exchanging of sodium for magnesium is known as water softening.*

phosphate detergents. Hard water contains dissolved calcium (Ca^{2+}) and magnesium (Mg^{2+}) ions from dissolved minerals, such as limestone ($CaCO_3$) and dolomite ($MgCO_3$). These ions cause the sodium ions in soaps and detergents to be replaced. The 2+ metal ions then cause the micelles to begin to clump together and fall out of solution as soap or detergent scum—the familiar bathtub ring. The loss of the soap or detergent micelles from solution reduces the cleaning action. Sodium tripolyphosphate ($Na_5P_3O_{10}$), strongly binds Mg^{2+} and Ca^{2+} ions in hard water, because one 2+ cation can replace two Na^+ cations. The 2+ cation gains extra stability by binding to two or more of the negative oxygen atoms on the phosphate (see Figure 6.6). This removes the Mg^{2+} and Ca^{2+} ions from solution, softens the water, and enhances the cleaning action of the detergent.

A serious problem can arise when phosphate detergent wastes flow into rivers and lakes. Sodium tripolyphosphate derives from phosphoric acid, an ingredient in fertilizers. In natural water systems, the phosphate detergent wastes act as a fertilizer for the growth of plant life. Microbes that degrade dead plant matter use up O_2 dissolved in the water. When excessive plant waste exists, this process reduces the amount of oxygen gas dissolved in the water (it may sink as low as 5 parts per mil-

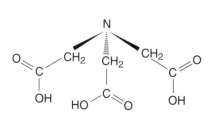

Sodium citrate, a metal ion chelator, is used as a builder in synthetic detergents.

Nitrilotriacetic acid (NTA) chelates a metal ion with the loss of three protons (hydrogen ions); it is used as a builder in synthetic detergents.

Figure 6.7. *The molecular structures of sodium citrate and nitrilotriacetic acid.*

lion). Then fish and other oxygen-dependent aquatic life die. Seven U.S. states and 27 cities had banned phosphates in laundry detergents by 1987. This ban was extended to most states by 1993, and most manufacturers removed phosphates from their formulations. Sodium carbonate (Na_2CO_3), or washing soda, is an alternative builder that has been used for many years. It precipitates Ca^{2+} and Mg^{2+} from solution as $CaCO_3$ and $MgCO_3$. Zeolites are porous, nontoxic minerals that exchange sodium ions for calcium ions in water. They are the solid builders used in most modern laundry detergents.

Alternative soluble builders for synthetic detergents, such as sodium citrate and nitrilotriacetate (NTA), shown in Figure 6.7, may be used instead of sodium tripolyphosphate to chelate the Ca^{2+} and Mg^{2+} ions. Some consumer groups oppose the use of NTA in the United States because of safety concerns. For sodium citrate, the salt of citric acid present in citrus fruits, toxicity is not an issue.

Atmospheric Gases

The common gases nitrogen and oxygen are extracted from liquid air. Water vapor and CO_2, which would freeze to form solids before the temperature needed to liquefy nitrogen ($-195.8°C$), are first removed by passing the air over CaO and NaOH to cause the reactions shown in equations (13) and (14).

$$CaO_{(s)} + H_2O_{(g)} \rightarrow Ca(OH)_{2(s)} \qquad (13)$$

Calcium oxide reacts with water in air to form calcium hydroxide.

$$NaOH_{(s)} + CO_{2(g)} \rightarrow NaHCO_{3(s)} \qquad (14)$$

*Sodium hydroxide reacts with carbon
dioxide in air to form sodium bicarbonate.*

Compression–expansion cycles cool the purified air until it liquefies; this is similar to the operation of an air conditioner compressor (Figure 6.8). Compressing a gas causes it to heat up, but expanding it causes it to cool. The components of liquid air (principally N_2 and O_2) are separated by distilling them at different temperatures. Liquid nitrogen boils at $-195.8°C$ and liquid oxygen boils at $-183.0°C$. Liquefaction of air requires energy to run the compressors. The prices of shares of stock in companies involved in gas separations often react strongly to changes in energy prices.

The largest application of gaseous nitrogen is to provide an inert atmosphere for making iron and steel. Molten iron and hot steel will combine with oxygen in the air to form iron oxide, but a blanket of pure nitrogen gas protects the hot, reactive metal from attack by oxygen. An inert nitrogen atmosphere is also used in the manufacture of chemicals that react with, or form explosive mixtures with oxygen in air. Liquid nitrogen finds wide use as a cooling agent because of its chemical inertness, low cost, and low temperature. It costs about $0.35 per liter, less than distilled water!

Gaseous oxygen is the oxidant of choice for the chemical industry because of its low cost. Its main use, however, is in the

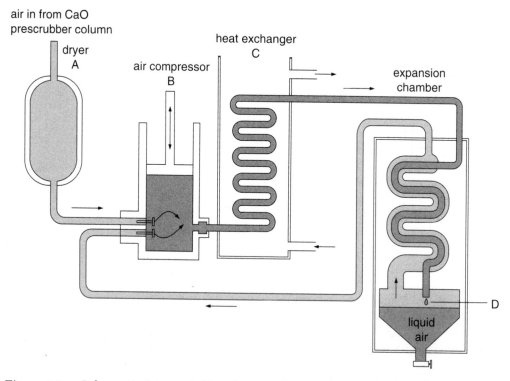

air in from CaO
prescrubber column

dryer
A

air compressor
B

heat exchanger
C

expansion
chamber

liquid
air

D

Figure 6.8. *Schematic for an air liquefaction plant. Entering air has water and carbon dioxide removed in a dryer (A) packed with NaOH. It is then compressed (B), the heat of compression is dissipated (C), and the gas is expanded (D) so that some of the air liquefies. The piston (B) is raised to begin another cycle. (Redrawn from* Chemical Principles with Qualitative Analysis, Sixth Edition *by William L. Masterton, Emil J. Slowinski and Conrad L. Stanitski, copyright © 1986 by Saunders College Publishing. Reprinted by permission of the publisher.)*

steel industry. Oxygen gas is blown through molten steel to oxidize carbon impurities to gaseous CO and CO_2. This oxidation step has to be monitored, because O_2 will oxidize iron to iron oxide once the carbon impurities are consumed. Furthermore, carbon steels require a small amount of residual carbon for their enhanced strength and hardness. In contrast to inert, colorless liquid nitrogen, blue liquid oxygen is a dangerous oxidant. It reacts explosively with most organic compounds. The central rocket motor of the space shuttle runs by combining liquid oxygen and liquid hydrogen.

Hydrogen and Other Gases Obtained from Natural Gas

Natural gas from North America is 60–80% methane. Natural gas fuels electric power plants and furnaces in homes, offices, and industries. Catalytic oxidation of methane gas with steam produces CO_2 and H_2 according to equation (15).

$$CH_{4(g)} + 2H_2O_{(g)} \overset{\text{catalyst}}{\rightleftharpoons} CO_{2(g)} + 4H_{2(g)} \qquad (15)$$

Methane and water partially react to form carbon dioxide and hydrogen in the presence of a nickel oxide catalyst.

Although carbon dioxide occurs in the atmosphere, its main commercial source is methane oxidation. Carbonated soft drinks are the largest market for gaseous carbon dioxide. Solid carbon dioxide (dry ice) is used as a refrigerant. Unlike most other common gases, when carbon dioxide is cooled at atmospheric pressure, it does not liquefy but instead directly freezes. And in the reverse process, dry ice does not melt but directly produces gaseous carbon dioxide in a process called sublimation.

Carbon dioxide gas liquefies only under pressure. The common CO_2 fire extinguisher contains liquid carbon dioxide under a pressure 60 times that of normal atmospheric pressure. When the valve of the extinguisher is opened, the carbon dioxide expands rapidly into the room as a gas. An expanding gas from a fire extinguisher cools rapidly and may freeze moisture in the air. Carbon dioxide fire extinguishers work by replacing the atmosphere around a burning object. Because pure CO_2 gas is more dense than air, it smothers the fire with a heavy blanket of unreactive CO_2 and prevents oxygen from reaching the fire.

An alternative source of hydrogen, which produces carbon monoxide instead of CO_2, is the high-temperature reaction between carbon and water given by equation (16).

$$C_{(s)} + H_2O_{(g)} \rightarrow CO_{(g)} + H_{2(g)} \qquad (16)$$

Carbon reacts with water to form carbon monoxide and hydrogen.

The composition of the resulting gas mixture can be further modified by the so-called water–gas shift reaction of equation (17). This reaction occurs over a solid iron oxide catalyst.

$$H_2O_{(g)} + CO_{(g)} \overset{catalyst}{\rightleftharpoons} H_2_{(g)} + CO_2_{(g)} \tag{17}$$

Water partially reacts with CO to form hydrogen and carbon dioxide in the presence of an iron oxide catalyst.

Some industrial reactions, such as methanol synthesis (equation 8), directly use the hydrogen and carbon monoxide mixtures from equation 16.

Fertilizers

Food consists of chemicals. Three elements—nitrogen (N), phosphorus (P), and potassium (K)—generally limit plant growth in soils. Protein synthesis in cells requires a source of nitrogen. And bone, DNA, and RNA synthesis requires phosphorus. Bone contains so much phosphorus that organic gardeners use bone meal as a natural source of phosphate fertilizer. Potassium ions ($K^+_{(aq)}$) in living systems help maintain membrane electrical potentials essential for the life processes of cells. Because sodium salts are so abundant in nature, they do not limit plant growth; however, the less abundant potassium must be replenished by addition of fertilizer. Carbon, oxygen, and hydrogen atoms, also necessary for synthesizing the molecules of life, are readily available from air (CO_2 and O_2) and rain (H_2O).

Fertilizers provide the means of supplementing soils with soluble forms of nitrogen, phosphorus, and potassium in areas where the soil lacks sufficient nutrients for agricultural needs. Fertilizers do not consist of the pure elements N, P, and K but instead contain inexpensive chemicals containing these elements, which plants can assimilate. A bag of fertilizer in the garden store usually carries a label that includes a designation such as 5–10–5 or 22–2–10. The first number stands for the percent total nitrogen as atomic N, the second for the percent phosphorus as phosphoric acid (H_3PO_4), and the third for the percent soluble potassium as potash (K_2CO_3). This historical practice is followed even though the chemical form of nitrogen and phosphorus in a modern fertilizer may occur in the single compound *ammonium hydrogen phosphate*, $(NH_4)_2HPO_4$, and the potassium may be present as KCl. The yearly U.S. production of fertilizers amounts to about 22 billion pounds of nitrogen, 18 billion pounds of phosphate, and 2 billion pounds of potash. That's about 200 pounds of fertilizer made yearly for each person in the country.

Ammonia is the starting point for the synthesis of all common bulk fertilizers. Ammonia can be injected as a gas directly into the soil to provide a nitrogen source for plant protein synthesis. Ammonia and nitric acid (made from ammonia oxidation) combine in a simple acid–base reaction to yield solid *ammonium nitrate*, NH_4NO_3. Ammonium nitrate dissolves in rainwater to furnish nitrogen to the soil as both a cation, NH_4^+, and an anion, NO_3^-, that plants assimilate. Solid fertilizers are convenient because of their ease of storage and application. Besides its use as a fertilizer, ammonium nitrate is a high explosive, as illustrated by the Texas City incident. A mixture of ammonium nitrate with 5–6% fuel oil serves as a blasting agent in the coal industry. Terrorists even began using ammonium nitrate fertilizer in the 1960s as an available explosive for homemade bombs. Some countries now permit the sale of ammonium nitrate fertilizer only when it is diluted with an inert substance (it may contain 20% limestone) so that it cannot explode. The alleged terrorists seized in the 1993 attempt to bomb New York City's Holland and Lincoln Tunnels were mixing ammonium nitrate fertilizer and fuel oil at the time of their arrest. A similar bomb was used in the 1993 bombing of the World Trade Center.

Two other common fertilizers are made by combining ammonia with inexpensive industrial acids. The reaction between ammonia and sulfuric acid yields *ammonium sulfate*, $(NH_4)_2SO_{4(s)}$. This salt is the 22–0–0 fertilizer commonly sold at garden stores to "green up" lawns. Phosphorus is added as dibasic *ammonium phosphate*, $(NH_4)_2(HPO_4)_{(s)}$. This compound forms by the reaction between ammonia and phosphoric acid. *Superphosphate*, $Ca(H_2PO_4)_2$, along with gypsum by-product from the reaction shown in equation (18), is an alternative phosphate source for fertilizers.

$$Ca_3(PO_4)_{2(s)} + 2H_2SO_{4(l)} + 4H_2O_{(l)} \rightarrow$$
$$Ca(H_2PO_4)_{2(s)} + 2(CaSO_4 \cdot 2H_2O)_{(s)} \qquad (18)$$

Phosphate rock reacts with sulfuric acid and
water to form superphosphate and gypsum.

Inexpensive salt fertilizers dissolve rapidly and enter the soil. For example, $(NH_4)_2SO_{4(s)}$ dissolves readily in water to form ammonium ions ($NH_4^+{}_{(aq)}$), and sulfate ions ($SO_4^{2-}{}_{(aq)}$). Excess fertilizer causes salt burn, often betrayed by patches of dead grass when a lawn is overfertilized. Salt fertilizers work best when applied in several small doses. Many high-priced lawn fertilizers are available. Several of these use urea as a slow-releasing nitrogen source that cannot cause salt burn. *Urea*, $NH_2C(O)NH_2$, is also a constituent in urine. The industrial synthesis of urea combines the base ammonia with the acid CO_2 at 185°C and 200 atm, as shown in equation (19).

$$CO_{2(g)} + 2NH_{3(g)} \rightleftharpoons H_2O_{(g)} + NH_2C(O)NH_{2(l)} \qquad (19)$$

Carbon dioxide reacts partially with ammonia to form water and urea.

The liquid urea produced in this reaction solidifies when cooled to room temperature. Urea decomposes slowly by reacting with water in the soil (the reverse of equation 19) to release ammonia as a source of nitrogen for plants.

The high solubility of the ionic forms of N, P, and K compounds (NH_4^+, NO_3^-, PO_4^{3-}, and K^+) in soil means that they wash away gradually when it rains. With heavy farming and irrigation, these minerals become depleted faster than they are replenished by natural means. In ancient times, the majority of the population engaged in farming because of the low efficiency of agricultural practices. Lands were often overfarmed, leading to the deterioration of the soil by erosion. Today only about 1.5% of the U.S. population engages in farming, yet food output has increased by orders of magnitude. Efficient farming with machinery and chemical fertilizers permits fields to be set aside for use by future generations. It also reduces the temptation to force marginal farmland into production and thereby curtails erosion. But there are negatives. Prolonged use of fertilizer salts and irrigation water causes salt buildup in the soil. A hard salt crust can form a foot or so below the surface, choking plant growth. And the economic advantages of large-scale farming have forced many small farms out of business, creating new social problems.

Pigments

Black and white pigments, the colors of most contrast, have the most applications. *Carbon black* pigment forms as soot by the incomplete burning of oil or natural gas. Carbon black is used as an additive to tire rubber, where it not only imparts a black color to automobile tires, but also strengthens and stabilizes the rubber. Carbon black finds other uses in printing inks, paints, paper, and plastics. Anyone whose hands have been stained with a newspaper appreciates the soot-like characteristics of carbon black.

White pigments generally contain *titanium dioxide* (TiO_2). Fine particles of TiO_2 appear white because they scatter visible light with high efficiency. Mineral sources of TiO_2, such as rutile, contain colored impurities and require purification. Heating rutile, carbon, and chlorine at 950°C (equation 20) produces gaseous $TiCl_4$, which separates from the solid mixture. At room temperature, $TiCl_4$ condenses to a liquid. Oxidation of $TiCl_4$ with oxygen at 1200°C, according to equation (21), produces white powdery TiO_2 suitable for use as a pigment.

$$2TiO_{2(s)} + 3C_{(s)} + 4Cl_{2(g)} \rightarrow 2TiCl_{4(g)} + CO_{2(g)} + 2CO_{(g)} \quad (20)$$

Impure rutile reacts with coke and chlorine to form titanium tetrachloride.

$$TiCl_{4(g)} + O_{2(g)} \rightarrow TiO_{2(s)} + 2Cl_{2(g)} \qquad (21)$$

Titanium tetrachloride reacts with oxygen
to form pure titanium dioxide and chlorine.

Titanium dioxide is used extensively in paints, as a surface whitener for paper, and as a filler in white rubber and plastic products.

Artists originally developed paint pigments for esthetic reasons with little concern for their toxicity. *White lead*, $PbCO_3 \cdot Pb(OH)_2$, was the pigment of choice for white paint before 1960. Because of its toxicity, it has been displaced by TiO_2, which is a nontoxic, insoluble mineral. The lead pigment in paint is also more reactive than TiO_2 and tarnishes to black lead sulfide (PbS) upon exposure to atmospheric hydrogen sulfide (H_2S).

Older paints that contained lead continue to pose an environmental threat. Lead-based paints hidden beneath a coat of nontoxic paint remain a hazard for small children, who, for example, may chew on wooden window sills. Those who sand away old paint create toxic dust that can be inhaled. Finely powdered paint dust that contains lead is especially dangerous. Brain damage caused by lead poisoning may be permanent. Chrome yellow or *lead chromate* ($PbCrO_4$) continues to be the pigment of choice for the yellow lane markers on highways. It contains toxic lead and oxidized chromium, which is a carcinogen. Finding an alternative pigment is difficult because few substitutes have the tolerance to mechanical wear, heat, and light that this application requires.

Reductants

Metal production often requires the reduction of metal ions present in mineral oxides, sulfides, or chlorides. The cheapest way to remove oxygen from metal oxides is to combine it with carbon. The carbon used in these reactions usually exists in a form called *coke*. Heating soft coal in the absence of oxygen distills away natural gas, ammonia, coal tar, and coal oil. The amorphous residue of carbon remaining is called coke. Blast furnaces require large quantities of coke to reduce iron oxides to metallic iron, as shown in equation (22).

$$Fe_2O_{3(s)} + 3C_{(s)} \rightarrow 2Fe_{(l)} + 3CO_{(g)} \qquad (22)$$

*Hematite or ferric oxide reacts with coke
to form pig iron and carbon monoxide.*

Figure 6.9. *Bicycles made with titanium alloy frames, in action at the Tour de France. (Reuters/ Bettman.)*

This reaction transfers the oxygen from iron to carbon, and the carbon monoxide gas separates easily from the solids and liquids involved. Temperatures in a blast furnace reach 1600°C (2912°F), where crude pig iron is molten. Pig iron is fed directly to a converter in making steel alloys.

When inexpensive coke cannot be employed as a reducing agent, then more exotic reductants, such as magnesium, are required. Titanium metal is made by the reduction of $TiCl_4$, as shown in equation (23). Coke is not a suitable reductant, because titanium reacts with carbon to form a stable compound, TiC. At 1000°C (1832°F), magnesium reduces $TiCl_4$.

$$TiCl_{4(g)} + Mg_{(l)} \rightarrow Ti_{(s)} + 2MgCl_{2(l)} \qquad (23)$$

At 1000 °C, titanium tetrachloride reacts with magnesium to form titanium and magnesium dichloride.

Titanium alloys exhibit the highest strength-to-weight ratio of all the commercial metals. They are used to make turbine blades, racing bike frames (Figure 6.9), and airplane engines and frames.

Figure 6.10.
Molecular structures of ethylene oxide and ethylene glycol.

Ethylene oxide

Ethylene glycol

Some metals, such as sodium, magnesium and aluminum, are too difficult to make by chemical reduction of their ores. In such cases, electrolysis of a high-temperature molten salt must be employed. In the manufacture of aluminum metal, electric power is used to force electrons onto the metal ions (Al^{3+}) present in a molten salt mixture containing dissolved ore, and pure metal deposits at the cathode of an electrolysis cell. For this reason, the price of stock of an aluminum company depends heavily on the cost of electrical energy.

Oxidants

Most industrial oxidation reactions involve the transfer of one or more oxygen atoms to the reactant. Oxygen gas is the least expensive oxidant and is used whenever possible. Air, which is 21% oxygen, often acts as the oxidizer itself, as in the burning of sulfur to make sulfuric acid (see equation 1). However, applications to the oxidation of organic substrates are limited by the tendency of oxygen to burn organic compounds completely to carbon dioxide and water.

Oxidation of ethylene by O_2 over a silver metal catalyst produces ethylene oxide. Even here, large amounts of CO_2 by-product are formed. The ethylene oxide obtained from this reaction goes into the manufacture of polymers and ethylene glycol (automotive antifreeze). The molecular structures of ethylene oxide and ethylene glycol are shown in Figure 6.10.

Chlorine is an oxidant used to make chlorinated organic chemicals, such as tri- and tetrachloroethylene solvents. Chlorine gas is made by passing electric current through molten sodium chloride. At the anode, chlorine gas (Cl_2) is generated. At the cathode, electrons are transferred to Na^+ ions to generate molten sodium metal. Chlorine is used in great quantity as a bleach for the pulp and paper industry and as a disinfectant for municipal water supplies (also see p. 196). The chemistry behind these applications stems from the reaction between chlorine and water, shown in equation (24).

$$Cl_{2(aq)} + H_2O_{(l)} \rightarrow 2H^+_{(aq)} + OCl^-_{(aq)} + Cl^-_{(aq)} \qquad (24)$$

Chlorine reacts with water to form protons
and hypochlorite and chloride anions.

Equation (24) shows that adding chlorine to water increases the acidity of the solution, so drinking water purified with chlorine must be neutralized before use. This can be done with salts such as NaOH and $Ca(OH)_2$.

The use of chlorine to disinfect drinking water suffers from reactions that occur between Cl_2 or OCl^- and organic chemicals (including those in bacteria) found in natural water sources. This may produce chlorinated hydrocarbons as by-products. Water disinfected with chlorine generally contains very small amounts of chloroform, which may increase the risk of liver cancer. The public consensus is that a slightly increased cancer risk from the presence of chloroform is outweighed by the benefit of eliminating serious infectious agents (giardia, cholera, typhoid, and amebic dysentery) from the water supply. Water purification poses difficult technological and chemical problems. It requires an inexpensive way to kill pathogenic bacteria without introducing human toxins in the process.

When chlorine is added to an aqueous sodium hydroxide solution, the acid produced (equation 24) is neutralized by hydroxide ($H^+ + OH^- \rightarrow H_2O$). This yields an aqueous bleach solution of NaOCl and NaCl. A 5% solution of NaOCl at pH 11 corresponds to liquid Clorox® bleach. Household bleach is a dangerous substance. A single drop of bleach can damage the human eye. Another serious hazard arises from the ignorant or inadvertent mixing of liquid bleach and ammonia cleaning solutions. Industry uses this reaction to make the liquid rocket fuel hydrazine (N_2H_4) by equation (25).

$$2NH_{3(aq)} + ClO^-_{(aq)} \rightarrow N_2H_{4(aq)} + H_2O + Cl^-_{(aq)} \qquad (25)$$

Ammonia reacts with hypochlorite to
form hydrazine, water, and chloride ion.

The initial product of the reaction, which forms rapidly by equation (26), is the highly toxic and volatile gas chloramine. Vapor evolved from this initial reaction poses a serious health hazard. Therefore, bleach and ammonia cleansers should never be mixed. Household scouring powders also may contain the OCl^- ion as a mixed salt that is about one-third $Ca(OCl)_{2(s)}$, calcium hypochlorite.

$$NH_{3(aq)} + ClO^-_{(aq)} \rightarrow NH_2Cl_{(aq)} + OH^-_{(aq)} \qquad (26)$$

Ammonia reacts with hypochlorite to form chloramine and hydroxide.

Adsorbents

Silica gel (SiO_2), alumina (Al_2O_3), and activated charcoals (C) make up a diverse class of materials used as adsorbents. Adsorbents are porous solids with a large surface area for their size. They are commonly used in chemical separation columns and as supports for catalysts in industrial reactors. They bind liquid or gaseous molecules to their surfaces. The adsorbed molecules can often be driven off the surface by heating to regenerate the adsorbent. Packets of silica gel are added as a water-absorbing agent in the storage of electronic equipment, vitamins, cameras and other goods that are sensitive to high humidity. Small amounts of silica gel additives in powdered foods (powdered coffee creamer, soups, and sugar) absorb moisture and prevent caking of the powder. Because silica gel is essentially sand with a large surface area, it is a nontoxic food additive.

Silica gel is made by adding acid to an aqueous solution of sodium silicate (Na_4SiO_4). The gelatinous precipitate contains a network of silicon atoms linked by oxygen bridges. This network forms around large pockets of the water solvent. The gel is dried after washing away the salt formed in the neutralization reaction. The dry solid, SiO_2, is called silica gel. When crushed to a free-flowing powder, this material has a large surface area because of its porous structure. Silica gel can absorb about 40% of its own weight of water. One gram of silica gel has a surface area of 800 square meters. Thus about 5.5 grams of silica (one heap-

Figure 6.11.
A fossilized diatom found in a sample of diatomaceous earth, as seen under a microscope. (Courtesy of the Celite Corporation.)

ing tablespoon) contains the same surface area (square footage) as a football field! Highly divided or porous solids have a very large surface area relative to a solid cube of the same substance.

A special porous form of silicon dioxide is found in diatomaceous earth. This consists of fossilized diatoms (single-celled plankton or algae) from ancient seas, whose structure is preserved as a SiO_2 shell. Figure 6.11 shows a micrograph of an individual particle in diatomaceous earth. The pores responsible for the large surface area are clearly delineated.

Often cobalt chloride is added to silica gel to act as a blue indicator. It is blue when the silica gel is dry and turns pink when the silica gel can't absorb anymore water. The blue form regenerates when the silica gel is heated to 300°C (570°F), which removes adsorbed water by vaporizing it.

Alumina (Al_2O_3) is a refractory material obtained from the mineral bauxite. Its melting point is 1900°C (3450°F). Many different crystalline forms of alumina exist, including sapphire. High-surface-area alumina finds use in industry as a catalyst and adsorbent. Automobile catalytic converters contain small amounts of a platinum, palladium, and rhodium alloy dispersed on the surface of an alumina support. Alumina is prepared by heating hydrated aluminum oxides, such as boehmite or diaspore, $Al_2O_3(H_2O)$, or gibbsite, $Al_2O_3(H_2O)_3$, below 1000°C. Powdered alumina absorbs water strongly to evolve heat. As in silica gel, the process reverses on heating (300–500°C) to drive away the adsorbed water as vapor.

Common barbecue charcoal is made by heating wood in the absence of oxygen. This process drives away water and other volatile chemicals. The residue contains carbon and the nonvolatile inorganic salts of the plant (NaCl, KCl, phosphate salts, and calcium salts). These salts leave behind a mineral ash when charcoal (or wood) is burned. Activated charcoal is formed by heating sawdust, animal bones (animal charcoal), and other plant or animal matter in the absence of oxygen. Additives (NaOH or H_2SO_4) help dehydrate the organic compounds in this process. Salts that remain wash away upon treatment with aqueous acids, so the product consists only of carbon. The manufacture of white cane sugar requires large amounts of purified activated carbon to remove colored impurities. Charcoals also help remove toxic chemicals in air pollution control, in gas masks, and in waste water treatment. Activated charcoal filters for purifying tap water are available in most hardware stores. One cubic inch of a high-quality activated charcoal contains surface area equivalent to that of five football fields.

Petrochemicals

Organic chemicals for the synthesis of plastics, paints, detergents, tires, and medicines come from oil. Although most oil is burned as an energy source, about 3% goes into the synthesis of organic chemicals, and half of this is used to make plastics. The processing of oil requires large-scale chemical plants (Figure 6.12).

Much effort is directed toward separating the various components. A barrel of Texas crude fresh from the ground contains a small (4%) amount of gaseous molecules (C_nH_{2n+2}, where

Figure 6.12. *A modern oil refinery. The high towers are for separating the components of oil by fractional distillation. (Courtesy of Amoco Photo Lab, Texas City Refinery.)*

$n = 1$ to 4—methane, ethane, propane, and butane) and larger fractions of liquid hydrocarbon products with the approximate compositions shown in Table 6.3. Each type of oil deposit differs somewhat in its composition. Oil refineries separate the various components in oil according to their boiling points (distillation). The percentage compositions in Table 6.3 represent those for a crude oil found in the midcontinental United States. Sometimes the gasoline fraction is further separated into light-boiling petroleum ether (boiling point 35–60°C) and ligroin (or naphtha, boiling point 60–100°C) fractions. At least 200 different organic chemicals occur in crude petroleum. Some of the paraffin molecules that are most abundant in oil appear in Figure 6.13. Paraffins are those molecules in which the carbon atoms are connected only by single bonds.

Another class of molecules found in oil are called cycloparaffins or naphthenes. They resemble paraffins, in that they

Table 6.3.   *Approximate Breakdown of Hydrocarbons Found in Oil*

Compound	Molecular Formula	Boiling-Point Range (°C)
Gases (4%)	CH_4–C_4H_{10}	< 40
Natural gasoline (33%)	C_5H_{12}–$C_{10}H_{22}$, aromatics	40–180
Kerosene (13%)	$C_{11}H_{24}$–$C_{12}H_{26}$, aromatics	175–230
Light gas oil (19%)	$C_{13}H_{28}$–$C_{17}H_{36}$	230–305
Heavy gas oil (14%)	$C_{18}H_{38}$–$C_{25}H_{52}$	305–405
Lubricating oil (10%)	$C_{26}H_{54}$–$C_{38}H_{78}$	405–515
Asphalt (7%)	Polycyclic structures	Nondistillable solid

contain no carbon–carbon double bonds, but the molecules consist of cyclic units, as shown in Figure 6.13. Crude oil from certain regions, such as Pennsylvania, contains predominantly paraffins. Such oil is referred to as paraffin-based crude. Other

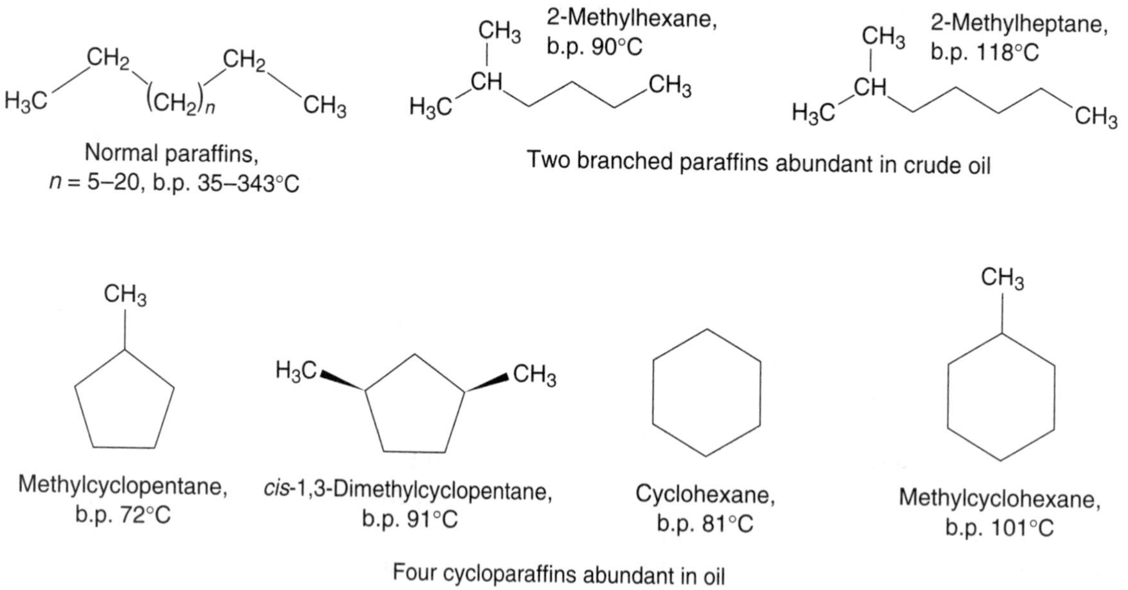

Figure 6.13.   *Some paraffin molecules found in crude oil.*

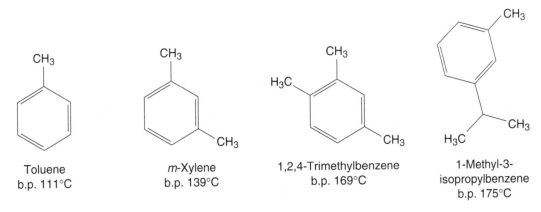

Toluene	m-Xylene	1,2,4-Trimethylbenzene	1-Methyl-3-
b.p. 111°C	b.p. 139°C	b.p. 169°C	isopropylbenzene
			b.p. 175°C

Figure 6.14. *Four abundant alkylbenzenes found in crude oil.*

petroleum, such as that from California, contains mainly cyclo-paraffins. It is called naphthene-based crude. Midcontinental oil from Texas and Oklahoma, called mixed-base crude, contains both components in comparable amounts.

The unsaturated carbon compounds (those containing car-bon–carbon double bonds) that are found in crude oil mostly contain aromatic rings, the bulk being alkyl substituted ben-zenes. Most aromatic compounds contain one or more six-mem-bered rings of carbon atoms with alternating double bonds in them. Some of the most abundant alkyl substituted benzenes found in oil are shown in Figure 6.14.

Petroleum ether and ligroin fractions of oil find application as solvents. Kerosene can be used as diesel or jet engine fuel. High-boiling gas oil is used for heating boilers. Of all the petro-leum fractions, gasoline is in the greatest demand and com-mands the highest price. Chemical treatment helps maximize the gasoline content before its separation from oil by fractional distillation. Catalytic cracking (the breaking of carbon–carbon bonds) of long-chain hydrocarbons in kerosene and gas oils into lower-boiling components improves the yield of gasoline. Crude oil also contains organic sulfur compounds that can destroy the activity of many of the catalysts used in its chemical treatment. Thus, desulfurization of crude oil is the first step in petroleum processing.

Figure 6.15. *The molecular structure of thiophene, an organic sulfur compound in oil.*

Catalytic Desulfurization

The removal of organic sulfur compounds, such as thiophene (Figure 6.15), occurs by treating oil with hydrogen over a molybdenum disulfide desulfurization catalyst.

Molybdenum disulfide (MoS_2) is an unusual solid in its own right. The structure of MoS_2, shown in Figure 6.16, resembles the layered structure of graphite. Only weak forces exist between the layers, so they slide easily past each other. This makes MoS_2 useful as an automotive lubricant. In the oil industry, molybdenum disulfide catalysts help remove sulfur from oil as gaseous hydrogen sulfide, $H_2S_{(g)}$. Gaseous hydrogen sulfide separates easily from the liquid hydrocarbons in oil. Thiophene undergoes desulfurization to form butadiene and hydrogen sulfide according to equation (27).

$$C_4H_4S_{(l)} + 2H_{2(g)} \xrightarrow{MoS_2} H_2S_{(g)} + C_4H_{6(g)} \qquad (27)$$

Thiophene reacts with hydrogen in the presence of the catalyst molybdenum disulfide to form hydrogen sulfide and butadiene.

Residual sulfur compounds in commercial gasoline (no process is ever 100% efficient) are often detectable by the smell of rotten eggs, which is characteristic of hydrogen sulfide. Catalytic converters in automobiles catalyze the reaction shown in equation (27). The small amount of $H_2S_{(g)}$ produced may be responsible for a foul-smelling automotive exhaust.

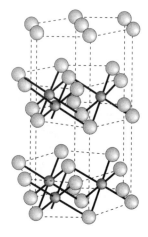

Figure 6.16. *Three-dimensional packing diagram of MoS_2, a desulfurization catalyst. (From Duward F. Shriver, P. W. Atkins, and Cooper H. Langford,* Inorganic Chemistry *© 1990, Oxford University Press. Redrawn by permission of Oxford University Press.)*

Catalytic Cracking

After desulfurization of oil, the heavier hydrocarbons (kerosene and gas oil) may be heated under pressure over a cracking catalyst. This breaks them down into gasoline. A mere 1% increase in the conversion efficiency of this process reduces U.S. oil imports by 22 million barrels per year. (And this reduces the trade deficit by $400 million.) A major advance in catalytic cracking technology occurred with the introduction of zeolites as catalysts for this process in the early 1960s. Zeolites are aluminosilicate minerals built of a network of aluminum, silicon, and oxygen atoms, with pores of molecular dimensions. For this

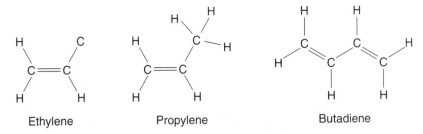

Figure 6.17. *The molecular structures of ethylene, propylene, and butadiene. These molecules are generated by catalytic cracking of larger molecules in crude oil.*

reason, chemists refer to them as molecular sieves. Besides separating molecules from one another according to size, zeolites contain acidic sites within their pores that help catalyze the cracking of long-chain hydrocarbons into shorter ones.

Unsaturated hydrocarbons, such as ethylene, also form under the harsh catalytic cracking conditions in reactions like equation (28).

$$CH_3-CH_{3(g)} \xrightarrow{\text{catalyst}} CH_2{=}CH_{2(g)} + H_{2(g)} \qquad (28)$$

Ethane cracks into ethylene and hydrogen.

A modification of temperature and catalyst maximizes formation of the unsaturated hydrocarbons ethylene, propylene, and butadiene. The structures of these three molecules are shown in Figure 6.17. These are especially valuable chemicals, because the reactivity of the double bond permits them to be modified to synthesize complex organic chemicals. Because of the tremendous volume of oil processed (55–60 million barrels a day), desulfurization and catalytic cracking amount to the largest application of catalysis in industry.

Catalytic cracking in oil refineries requires stringent engineering safety procedures, because hot, pressurized hydrocarbon gases will explode in the presence of oxygen. For this reason, oil refineries were originally situated several miles from population centers. Although accidents are rare, the results are devastating when they occur, The tremendous economic damage and loss of life suffered in refinery accidents motivate both management and employees to adopt stringent safety measures.

220

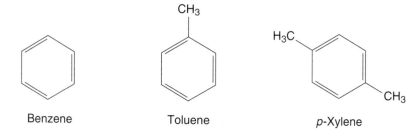

Figure 6.18. *The molecular structures of three aromatic molecules produced by catalytic reforming of crude oil.*

Benzene

Toluene

p-Xylene

Catalytic Reforming

Catalytic reforming is important for the production of aromatic hydrocarbons, such as benzene, xylene, and toluene. The molecular structures of these aromatics appear in Figure 6.18. Catalytic reforming reactions use metal catalysts dispersed on solid oxides of aluminum and silicon. In this reaction, a petroleum fraction and hydrogen gas are heated under pressure in the presence of a catalyst.

Catalytic reforming reactions also transform normal paraffins with a low octane rating into branched paraffins with a higher octane value. The octane number of a fuel is its rating on an arbitrary scale; it reflects the amount of engine knocking. Isooctane is assigned an octane number of 100, and n-heptane, a fuel that causes severe engine knocking, is assigned an octane number of

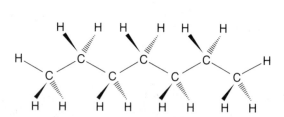

n-Heptane, a poor engine fuel with an octane number of zero

Isooctane, a good engine fuel with an octane number of 100

Figure 6.19. *The molecular structures of n-heptane and isooctane. These two chemicals are used to define the octane number used to characterize gasoline.*

zero. The molecular structures of these two compounds appear in Figure 6.19. Automobiles require gasoline fuel with an octane rating of at least 87 to run smoothly. In the past, low-octane gasoline could be boosted to a higher rating by the addition of *tetraethyl lead*, $Pb(C_2H_5)_4$ (Figure 6.20). Hence the name *leaded gasoline*. Given the high toxicity of lead, and given the millions of tons introduced into the environment by the combustion of leaded gasoline, the switch to unleaded gasoline was sensible. Catalytic reforming grew even more important when unleaded gasoline became the standard for new automobiles, because this process produces molecules with high octane ratings. For example, toluene has an octane rating of 110.

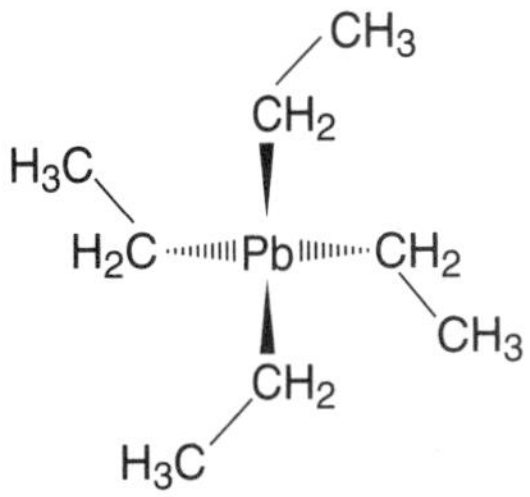

Figure 6.20. *The molecular structure of tetraethyl lead, the octane enhancer of leaded gasoline.*

Unleaded gasoline also contains benzene. Benzene is a weak human carcinogen. Toluene, on the other hand, is not carcinogenic. For many individuals, their chief exposure to carcinogenic chemicals such as benzene occurs when they pump gasoline or change the oil in their cars. The amendment to the U.S. Clean Air Act, passed in 1990 and requiring a reduction in the levels of benzene in gasoline, stimulated the development of new processes for benzene removal. Alternative octane enhancers, such as ethanol and *MTBE* (methyl–*t*–butyl ether, $CH_3OC(CH_3)_3$), shown in Figure 6.21, have been introduced to make high-octane gasoline that has a reduced content of aromatics and reduced emissions of pollutants.

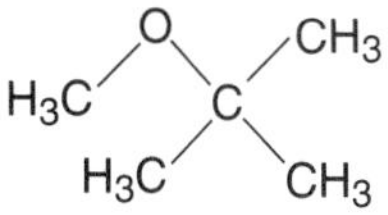

Figure 6.21. *The molecular structure of MTBE, an additive in high-performance, low-emission unleaded gasolines.*

Catalytic Hydrogenation of Benzene

Benzene can be removed from gasoline by a reaction in which it is hydrogenated to cyclohexane (noncarcinogenic) as shown in equation (29).

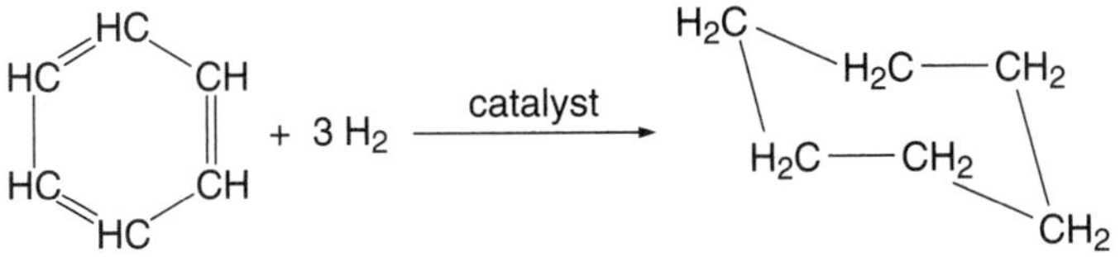

(29)

Benzene reacts with hydrogen to form cyclohexane in the presence of a catalyst.

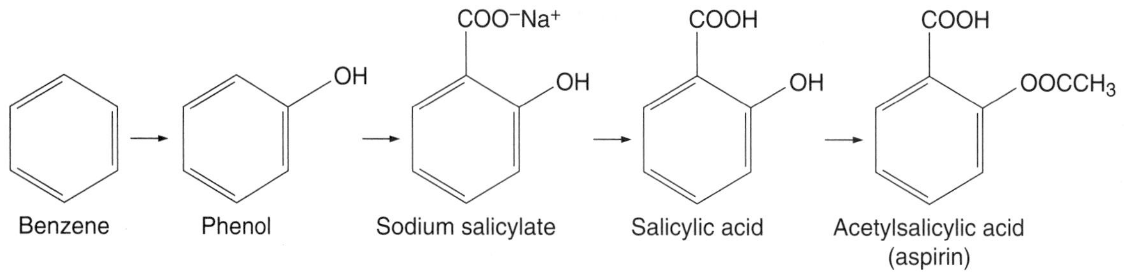

Figure 6.22. *The synthesis of aspirin from benzene, a carcinogenic aromatic chemical obtained from oil.*

In general, the hydrogenated product has a lower octane number than the aromatic precursor. Thus, in manufacturing high octane gasoline, it is desirable to selectively hydrogenate benzene, leaving toluene unreacted. Chemists in oil companies are actively seeking selective catalysts for this process. An alternative approach, converting benzene to less noxious alkylbenzenes by existing technology, is in use.

The oil-based economy extends far beyond the gasoline that is produced for automobiles. Aspirin synthesis (Figure 6.22) begins with benzene, which comes from an oil refinery. Various other chemicals are used to bring about the changes indicated in each step of the aspirin synthesis. Many other substances in the home and workplace—clothes, tires, plastic casings on telephones, plastic coatings on computer chips, and cough medicine—come from petrochemicals. These and other materials of commerce are considered in the following chapter.

CHAPTER 7

Synthetic Materials

Wallace H. Carothers

Charles M. A. Stine, the head of E. I. du Pont de Nemours's Central Research Department, had put his career on the line before the company's executive committee. He wanted them to back an effort in fundamental research, as opposed to the applied research already supported at du Pont. In 1927 Stine was given money to build and staff a new laboratory along the banks of the Brandywine River at du Pont's Experimental Station in Wilmington, Delaware. Organic chemistry was an area crucial to the company's business, and an outstanding individual was needed to head research in that area. Stine successively offered the job to two prominent academicians, but they declined. Then Stine approached several young professors whose stars were on the rise—but again, no luck. In desperation, the job was offered to a bright 31-year-old chemistry instructor at Harvard, Wallace H. Carothers. The young chemist with fanciful

Wallace Carothers (Courtesy of the Hagley Museum and Library, Wilmington, DE.)

ideas about polymers joined du Pont in February 1928.

Carothers was an academic idealist throughout his career. He regarded any business gain he realized from basic

223

research as incidental to the pursuit of knowledge. He frequently found himself at odds with management, especially after Stine's promotion to the executive committee. Nevertheless, in 1930 Carothers's research group discovered (by chance) the first commercially successful synthetic rubber, neoprene. In other experiments, Carothers proved that polymers are very large molecules made by successively joining small molecules with covalent bonds. He showed that polymers form by two fundamental reactions, which he called addition and condensation. He recognized that the successful construction of certain high-molecular-weight polymers requires chemicals of exceptional purity (99.5% isn't nearly good enough). Carothers, the chemical idealist, can be credited with launching the most important and lucrative application of organic chemistry: synthetic polymers.

Wallace Carothers never lived to receive a Nobel prize. Since his days at Harvard, he had suffered from bouts of nervous depression. In a letter to a friend, he expressed reservations about moving to du Pont owing to "neurotic spells of diminished capacity which might consti-

tute a much more serious handicap there than here." Although Carothers saw his discovery of neoprene rubber put into commercial production by 1932, he wrestled with serious self-doubt about his worth as a scientist. Under continued pressure from management, his research group focused on making their newly discovered polyester and polyamide (nylon) condensation polymers suitable for commercial application. Key results they obtained in 1935 showed that nylon polymers could be drawn into strong synthetic fibers that behaved like artificial silk. Shortly thereafter, Carothers succumbed to severe clinical depression. Neither his marriage nor his election to the National Academy of Sciences in 1936 could halt the deepening despair. In January 1937, Wallace's beloved sister Isobel (a nationally known radio entertainer) died at age 31 from pneumonia secondary to a strep infection. She left behind a husband, a young adopted child, and a tortured older brother. Wallace H. Carothers drank a cyanide solution on April 29, three weeks after the patent for nylon was filed and six months before the birth of his daughter Jane.

Petrochemical–Based Polymers: Rubber, Textiles, and Plastics

Many useful polymeric solids derive from oil. Clothing, non-stick cookware, and bulletproof vests all rely on polymer technology. Lightweight plastics are essential for the manufacture of fuel-efficient cars. Polymer water pipes are rapidly replacing metal plumbing as a means of reducing the release of toxic heavy-metal ions by corrosion of metal-plumbed systems. If you stop and look around, you will be surprised at the amount of plastics and synthetic fibers in the average home or office. The odds are that you have synthetic rubber soles on your shoes and are reading this book through glasses with plastic lenses and frames (or perhaps polymer contact lenses). Polymers are long molecules built by linking many similar small molecules called monomers. Proteins, rubber, and plastics are all examples of polymers.

Addition Polymerization

Polyethylene is the simplest polymer. It forms by making carbon–carbon bonds between adjacent molecules of gaseous ethylene (C_2H_4). Polymerization is initiated by oxygen at high temperature and pressure, by supported metal catalysts (such as chromium on alumina), or by titanium-based catalysts in solution. Figure 7.1 shows how polyethylene forms after an initiator, such as O_2, adds to an ethylene molecule and leaves an unpaired electron, signified by the dot, on one carbon atom. Recall that such a species is a free radical. The free radical can add to the double bond of another ethylene molecule to form a new carbon–carbon bond and generate an unpaired electron on the end of the growing chain. Because the polymer forms by adding ethylene molecules to the growing chain, the process is termed *addition polymerization*. Polyethylene's mechanical strength depends on the conditions (temperature, pressure, and catalyst) of polymerization. As one of the least expensive polymers, it is used in plastic bags for packaging (sandwich and food storage bags are polyethylene), in electrical insulation, and as a plastic material (for assorted kitchenware containers). Its low melting point (85–110°C), near that of boiling water, limits the possible applications. This is why polyethylene food containers often deform when used in microwave ovens.

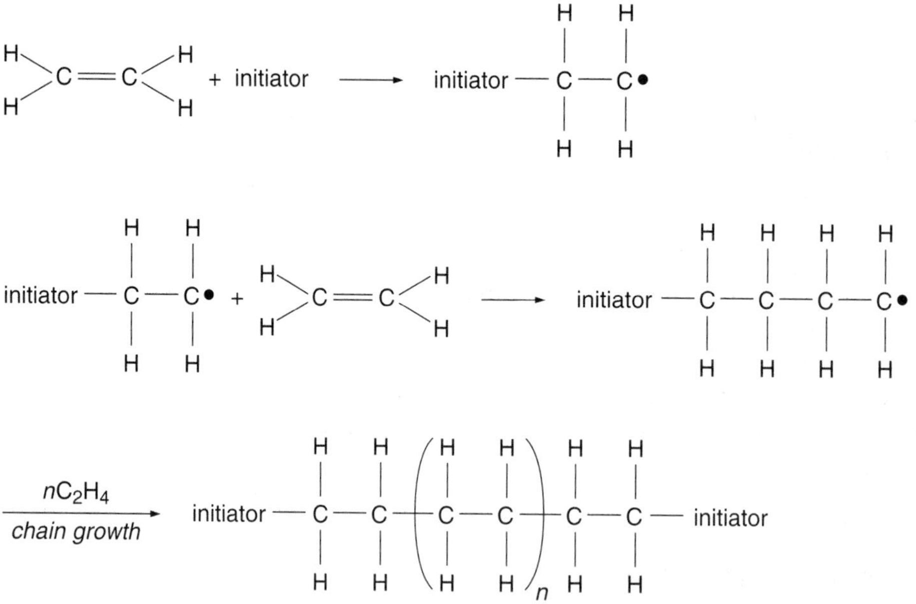

Figure 7.1. *Addition polymerization of ethylene forms solid polyethylene. Typical values of* n *are 500–1000 for the polymers.*

A stronger and higher-melting (165°C) plastic forms by addition polymerization of propylene. Polypropylene's major uses are in the automobile industry and as an abrasion-resistant fiber for the manufacture of ropes, carpets, shoe soles, and clothing. The low flammability of polypropylene, and its moisture permeability, also make it suitable for athletic apparel. Polypropylene long underwear for outdoor winter sports is especially popular. Many high-quality loudspeakers use polypropylene cones. The high thermal stability of polypropylene makes it especially useful for automotive speakers exposed to direct sunlight. And its high strength makes it suitable for automobile bumpers.

Other common polymers derived from polymerization of a substituted ethylene are listed in Table 7.1. Many are called vinyl addition polymers, because they originate with monomers that contain a vinyl (CH_2=CH−) group. Enhancing the thermal stability, flow properties, color, adhesion, and light stability of polymers frequently requires the addition of other compounds.

Table 7.1 *Structures and Uses of Addition Polymers of Substituted Ethylenes*

Polymer Name	Structure of Repeat Unit	Applications
Polyethylene		Plastic bags and containers, toys
Polypropylene		High-strength plastic bags, structural plastic in automobile interiors, fiber in durable clothing and carpets
Polyvinyl chloride (PVC)		Durable plastic in airline trays, structural plastics, PVC pipe for plumbing, vinyl fake leather upholstery
Polyvinyl alcohol (PVA)		Water-soluble polymer used in eye drops, shampoos, cosmetics, inks, textile sizes, coatings, and plastics
Polyacrylonitrile (acrylics, Orlon® fibers)		Super glues, adhesives, carpet and clothing fibers
Polymethylmethacrylate (Lucite® or Plexiglas®)		Automobile tail light covers, lenses, plastic windows
Polytetrafluoroethylene (Teflon®)		Nonstick cookware, low-friction gears, corrosion-resistant containers, parts, and coatings
Polystyrene		Plastics, thermal foam and electrical insulation, foam cups, picnic plates, and foam ice chests

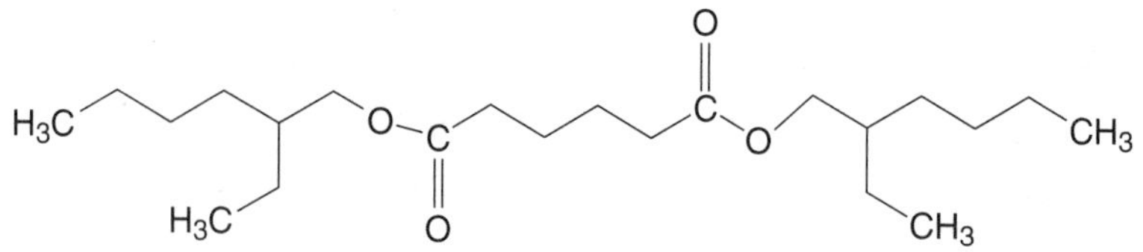

228

Figure 7.2. *Di-2-ethylhexyl adipate, an additive that imparts plasticity to PVC film.*

The development of Teflon® cookware forced chemists to discover glues to bind the slippery polymer (often used as a lubricant) to a metal surface. Strong polyvinyl chloride film, used as a food wrap, requires the addition of a plasticizer, such as DEHA (di-2-ethylhexyl adipate), to impart the flexibility necessary for the film to cling. The molecular structure of DEHA is given in Figure 7.2. There has been concern about plasticizer leakage into food. Foods that contain fats or oils can slowly dissolve organic plasticizers. This process speeds up with heated foods. The British Committee on Toxicity of Chemicals in Food, Consumer Products, and the Environment discourages the use of plastic food wrap in microwave ovens.

One important trick in polymer chemistry is the use of *cross-linking*, which adds rigidity to a polymer. Polystyrene is a plas-

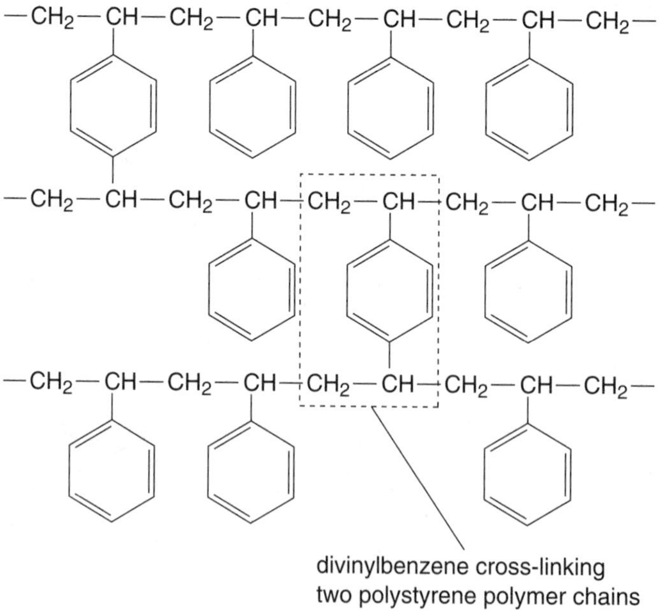

Figure 7.3. *Illustration of how the polymer chains in polystyrene can be cross-linked with divinylbenzene to produce a rigid plastic.*

divinylbenzene cross-linking two polystyrene polymer chains

tic material that will dissolve in organic solvents such as benzene to give a thick syrup of long polymer molecules. If a small
amount (1–20%) of 1,4-divinylbenzene is added to styrene during its polymerization, a rigid, insoluble polymer forms. Divinylbenzene is called a cross-linking agent because it stitches together the individual polystyrene chains to produce a strong three-
dimensional cross-linked network (Figure 7.3).

It is also possible to polymerize a mixture of ethylene and
propylene. This yields a copolymer that contains a random distribution of ethylene and propylene units in the polymer chain.
Unlike either polyethylene or polypropylene, the ethylene–
propylene copolymer is a nonrigid elastic material that can be
modified for use in synthetic rubber tires.

Natural and Synthetic Rubber

Natural rubber, or latex rubber, is a sticky polymer that consists
of 93% polyisoprene, as depicted in equation (1).

Isoprene polymerizes into polyisoprene.

Raw latex from the sap of the rubber tree consists of a milky suspension of rubber particles in water. Much of the water is spun
out in a centrifuge before further processing. In 1846 the U.S.
chemist Charles Goodyear discovered that this sticky substance
could be heated with sulfur (a process called vulcanization) to
prepare a useful, nontacky rubber. Vulcanization cross-links the
chains of natural rubber by forming disulfide bonds (–S–S–)
between them, much the same way that divinylbenzene cross-
links polystyrene. By the early 1900s, vulcanized latex rubber
had become a widely used material. It continues to be used for
the manufacture of condoms.

Access to natural-rubber plantations in Malaysia was limited during World War II. Because of the strategic importance of automobile and truck tires to the war effort, chemists in the Allied countries had to develop synthetic substitutes. One obvious strategy was to make isoprene and polymerize it, but isoprene was too expensive for this to be practical. The related chemical butadiene, obtained inexpensively from oil, polymerizes to make the polybutadiene elastomer shown in equation (2). (A polymer with elastic, rubbery properties is called an elastomer.) The first polybutadiene polymers were unsatisfactory as a substitute for natural rubber.

$$n \ H_2C\!\!=\!\!CH\!-\!CH\!\!=\!\!CH_2 \quad \rightarrow \quad \left[CH_2\,\overset{\displaystyle H}{\diagdown}C\!=\!C\overset{\displaystyle H}{\diagup}\,CH_2 \right]_n \qquad (2)$$

Butadiene polymerizes in polybutadiene.

Chlorobutadiene had been polymerized earlier by Carothers to make neoprene rubber (employed in oil-resistant gaskets and hoses), but its cost made it unsuitable as a natural-rubber substitute.

Through additional experimentation, chemists discovered a copolymer between butadiene (~75%) and styrene (~25%) that gave a SBR (styrene–butadiene rubber) suitable for use in automobile tires. The process for making polybutadiene was later improved enough so that it too could be used as a substitute for natural rubber. This primarily involved learning how to ensure that the double bonds in polybutadiene had the two hydrogen substituents on the same side of the double bond (called the *cis* configuration, shown in Figure 7.4). A polymer that contains a mixture of *cis* and *trans* double bonds produces a rubber with poor properties. Natural isoprene rubber also contains a *cis* configuration about the C=C double bond in the polymer. Modern radial tires employ polybutadiene and natural rubber, whereas bias ply tires use SBR copolymer. The copolymer between ethylene and propylene is also finding increasing use as a synthetic rubber.

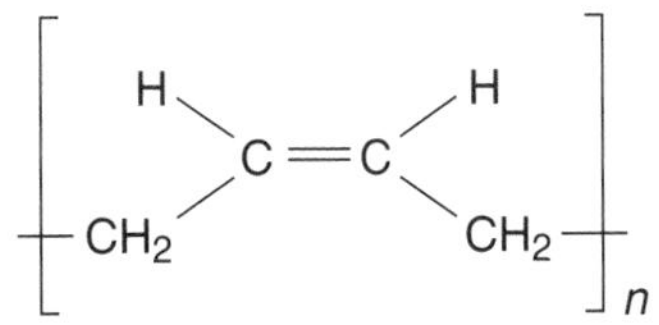

cis-Polybutadiene, a
useful synthetic rubber

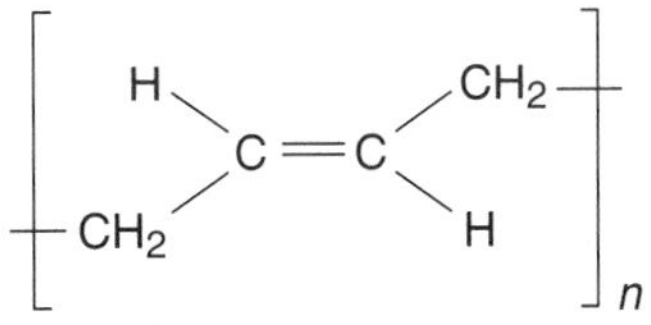

trans-Polybutadiene, an
inferior synthetic rubber

Figure 7.4. *The two possible geometries of the butadiene molecule in a polybutadiene polymer.*

The presence of the carbon–carbon double bonds in natural and synthetic rubber provides reactive sites where atmospheric oxidants, such as oxygen and ozone, can attack the structure. This breaks the polymer chain and causes the rubber to fragment. That is why old rubber tends to crumble. Ozone, a component of photochemical smog, is exceptionally reactive. Rubber objects age rapidly in cities with severe ozone pollution.

The main problem facing rubber manufacturers is difficulty in recycling used tires. Of the 200 million scrap tires produced each year in the United States, only 20 million are recycled. The others find their way to landfills or accumulate in above-ground storage areas. Even incineration of tires is problematic, because the cost of transporting the tires must be deducted from the value of the energy recovered when they are burned. An economical solution remains to be found.

Condensation Polymerization: Synthetic Fibers

Addition polymers form by the stepwise addition of a single monomer to a growing polymer chain. All atoms of the monomer molecules are retained in the polymer. This is not the case in a *condensation polymerization* reaction. Two different monomers are generally used to grow condensation polymers. Each monomer has two identical reactive groups, represented by the Xs and Ys in Figure 7.5. Condensation polymers form by a chemical reaction wherein each monomer combines with a reactive end of the growing polymer chain, eliminating a small molecule, such as water. Wallace Carothers was the first to propose and make practical the reaction sequence shown in Figure 7.5. His fundamental research on polyester and polyamide fibers paved the way for the commercial development of synthetic fibers.

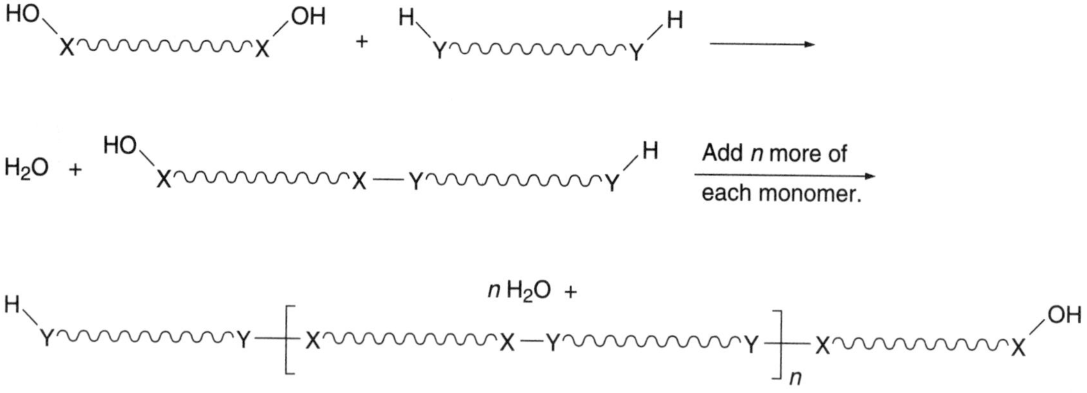

Figure 7.5. *Schematic of the formation of a condensation polymer from two different monomers with reactive ends X and Y. The attachment of monomers to the polymer occurs with the loss of water.*

Several examples of such condensation polymers appear in Table 7.2. Epoxy polymer glue works by forming a condensation polymer. Epoxy glue is sold as two separate ingredients, which form an adhesive when mixed. The two components react to make a condensation polymer on mixing, which makes the glue set. This contrasts with other polymer glues, such as a cyanoacrylate super glue, which comes in a single tube. Here the exposure to air initiates an addition polymerization reaction that causes the glue to set. Ultraviolet light can also initiate addition polymerization of cyanoacrylate glues. When mixed with a white pigment, cyanoacrylate polymers are used by dentists as tough plastic caps for damaged teeth.

Nylon-6,6 (the 6,6 denotes condensation of two 6-carbon monomers) melts near 250°C (482°F), and it is an important commercial condensation polymer. The two monomer components for nylon-6,6 are adipic acid (1,6-dihexanoic acid) and hexamethylene diamine (1,6-diaminohexane). When combined, the acid (COOH) groups transfer protons to the amine (NH_2) groups to form a salt, as shown in Figure 7.6. Subsequent heating of the salt to 275°C eliminates water as steam. At the same time, an amide bond (−NHC(O)−) forms between the carboxylic acid (−COOH) end groups and the amine (−NH_2) end groups, which yields the polymer shown in Table 7.2. Proteins in the body are put together by this same process of amide bond for-

Table 7.2 *Examples of Condensation Polymers*

Polymer Name	Structure of Repeat Unit	Applications
Nylon-6,6	$\left[\!-N(H)-(CH_2)_6-N(H)-C(=O)-(CH_2)_4-C(=O)-\!\right]$	Nylon clothing, ropes, carpets, tents, and injection-molded machine parts
Polyesters (Dacron® fiber, Mylar® film)	$\left[\!-C(=O)-C_6H_4-C(=O)-O-CH(H)-CH(H)-O-\!\right]$	Polyester clothing and carpet fibers (Dacron®) and Mylar® plastic film
Polycarbonate (Lexan® plastic)	$\left[\!-O-C_6H_4-C(CH_3)_2-C_6H_4-O-C(=O)-\!\right]$	High-impact plastics and electrical insulation
Polyurethane	$\left[\!-C_6H_3(CH_3)-N(H)-C(=O)-OCH_2CH_2(OC(CH_2)_4COCH_2CH_2)_{5-10}O-C(=O)-N(H)-\!\right]$	Foam padding and insulation, varnishes and coatings
Epoxy resins	$\left[\!-O-C_6H_4-C(CH_3)_2-C_6H_4-O-CH_2-CH(OH)-CH_2-\!\right]^{*}$	Adhesives, molding plastics, fiberglass resins, and coatings

*Curing of epoxy resins requires cross-linking the polymer chains terminated with epoxide groups by the reaction with polyamines such as $NH(CH_2CH_2NH_2)_2$.

mation. The making of an amide bond (equation 3) results in the elimination of water, and it is important to recognize that this reaction can be reversed. This represents a weakness of nylon polymers. Acids can catalyze the reverse reactions of those shown in Figure 7.6 and decompose the nylon polymer back into monomers.

$$RCOOH + H_2NR' \rightarrow RC(O)NHR' + H_2O \qquad (3)$$

A carboxylic acid reacts with an amine to form an
amide and water. (R and R' represent organic groups.)

Thus nylon and polyester polymers (Table 7.2) show much less chemical resistance than most addition polymers. Concentrated sulfuric acid attacks the addition polymer polyethylene very slowly, but nylon and polyesters decompose instantly.

The du Pont company built the first large-scale plant to make nylon in Seaford, Delaware, in 1938. It cost $8 million to build. About 11.6 billion pounds of nylon polymers are now produced worldwide each year. That's more than two pounds for every person on the planet! What makes nylon an especially useful polymer is its ability to undergo melt-spinning into a fiber of high strength. This involves rapidly pushing the molten polymer through a small hole, whereupon it cools to yield threads as shown in Figure 7.7. Nylon fibers are widely used to make carpets, clothing, automotive air bags, ropes, and the reinforcing cord in tires. One square inch of nylon rope will support

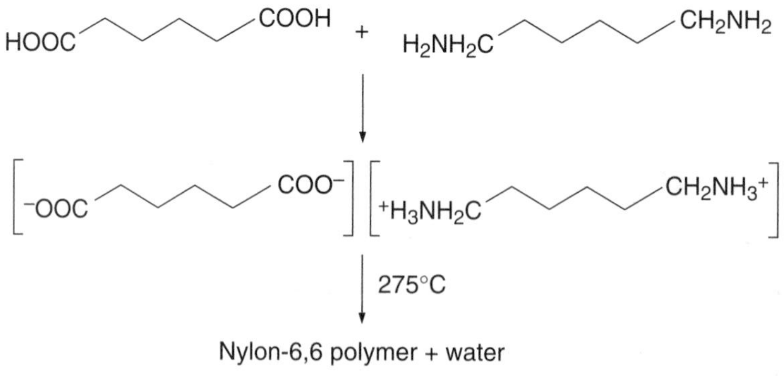

Figure 7.6.   *The synthesis of nylon from the monomers adipic acid and hexamethylene diamine*

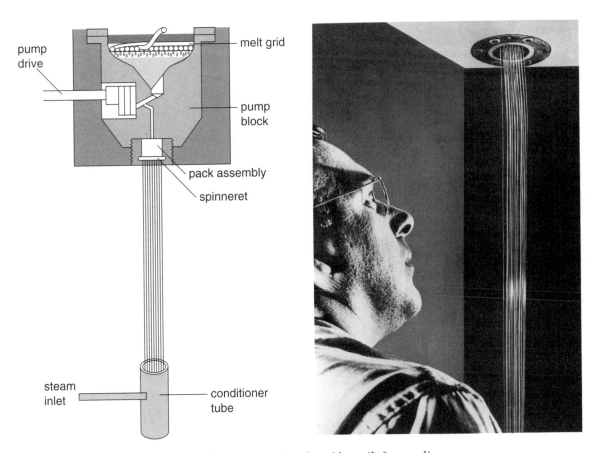

Figure 7.7. *Diagram of the melt-spinning of nylon fibers (left panel) and the extrusion of individual fibers from holes in the spinneret (right panel). (Figure redrawn from H. F. Mark, S. M. Atlas, and E. Cernin, Man-Made Fibers, Volume 2, copyright © 1968 by John Wiley & Sons, Inc. Reprinted by permission of John Wiley & Sons, Inc. Photo courtesy Hagley Museum and Library, Wilmington, DE.)*

60,000 pounds before it breaks. The molten polymer can also be injection molded into machine parts.

Kevlar® is a more recently discovered type of nylon condensation polymer made by the reaction between *para*–phthalic acid and *para*–diaminobenzene, as shown in Figure 7.8. Like nylon-6,6, Kevlar® forms by the elimination of water, and its two components are joined by amide bonds. It exhibits exceptional strength as a fiber, which makes it useful in high-strength truck tires and in bulletproof vests. Kevlar® also has better thermal stability than other nylons.

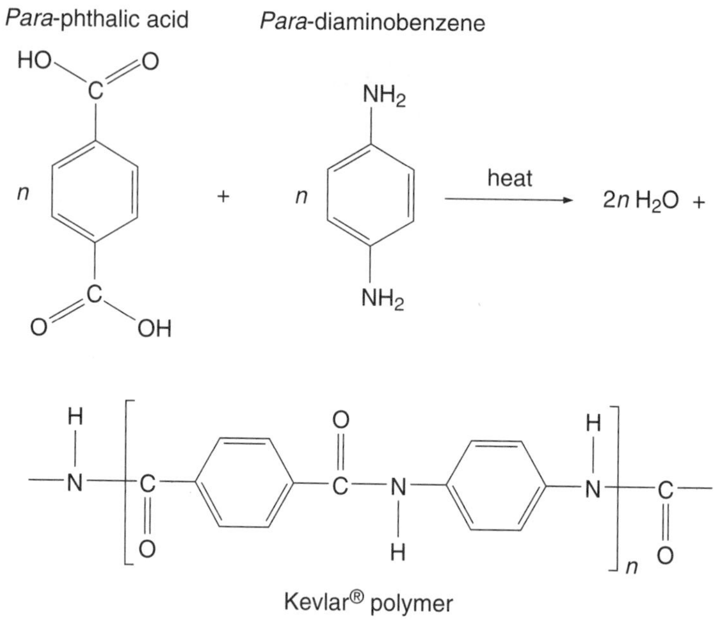

Figure 7.8. *The synthesis of Kevlar®* *from* para-*phthalic* *acid and* para-*diaminobenzene.*

Stability of Organic Polymers

Organic plastics exhibit limited stability in air at elevated temperatures. Heating may break a carbon–carbon bond in the polymer backbone, and oxygen in the air can attack reactive atoms at the break and initiate combustion. But the instability of certain organic polymers may be advantageous. Much research focuses on making biodegradable polymers for use in disposable products. Products made from biodegradable polymers would have minimal long-term effects on the environment.

Because an organic polymer is a solid derived from oil, it burns just like a hydrocarbon fuel to produce carbon dioxide and water. Polyethylene has about the same fuel content as an equivalent weight of gasoline. Replacing all the hydrogen in a polymer with an element that cannot be oxidized, such as the fluorine in Teflon® (see Table 7.1), produces an organic polymer with better thermal stability. Even partial substitution of hydrogen with chlorine, as in polyvinyl chloride (see Table 7.1), imparts fire resistance to an organic polymer. Oxidation-resistant polymers attracted public attention in the 1970s when it was realized that they could generate toxic gases in fires. Polyvinyl chloride, a fire-resistant plastic used in airplanes and in vinyl upholstery for cars, generates phosgene gas (Cl_2CO) when it burns. Phosgene gas was used as a chemical-warfare agent in World War I. Intense, irritating, black smoke made it impossible to control a fire on the British destroyer *Sheffield* when it was hit by a missile in the Falklands war between Great Britain and Argentina. Ultimately the ship sank. The PVC insulation used to coat the ship's 4 miles of electrical wiring appeared to be the source of the problem.

Inorganic Polymers, Glasses, and Ceramics

Silicon Rubber or Silicones

One way to make materials with greater thermal stability and resistance to air oxidation is to reduce their flammable hydrocarbon content. Inorganic compounds that don't contain carbon offer one possible solution. Hybrid inorganic–organic polymers, such as silicones, retain the desirable elastic properties of organic polymers but have increased resistance to air oxidation. Silicones contain silicon, an element prone to making bonds to oxygen. Unlike carbon, silicon rarely forms double bonds to other elements. Oxygen compounds of silicon primarily contain Si–O–Si single-bond linkages. Silicon–oxygen and silicon–carbon bonds are quite strong.

In 1940 the chemist Eugene Rochow at the General Electric research laboratories discovered an efficient synthesis of organosilicon compounds (equation 4).

$$Si + 2CH_3Cl \xrightarrow{Cu} Si(CH_3)_2Cl_2 \qquad (4)$$

Silicon reacts with chloromethane in the presence of copper catalyst to form dichlorodimethylsilane.

This reaction proceeds at 300°C (570°C) with copper powder as a catalyst. The primary product, dichlorodimethylsilane, can undergo hydrolysis to form a silicone polymer, as shown in

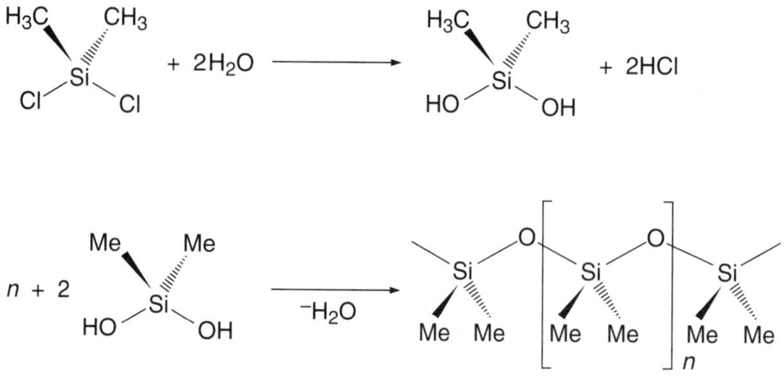

Figure 7.9. *The synthesis of an inorganic silicone polymer by adding water to the monomer dichlorodimethylsilane.*

Figure 7.9. The length of a silicone polymer chain, n, can be controlled to produce temperature-resistant oils, high-temperature hydraulic fluids (automotive transmission fluid), and nontoxic additives for sun tan oils, cooking oils, and cosmetics. The $n = 200$–350 polymer, known as simethicone, behaves as a viscous liquid in Pepto Bismol® to soothe and coat the stomach. Silicone transformer oils serve as nontoxic replacements for PCBs (polychlorinated biphenyls).

Long-chain silicone polymers assume a rubbery consistency if finely divided silica (SiO_2) is added. Silicone rubber is used to make nonflammable suits for fighting intense fires or rubber gaskets for use at high temperatures. Neal Armstrong stepped onto the moon in 1969 wearing silicone rubber boots! Depending on the specific application, the methyl (CH_3) groups in the polymer are replaced with different organic groups. Self-curing RTV (room-temperature-vulcanized) sealant is also a silicone polymer. Most children have had hands-on experience with another silicone polymer called "Silly Putty®."

Silicone gel breast implants came under scrutiny in 1991 because of adverse effects from implants that leaked. The implants consist of a sac of silicone polymer mixed with silicone oil. The polyurethane (see Table 7.2) foam coating on some implants breaks down in the body. Decomposition products include toluene-2,4-diamine (shown in Figure 7.10), a known animal carcinogen. Other concerns about leakage of the silicone oil and resulting autoimmune disorders prompted the U.S. FDA to limit breast implants in 1992 until safety studies were completed. The ban was reaffirmed in 1993. Problems with cosmetic implants raise sobering questions about the chemical compatibility of synthetic materials and body chemicals. Heart valves, joint replacements, limb prostheses, tooth substitutes, and contact lenses all require biocompatible materials. Research initiatives to study biocompatibility problems are in their infancy.

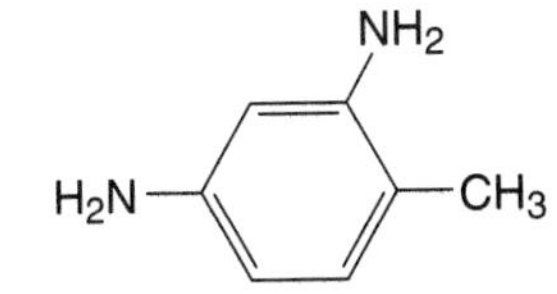

An extremely serious problem is the illegal use of silicone fluid injections to remove skin wrinkles. In this inexpensive procedure, an injected silicone fluid fills out a wrinkle. Later the oil may diffuse slowly through surrounding tissues. Manufacturers of silicone fluids forbid purchase of their product for this use, because industry safety studies show serious complications may arise from silicone injections. But in spite of an FDA ban, some dermatologists inject industrial silicone fluids into patients. Skin ulcers, gangrene, nerve damage, permanent disfigurement, and other complications occur in about 30% of the unsuspecting victims.

Figure 7.10.
Toluene-2,4-diamine, a potentially carcinogenic chemical released by the breakdown of polyurethane foam in the human body.

Fiberglass, Graphite, and Boron Composites

Particulate fillers, such as the carbon black in automobile tires and the SiO_2 in silicone polymers, often increase the stiffness and toughness of a polymer. Materials scientists have discovered that a fibrous filling agent confers remarkable mechanical strength to *composite materials*. The general structure of a composite material is given in Figure 7.11. Fiberglass is a common example of such a material. Fiberglass contains fibers of glass (approximately 20% by volume) embedded in an epoxy resin. This combination produces a tougher and less expensive structural material than the pure epoxy resin. Fiberglass–epoxy laminates are used to make automobile bodies (General Motors Corvette automobiles), boat hulls, fishing rods, and skis.

The specific fiber used in a composite material may be glass, boron, graphite, or even another polymer (fibers of Kevlar®). Epoxy resins reinforced with boron, Kevlar®, and graphite fiber exhibit tensile strengths nearly equal to those of steel and titanium. Because the polymer composites are one-third to one-fifth as dense as metals, the composites are stronger than steel on a

Figure 7.11.
Schematic of a composite material wherein fibers of glass, plastic, graphite, or boron are imbedded in a polymer matrix—often epoxy—to produce a strong, lightweight structure.

Figure 7.12. *Martina Navratilova prepares her composite tennis racquet to volley a low return en route to a Wimbledon championship. (UPI/Bettmann.)*

per-weight basis. Expensive composites are reserved for premium applications, such as the production of airplane parts (helicopter rotor blades), golf clubs, skis, and tennis racquets.

Glass

Glass making is an ancient art. The chemistry relies on the ability of silicon dioxide (SiO_2) to accept oxide (O^{2-}) ions when heated with a metal oxide or metal carbonate. Metal carbonates can evolve CO_2 gas and form a metal oxide upon heating. Thus it is possible to substitute metal carbonates, such as $CaCO_3$, for metal oxides in glass making if the carbonate is less expensive. As an example, simple calcium silicate glass can be prepared by either equation (5) or equation (6).

$$CaO_{(s)} + SiO_{2(s)} \rightarrow CaSiO_{3(s)} \qquad (5)$$

Lime reacts with silica to form calcium silicate glass.

$$CaCO_{3(s)} + SiO_{2(s)} \rightarrow CaSiO_{3(s)} + CO_{2(g)} \tag{6}$$

Limestone reacts with silica to form
calcium silicate glass and carbon dioxide.

The reaction of equation (5) occurs in a blast furnace when pig iron is made. Silica, an impurity in iron ores, does not melt at the temperature of molten iron. When lime is added, the silica reacts to form calcium silicate glass (called slag). The glass melts and floats on top of the liquid iron, where it is easily separated.

Glasses have an ill-defined structure in the solid state—like a suddenly frozen solution. Solids with no crystalline order are called amorphous solids. Chemical compositions of glasses vary, but they usually show an excess of SiO_2. Commercial glass, a hard, chemically resistant substance, may be colored, translucent, or opaque. The few substances that react with glass are those able to cleave the strong Si–O bonds in the glass lattice. Only fluorine forms a stronger bond to silicon than oxygen. For that reason, hydrofluoric acid, $HF_{(aq)}$, is one of the few substances that can dissolve glass. Concentrated alkali solutions (NaOH) slowly dissolve glass by breaking Si–O–Si bonds. The attack of OH^- groups forms new Si–O bonds, as shown in equation (7).

Hydroxide ion attacks a Si–O bond to break apart the structure of glass.

The Egyptians made glass in 1400 B.C. by heating to 800°C (1450°F) mixtures of sand, limestone (dolomite, $CaCO_3 \cdot MgCO_3$), and sodium carbonate. These natural minerals often contain small amounts (0.5–2%) of Al_2O_3, Fe_2O_3, and potassium carbonate. Table 7.3 gives the compositions of several glasses. Unlike most chemical compounds, common glass has a variable composition, and the percentages of the various components can be altered to make glasses with different properties. Pure SiO_2 can itself be melted and cooled to make a superior silica (quartz) glass; however, this was unknown to early societies because it is difficult to obtain the temperatures above 1700°C (3100°F) necessary to melt quartz sand. The composition of window glass has changed very little from the early days, as Table 7.3 shows.

The high chemical stability of glass is evident in lead crystal (see Table 7.3). Lead is an extremely toxic heavy metal, yet crystal goblets have been used for ages with no reported ill effects. Glass binds the Pb^{2+} ion so effectively that little of it escapes. Heavy lead ions cause leaded glass to bend light more strongly than ordinary glass, which gives cut crystal a striking appearance. Recent experiments suggest that with time, small amounts of lead may leach out of crystal storage flasks. A Columbia University study showed that lead levels in wine stored 4 months in a crystal decanter increased 60-fold. A sample of brandy showed as much as 21,500 micrograms of lead per liter of liquid. A lead concentration above 50 micrograms per liter in a person's blood is considered the threshold for lead poisoning.

Table 7.3. *Composition of Various Glasses*

Type	% SiO_2	% Na_2O	% CaO	% MgO	% Al_2O_3	% Other
Egyptian	64	21	8	5	1.5	0.5 K_2O
Roman (A.D. 100)	70	17	9	1	3	0
Venetian	73	19	5	0	3	0
Window	72	12	10	2	1	3 K_2O
Pyrex	80	4	1	0	2	11 B_2O_3
Optical	69	6	12	0	0	12 K_2O
Crystal	55	0	0	0	1	11 K_2O, 33 PbO
Quartz (silica)	100					

Fortunately, no problem occurs with the short-term use of crystal wine goblets.

Molten glass must be cooled slowly (called annealing) or strain develops. Strain patterns can be visualized in a piece of glass held between crossed polarizing lenses. Because glass does not bend like metal or plastic, the presence of strain can cause glass objects to shatter spontaneously. Thermal strain develops upon the heating of a glass object because most materials expand upon heating. For glass, the effect of strain is especially bad because it is a brittle thermal insulator. Thus the hot surface of a piece of glass expands more than the cool inner region, and strain develops between the two regions. This strain can pull the glass apart. For metals, the problem of thermal strain isn't so serious. Heat conducts rapidly from the surface of a metal to its interior, which minimizes the thermal strain. Metals are also intrinsically less brittle than glass. Safety glass, used in automotive windshields, is purposely designed to incorporate strain that causes it to fragment into many small pieces when broken.

Thermal strain in glass is diminished by making the amount of expansion for glass small. In Pyrex® glass (see Table 7.3), B_2O_3 substitutes for some of the CaO and Na_2O. This greatly reduces the expansion on heating and the contraction on cooling. Chemists routinely use Pyrex® glassware in reaction vessels that must be heated and cooled. Glass containers designed for cooking in the home also use Pyrex® glass. Because Pyrex® is more expensive, bottles and drinking glasses usually are not made of it, which is why they shatter if put in an oven or flame. Cooling with ice poses less of a problem for glass objects. The temperature change involved (20–0°C) is much smaller than the (100–500°C) change involved in heating a glass object in an oven or over an open flame. However, for low-temperature applications (such as handling liquid nitrogen at –196°C), Pyrex® must be used instead of regular glass.

Iron oxide impurities in the materials used to make glass would color it yellow or brown. Purified components are necessary to make colorless glass. Conversely, the addition of small amounts of colored metal oxides gives glass color. A trace of cobalt oxide colors glass blue, chromium or copper oxide imparts

Figure 7.13. *Photograph of a glass blower crafting a wine goblet from molten glass. (Courtesy of Corning Incorporated.)*

a green color, and manganese oxide gives a violet color. The green color of bottle glass arises from the presence of added ferrous oxide (FeO). The faint green color of window glass, which is evident when you examine its edge, arises from traces of iron impurities. Either fine gold particles or a mixture of Se and CdS particles can give glass a red color. Milky white glass contains small, white particles of calcium sulfate ($CaSO_4$) additive. The next time you encounter an artisan blowing glass at a fair, note the varying consistency of the glass as it changes from a heavy syrup when white hot (Figure 7.13) to the consistency of stiff taffy when red hot, to a solid breakable object when cold.

Ceramics

Ceramic materials maintain their structural integrity at much higher temperatures than ordinary glass. Ceramics may be crystalline substances or amorphous (so-called glass ceramics). The earliest ceramic pottery made by humans contained clay, at first simply dried in the sun, later hardened by firing in an oven. Pure clay, or kaolin, is a compound of formula $Al_2O_3 \cdot 2SiO_2 \cdot 2H_2O$. Heating drives out the water and forms a silica–alumina glass, which melts at 1755°C. Pure clay, which is free from iron, is fired to make white china. The familiar brown or red clay pottery owes its color to the presence of iron oxide (rust) impurities. One can color white clay just like glass, by adding colored metal oxides to it. Porcelain, a glass ceramic, is made by adding a high percentage of Al_2O_3 to a SiO_2-based glass. White porcelain cookware consists of about 55% SiO_2, 20% Al_2O_3, and the remainder is CaO and MgO.

Other ceramics besides those based on clay play an important part in current technology. One area of interest is the development of ceramics that can be machined into parts for high-temperature diesel engines and turbines. These materials could result in the next generation of fuel-efficient engines. They allow higher operating temperatures than are possible with metals and a greatly reduced weight (the heaviest part of a car is the engine). A candidate for structural ceramics includes zirconia (ZrO_2), stabilized by addition of CaO, which is usable up to 800°C (1470°F). Its melting temperature lies above 1700°C (3100°F). Silicon carbide (SiC) and silicon nitride (Si_3N_4) engines can be operated at 1500°C (2700°F) and 1300°C (2400°F), respectively. Besides the savings in weight, ceramics eliminate the need for cooling the engines, because they can tolerate higher operating temperatures than metals.

The virtue of ceramics, their refractory nature, poses the greatest problem for their manufacture into useful shapes. Silicon carbide remains a solid to 3000°C (5400°F). It can't be melted and poured into a mold to make a useful structure. The difficulty in pressing ceramic materials into usable parts is the main barrier to their application. The conventional method of molding ceramics is to heat a fine powder of the ceramic material under pressure (called sintering).

The chemical inertness (similar to that of glass), low density, and mechanical strength of ceramics are properties that have made them useful in surgical implants. Bioceramic aluminum oxide (Al_2O_3), made by compressing powdered Al_2O_3 and firing at 1500–1700°C, has been used in knee and hip prostheses, dental implants, and porcelain dental crowns. Chemical and mechanical (porosity) modifications to ceramics help encourage the ingrowth of bone or tissue when fixation to bone structure in the body is necessary. These applications require the collaborative efforts of physicians, biologists, chemists, biomedical engineers, and materials scientists.

A remarkable "high-temperature" ceramic superconductor was discovered in 1986 by Alex Müller and Georg Bednorz at the IBM research laboratories in Zurich. They received a Nobel prize one year after their discovery. The compound, a ceramic metal oxide of approximate composition $YBa_2Cu_3O_7$, is often called the 123 superconductor from the subscripts for the metallic elements in its formula, $Y_1Ba_2Cu_3$.

Superconductors are materials that conduct electricity with no resistance. Most must be cooled to extremely low temperatures before they superconduct. Strong interest in the "high-temperature superconductors" arose from their superconductivity in the −195° to −180°C "high-temperature" range. Previous known superconductors worked only below −253°C, and expensive liquid helium (boiling point −269°C) was needed to cool them to very low temperatures. High-temperature superconductors can be cooled with relatively inexpensive liquid nitrogen. It is possible that superconductors could be used in electric power transmission. Normal conductors convert some electric power to heat when they conduct electricity. Energy losses in normal wires can exceed 30% for electric power transmission over long distances. Superconductors also could be used to construct strong electromagnets for magnetically levitated trains that run on frictionless tracks. A significant barrier to the practical use of these materials remains the difficulty in processing the refractory ceramic into wires and other useful forms.

CHAPTER 8

Biochemistry

The Chemical Composition of Life

A human body consists of approximately 10 trillion cells. The construction of cells, the functions that they carry out, and the communication between them involve molecules. Molecules transmit pleasure and pain. Molecules metabolize and extract needed ingredients from foods and beverages. Molecules transport oxygen from the lungs to the muscles and brain. Molecules impart strength to skin and muscles. Molecules are responsible for the senses of sight, smell, hearing, taste, and feeling. Tears, which lubricate the region between the eyeball and the eyelid, are salt solutions excreted by the lachrymal gland of the eye. Living beings consist largely of the molecules known as carbohydrates, polysaccharides, proteins, nucleic acids, and water.

Surprisingly, only about 20% of the elements are essential building blocks for molecules found in plants and animals. Of these, only 11 elements are common to most living systems. Four elements—carbon, hydrogen, nitrogen, and oxygen—account for 99% of the total elemental makeup of humans. Oxygen and hydrogen are the most abundant, reflecting the high water content of the human body. Carbon and nitrogen, together with hydrogen and oxygen, are the basic elements of organic structures and metabolites. The other 7 elements, which represent 0.9% of the total atoms in the human body, are sodium, potassium, calcium, magnesium, phosphorus, sulfur, and chlorine. Although they are present to a lesser extent than C, H, N, and O, these atoms play important roles in many life processes. In addition to these elements, many metals and nonmetals are required in trace amounts. Ironically, some essential trace elements are toxic at slightly higher concentrations (examples include arsenic, chromium, cadmium, copper, and nickel). Exposure to the environment also results in the incorporation of trace concentrations of many nonessential elements (such as uranium).

Describing the chemistry of living systems in terms of their elemental makeup is a formidable task, but fortunately, a large number of biological molecules can be described in terms of a small number of molecular building blocks. The bulk of the organic matter found in human bodies can be built from a collection of approximately 50 different molecular units. Amino acids are the molecular units used to construct the thousands of proteins found in the human body. Likewise, the nucleic acid DNA is built from a set of four different molecular building blocks called nucleotides. The great diversity of life on Earth arises from the seemingly unlimited ways in which these molecular units can be joined.

Amino Acids

Viewing the three-dimensional structure of a protein can be an intimidating experience. Consider the protein chymotrypsin, shown in Figure 8.1. This protein catalyzes the digestion of proteins from food in the small intestine. Proteins that catalyze chemical reactions are called enzymes. On the atomic level, chymotrypsin is composed of hydrogen (H), carbon (C), oxygen (O), nitrogen (N), sulfur (S), and zinc (Zn). Its three-dimensional structure results from the bonding interactions between these atoms. The diameter of this protein is less than one ten-thousandth of the diameter of a human hair.

The molecular building blocks of proteins are called *amino acids*. Of the 20 common amino acids, 19 have the general structure shown in Figure 8.2. Amino acids contain a basic amino (NH_2) end group and an acidic carboxyl (CO_2H) end group. These two groups are connected by covalent bonds to a central carbon atom (see Figure 8.2). The uniqueness of a particular

Figure 8.1. *Three-dimensional structure of chymotrypsin, a digestive enzyme found in the small intestine. (Image courtesy of Molecular Simulations, Inc., Burlington, MA.)*

amino acid lies in the chemical composition of the side chain, represented by R, that is also connected to the central carbon atom. Figure 8.3 gives the structures, names, and abbreviations of the amino acids found in proteins. The structure of proline differs slightly from that of other amino acids. Amino acids can exist as either D- or L- mirror image isomers. Only the L-amino acids are used by living systems.

A large supply of amino acids is needed to synthesize proteins. The human body carries out chemical reactions to synthesize many of the amino acids. Others must be supplied by foods. Humans need to eat plant or animal proteins that contain the following essential amino acids: lysine, tryptophan, threonine, methionine, phenylalanine, leucine, valine, and isoleucine. Meat protein usually contains all of these. Vegetarians need to plan their diets carefully to include all the essential amino acids. For example, beans contain isoleucine and lysine but are deficient in tryptophan and the sulfur amino acids (cysteine and methionine). Rice is deficient in isoleucine and lysine but contains enough of the others.

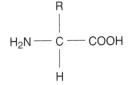

Figure 8.2. *General structure of an amino acid.*

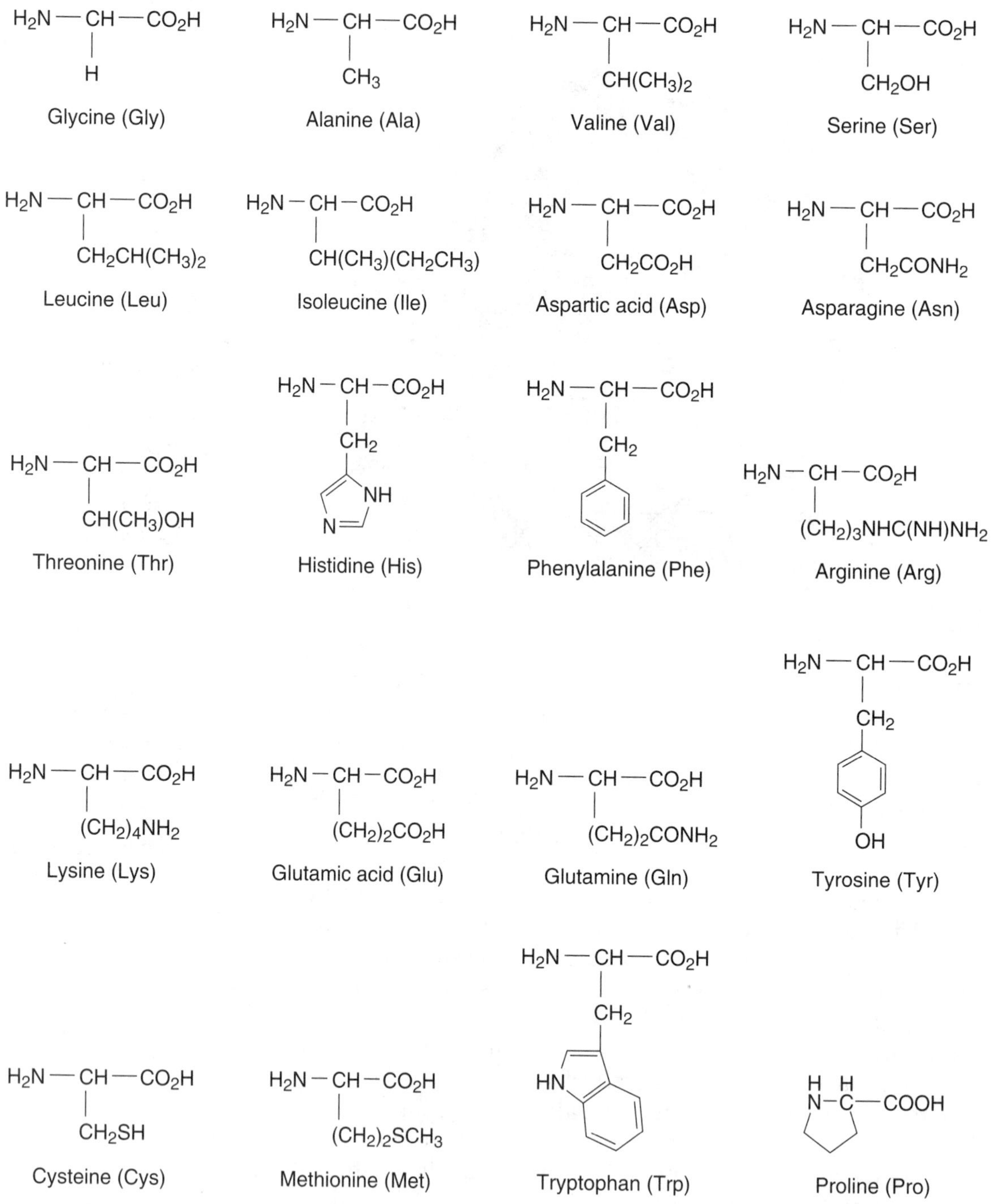

H₂N — CH — CO₂H
|
H

Glycine (Gly)

H₂N — CH — CO₂H
|
CH₃

Alanine (Ala)

H₂N — CH — CO₂H
|
CH(CH₃)₂

Valine (Val)

H₂N — CH — CO₂H
|
CH₂OH

Serine (Ser)

H₂N — CH — CO₂H
|
CH₂CH(CH₃)₂

Leucine (Leu)

H₂N — CH — CO₂H
|
CH(CH₃)(CH₂CH₃)

Isoleucine (Ile)

H₂N — CH — CO₂H
|
CH₂CO₂H

Aspartic acid (Asp)

H₂N — CH — CO₂H
|
CH₂CONH₂

Asparagine (Asn)

H₂N — CH — CO₂H
|
CH(CH₃)OH

Threonine (Thr)

H₂N — CH — CO₂H
|
CH₂
NH
N

Histidine (His)

H₂N — CH — CO₂H
|
CH₂

Phenylalanine (Phe)

H₂N — CH — CO₂H
|
(CH₂)₃NHC(NH)NH₂

Arginine (Arg)

H₂N — CH — CO₂H
|
(CH₂)₄NH₂

Lysine (Lys)

H₂N — CH — CO₂H
|
(CH₂)₂CO₂H

Glutamic acid (Glu)

H₂N — CH — CO₂H
|
(CH₂)₂CONH₂

Glutamine (Gln)

H₂N — CH — CO₂H
|
CH₂
OH

Tyrosine (Tyr)

H₂N — CH — CO₂H
|
CH₂SH

Cysteine (Cys)

H₂N — CH — CO₂H
|
(CH₂)₂SCH₃

Methionine (Met)

H₂N — CH — CO₂H
|
CH₂
HN

Tryptophan (Trp)

H H
N — C — COOH

Proline (Pro)

Figure 8.3. *The 20 common amino acids found in proteins.*

Amino acid supplements are available in health food stores. Caution should be exercised in taking these products. In the 1970s, L-tryptophan (L-Trp) was promoted as a diet supplement and as a "cure" for insomnia and stress. A contaminant from the manufacturing process used by a Japanese supplier of L-Trp caused a rare blood disease, which led to 31 deaths and numerous cases of paralysis. As a result, the United States government banned the sale of L-Trp diet supplement pills in 1973. Despite the ban, as of 1993 a loophole in the law allowed L-Trp to be sold in combination formulas marketed as nutrient supplements.

Building Polypeptides

Amino acids polymerize when the terminal acid group (CO_2H) of one amino acid reacts with the terminal amino group (NH_2) of another. This chemical reaction, given by equation (1), forms water and a carbon–nitrogen bond between the two amino acids; this C–N bond is called a peptide bond.

$$
H_2N-\underset{\overset{|}{R_1}}{CH}-CO_2H \quad + \quad H_2N-\underset{\overset{|}{R_2}}{CH}-CO_2H \quad \rightleftharpoons
$$

$$
H_2N-\underset{\overset{|}{R_1}}{CH}-\underset{\overset{\|}{O}}{C}-NH-\underset{\overset{|}{R_1}}{CH}-CO_2H \; + \; H_2O \qquad (1)
$$

peptide bond

The formation of a peptide bond between two amino acids.

The energetics involved dictate that an amino acid polymer should decompose into individual amino acids in water solution. Fortunately, the rate of this reaction is very slow. The reaction can be catalyzed by acids or bases, however. Drain cleaners such as Drano® work by catalyzing the decomposition of proteins with base.

Multiple reactions of the type shown in equation (1) generate a polymer chain that consists of many amino acid groups. Such a polymer is called a *polypeptide.* The order of the amino acids

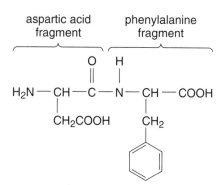

Figure 8.4. *The molecular structure of the dipeptide Nutra-Sweet® (Asp-Phe).*

in the polypeptide chain is written beginning with the amino end. The smallest polypeptide consists of two amino acids and is called a *dipeptide.*

The dipeptide aspartame, shown in Figure 8.4, demonstrates that the chemical properties of peptides depend on the order in which the amino acids appear. Marketed as a low-calorie sweetener, NutraSweet®, it is composed of the amino acids phenylalanine and aspartic acid. The amino terminus is on the aspartic acid group. The phenyalanine has the terminal CO_2H group. By convention, this dipeptide is abbreviated Asp-Phe. By contrast, the dipeptide Phe-Asp, in which the phenylalanine group contains the amino terminus, is not sweet. The chemical receptors in the taste buds are sensitive to the order of the amino acids.

Proteins

Proteins differ from the polymers discussed in Chapter 4 in two important ways. In synthetic copolymers such as styrene-butadiene rubber, there is a wide distribution of molecules with different lengths. And the "molecular units" that make up copolymers occur in a random sequence. In contrast, a protein has a precise length and the amino acids occur in a precise order. It is also important to realize that the relative amounts of any one particular amino acid vary significantly between different proteins. Figure 8.5 shows the relative amounts of the 20 amino acids in myoglobin (a protein that stores and transports oxygen

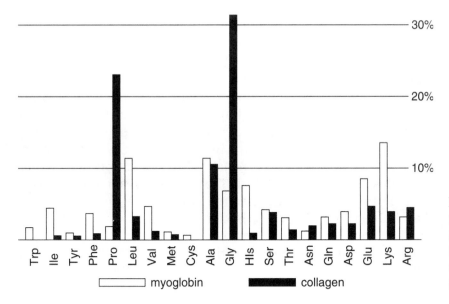

Figure 8.5. *The relative amounts of the 20 amino acids in the polypeptide chains of myoglobin and collagen.*

in muscle tissue) and collagen (the fibrous protein that makes up skin). Glycine (Gly) and proline (Pro) account for over half the amino acids in collagen. Myoglobin, on the other hand, contains less than 10% of these two amino acids.

Like the polymers used in making synthetic polymers, polypeptide chains can be cross-linked to make the structure more rigid. Cross-linking occurs between cysteine groups in the individual chains. The S–S bond is called a disulfide bond; Figure 8.6 shows the disulfide bond linkage. Disulfide bonds play an important role in stabilizing the three-dimensional folded structures of many proteins, including insulin, antibodies, lysozyme, and hair proteins.

The fibrous strength of hair comes from intertwined polypeptide chains. Over 15% of the amino acid groups in these polymers are cysteines. The natural texture of hair arises from the disulfide bonds that exist between the intertwined polypeptide chains. This texture can be artificially modified in a process called a permanent wave, which is available in most hair salons. This technique is based on breaking and reforming disulfide bonds. The first step in giving a perm is to constrain the hair to the desired shape with curlers. On the molecular level, this has the effect of twisting, stretching, or curling the polypeptide

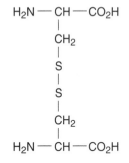

Figure 8.6. *The disulfide bond between two cysteine molecules. This bond can be used to link protein chains and is important in determining the three-dimensional structures of proteins.*

chains. The hair is then treated with keratin, a natural component of the outer layer of hair, which breaks many of the existing disulfide bonds. The polypeptide chains unwind, and the limp hair conforms to the shape imposed by the curlers. The new three-dimensional structure of the intertwined protein chains will have some cysteine groups in close proximity to one another. At this point the hair is treated with a basic solution, which promotes disulfide bond formation between neighboring cysteine groups. This sets the structure of the polypeptide chain as that defined by the curlers. Because of the new bonds formed, the hair remains in the desired shape after the curlers are removed. Over time, however, these bonds break and the hair returns to its natural shape. A permanent, in other words, isn't.

Protein Structure

The three-dimensional structures of proteins determine to a large extent their chemical reactivity and physical properties. Many proteins consist of a single polymer chain; others have two or more interwoven or interacting chains. And some proteins are made of individual polypeptide chains connected through one or more disulfide bonds. In addition to amino acids, many proteins also contain small molecules and metal ions.

The ordering of amino acids in the chain defines the *primary structure* of the protein. For demonstration purposes, consider the single-chain protein myoglobin. Myoglobin is responsible for the transport and storage of oxygen in muscle tissue. There

Figure 8.7. *The sequence of amino acids in myoglobin.*

Val-Leu-Ser-Glu-Gly-Glu-Trp-Gln-Leu-Val-Leu-His-Val-Trp-Ala-Lys-Val-
Glu-Ala-Asp-Val-Ala-Gly-His-Gly-Gln-Asp-Ile-Leu-Ile-Arg-Leu-Phe-Lys-
Ser-His-Pro-Glu-Thr-Leu-Glu-Lys-Phe-Asp-Arg-Phe-Lys-His-Leu-Lys-
Thr-Glu-Ala-Glu-Met-Lys-Ala-Ser-Glu-Asp-Leu-Lys-Lys-His-Gly-Val-Thr-
Val-Leu-Thr-Ala-Leu-Gly-Ala-Ile-Leu-Lys-Lys-Lys-Gly-His-His-Glu-Ala-
Glu-Leu-Lys-Pro-Leu-Ala-Gln-Ser-His-Ala-Thr-Lys-His-Lys-Ile-Pro-Ile-Lys-
Tyr-Leu-Glu-Phe-Ile-Ser-Glu-Ala-Ile-Ile-His-Val-Leu-His-Ser-Arg-His-
Pro-Gly-Asn-Phe-Gly-Ala-Asp-Ala-Gln-Gly-Ala-Met-Asn-Lys-Ala-Leu-
Glu-Leu-Phe-Arg-Lys-Asp-Ile-Ala-Ala-Lys-Tyr-Lys-Glu-Leu-Gly-Tyr-Gln-Gly

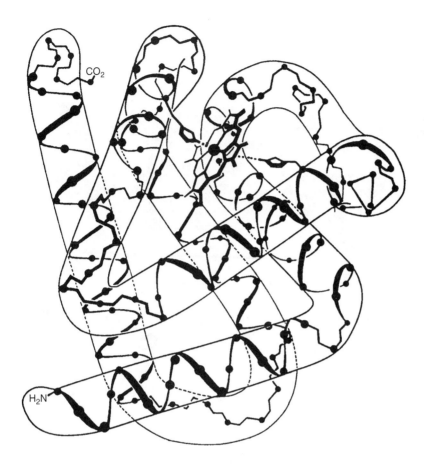

Figure 8.8. *Three-dimensional structure of myoglobin, the oxygen-storage protein in muscles. (Adapted from Hans Neurath ed.,* The Proteins, *2nd ed., Vol. II, copyright © 1964 by Academic Press. Redrawn by permission of Academic Press and Richard E. Dickerson.)*

are 153 amino acid groups in myoglobin. The linear polypeptide chain, whose sequence is shown in Figure 8.7, folds into the three-dimensional structure shown in Figure 8.8. How proteins fold into their unique three-dimensional structures is currently not understood. The human body initially makes proteins in the linear form, but before a protein becomes biologically active, it must fold into a particular three-dimensional structure.

Close examination of a folded protein (Figure 8.8) often reveals that contiguous segments of the polypeptide chain adopt regular coiling patterns. These localized patterns are called *secondary structures* of proteins. There are three types of secondary structures: random coils, α-helices, and β-pleated sheets. The simplest to envision is the random coil. This term describes seg-

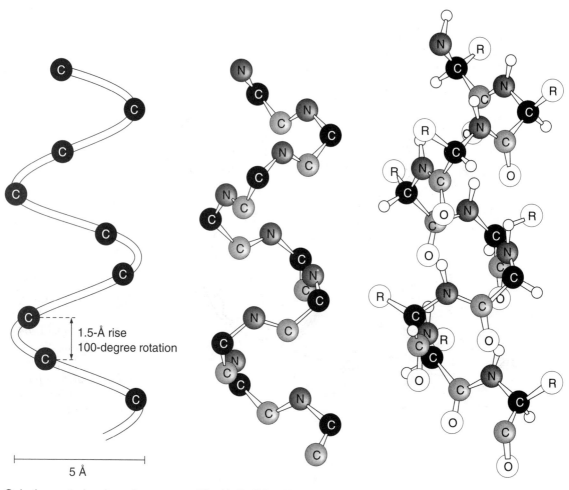

Only the central carbon atom on each amino acid is indicated.

The N–C–C backbones of the amino acids are indicated.

The entire helix and the side chains are labeled.

A

Figure 8.9. *Three-dimensional structures of α-helices and β-pleated sheets.*

(**A**) *The α-helix is a rod-like protein structure. A complete revolution around the axis of the helix takes 3.6 amino acid groups and results in a linear displacement of 5.4 Å. The structure is stabilized by hydrogen bonds. (Adapted from* Biochemistry, 3rd ed., *by Lubert Stryer. Copyright © 1988 by Lubert Stryer. Used with permission of W. H. Freeman and Co.)*

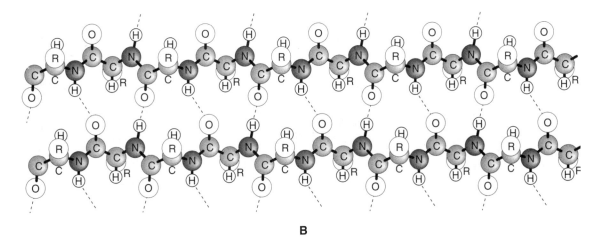

B

Figure 8.9 *(continued)*

(B) *A polypeptide in a β-pleated sheet is nearly fully extended. Adjacent strands run in opposite directions and are stabilized by hydrogen bonds between NH and C=O groups in the different strands. (Redrawn from* Genes and Genomes: A Changing Perspective *by Maxine Singer and Paul Berg. Copyright © 1991. Used with permission of University Science Books.)*

ments of amino acids that appear to be randomly oriented. On the other hand, α-helices and β-pleated sheets are two orderly three-dimensional arrangements of a chain. Figure 8.9 shows the arrangements of neighboring amino acids in an α-helix **(A)** and a β-pleated sheet **(B)**. Completely α-helical proteins occur in hair and wool fibers and help give them elasticity. The β-pleated sheet derives its name from its more extended chain structure, which allows neighboring chains to associate loosely in a sheet-like fashion. When a fibrous α-helical protein is subject to moist heat and stretched to about twice its initial length, its structure transforms to that of a β-pleated sheet. The relative percentages of α-helix, β-pleated sheet, and random coil vary significantly among proteins. Myoglobin (Figure 8.8) contains 75% α-helix and 25% random coil. Carboxypeptidase A, a digestive enzyme, contains 38% α-helix, 17% β-pleated sheet, and 45% random coil. Chymotrypsin (see Figure 8.1) contains several β-pleated sheets and very little α-helix. Both the α-helix and the β-pleated sheet structures are stabilized by strong dipolar forces between N–H and C=O groups wherein the positive hydrogen end of the N–H dipole is

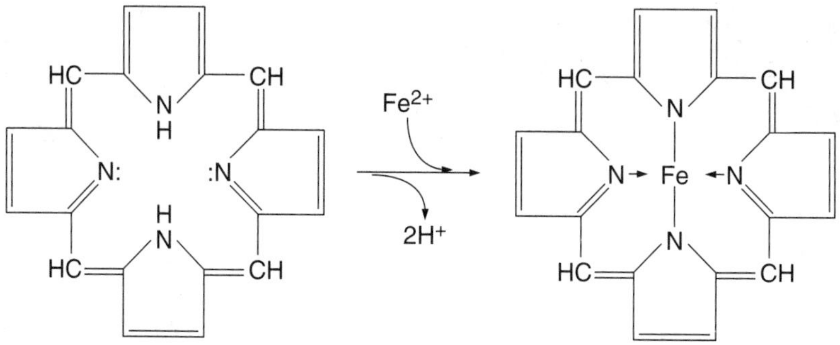

Figure 8.10. *The binding of positive ions by a porphyrin molecule.*

attracted to the negative oxygen end of the C=O dipole. This particular type of dipole–dipole interaction (where the bonding involves a hydrogen atom) is commonly referred to as a hydrogen bond.

Metal ions and organic molecules are often found embedded in proteins and are required for their chemical functions. One example is the porphyrin molecule, whose general structure is shown in Figure 8.10. This molecule is neutral in charge but contains two acidic hydrogens on the central nitrogen atoms. Under basic conditions, these hydrogen atoms dissociate; the resulting negatively charged nitrogens stabilize the binding of a metal cation in the center of the molecule.

Figure 8.11 shows two important porphyrin molecules. Fe-protoporphyrin IX (also called a heme group) is a porphyrin derivative found in myoglobin, hemoglobin, and cytochrome *c*. In myoglobin and hemoglobin, Fe-protoporphyrin IX serves as the binding site for a molecule of oxygen. In cytochrome *c*, the heme group participates in electron transfer chemistry that is crucial in respiration. Chlorophyll *a* is a different porphyrin derivative, which is the chief green pigment found in plants. It serves as an electron carrier in plant photosynthesis.

Protein Function

The protein collagen is responsible for the strength exhibited by hair, skin, claws, feathers, cartilage, and tendons. Collagen forms water-insoluble fibers of high tensile strength. It does not adopt

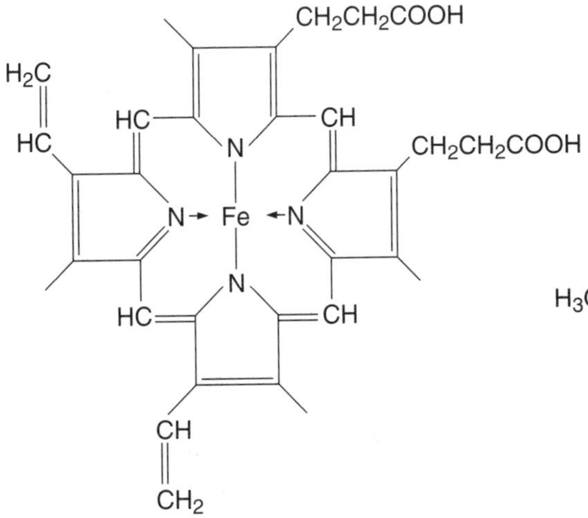

Protoporphyrin IX

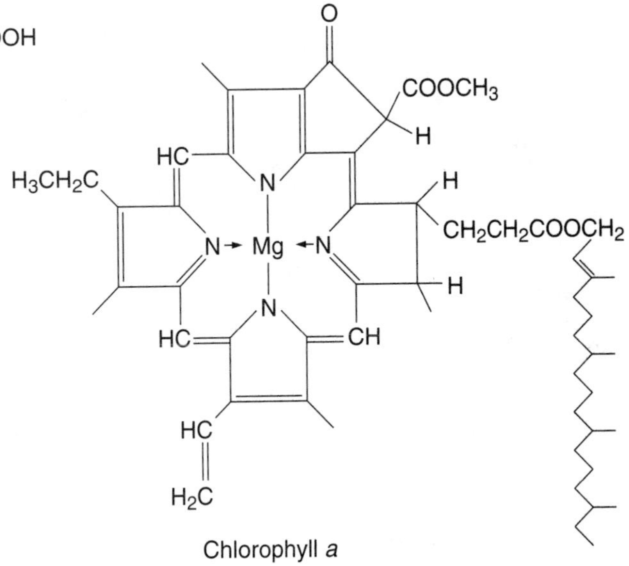

Chlorophyll *a*

Figure 8.11. *Examples of important porphyrin molecules. Protoporphyrin IX is the oxygen-binding site in hemoglobin and myoglobin and the electron carrier in cytochrome* c. *Chlorophyll is the electron carrier in plant photosynthetic systems.*

an α-helical structure but rather forms a special triple helix from three twisted polypeptide chains. This structure (Figure 8.12), which resembles the twisted fibers in a rope, gives the protein fibers great strength. Collagen is characterized by an unusually high level of the amino acids glycine (about 33%) and proline (24%); see Figure 8.6. Collagen forms rigid rods about 1/100 of an inch long. It is one of the longest proteins known.

Proteins protect living systems from invading molecules. The immune system manufactures proteins called antibodies. Antibodies target specific chemical structures on the surfaces of viruses and bacteria and remove them from the human body. Some chemicals can also trigger an immune response, more commonly called an allergic reaction. For example, *Toxicoden-*

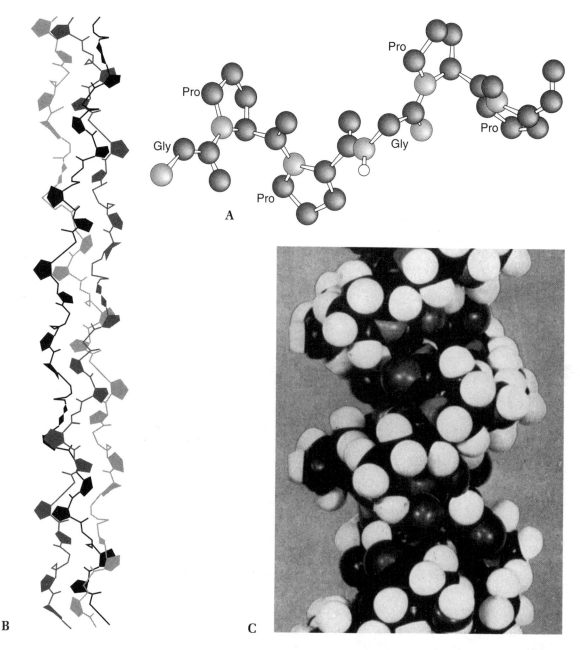

Figure 8.12.   *Molecular details of the structure of collagen. (A) The long chains are formed from a basic repeating structure Gly-Pro-Pro. (B) A model of how the three individual chains interact to form the triple-strand collagen helix. (C) A space-filling model of the collagen structure. (A and B: From* Biochemistry 3rd ed. *by Lubert Stryer. Copyright © 1988 by Lubert Stryer. Used with permission of W. H. Freeman and Company. C: Courtesy of Alexander Rich. From* Journal of Molecular Biology, Vol. III. *Copyright © 1961 by Academic Press. Used by permission of Academic Press and Alexander Rich.)*

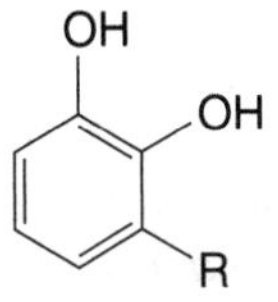

R-groups for the four components of urushiol

$(CH_2)_{14}CH_3$

$(CH_2)_7CH{=}CH(CH_2)_5CH_3$

$(CH_2)_7CH{=}CHCH_2CH{=}CH(CH_2)_2CH_3$

$(CH_2)_7CH{=}CHCH_2CH{=}CHCH_2CH{=}CH_2$

Figure 8.13. *Urushiol, the irritant in the oil from poison ivy.*

dron radicans (poison ivy) triggers an immune response in the skin of many people. The chemicals that cause the allergic reaction are contained in an oil on the plant leaves. The active agent, urushiol, is a mixture of the four organic compounds shown in Figure 8.13.

Proteins mediate the generation and transmission of nerve responses. On a molecular level, the message of pain is carried between nerve cells by chemical reactions between acetylcholine (Figure 8.14) and the acetylcholine receptor proteins. A complete breakdown of the neural transmission process leads to death. Nerve gases such as tabun and sarin (Figure 8.15) cause respiratory paralysis by interfering with the protein responsible for breaking down acetylcholine. The resulting buildup of acetylcholine in the body overstimulates the nerves until exhaustion and paralysis occur.

The human body has many proteins that transport and store specific molecules. Myoglobin transports and stores oxygen in muscle tissue. Hemoglobin transports and stores oxygen in red blood cells. Transferrin transports iron in the blood plasma, and ferritin stores iron in the liver.

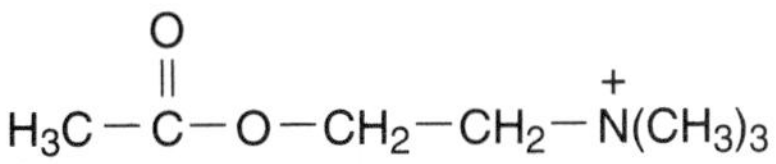

Figure 8.14. *The structure of acetylcholine, an ion involved in the transmission of nerve impulses.*

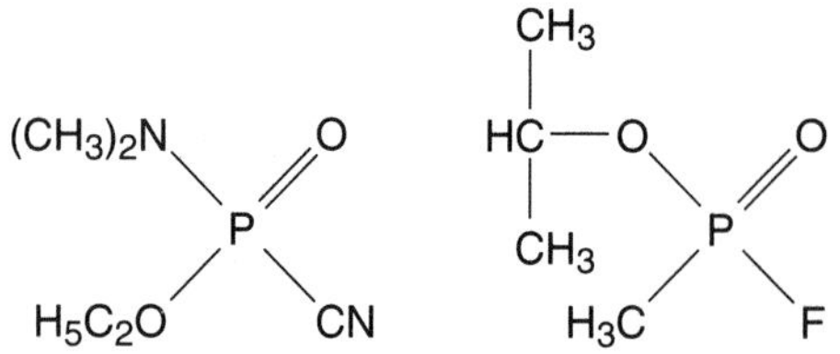

Figure 8.15. *Molecular structures of the nerve gases tabun (left) and sarin (right). These molecules inactivate the enzyme acetylcholinesterase, which is responsible for the breakdown of acetylcholine.*

Enzymes

Some proteins speed up reactions that are essential for life. A protein that acts as a catalyst is called an enzyme. An *enzyme* is generally designed to promote a specific reaction. The reactant molecule acted on by an enzyme is called the *substrate* and the region of the enzyme that interacts with the substrate is called the *active site.* Enzymes, being proteins, are composed of amino acids. The number of amino acid groups in the active site where the chemistry occurs is extremely small. The specificity of the enzyme depends on the geometry of the active site and on the spatial constraints imposed by the entire polypeptide chain. This specificity is often likened to a lock-and-key mechanism. The enzyme behaves as a lock, and the substrate is the unique key that fits into the lock. After the substrate has been transformed into the reaction product, it is released from the enzyme and a new substrate molecule can bind. The power of enzymes can be appreciated by considering their effect on several chemical reactions performed by the human body.

Alcohol Dehydrogenase

The enzyme alcohol dehydrogenase binds ethanol and converts it to acetaldehyde. The reaction involves oxidizing ethanol to acetaldehyde. The enzyme accelerates the rate of this reaction by a factor greater than 20 billion! The structure of alcohol dehydrogenase is optimal for binding with ethanol. It can react with other alcohol molecules, but those reactions occur at much slower rates. If methanol, or wood alcohol, is ingested accidentally, the oxidation reaction that is mediated by alcohol dehydrogenase occurs more slowly. This decrease in the rate of metabolism enables methanol to participate in other unwanted chemical reactions in the body and to cause serious complications such as blindness and death. Ethanol is less toxic because alcohol dehydrogenase metabolizes it quickly .

The first step in the oxidation of ethylene glycol in the body also involves the enzyme alcohol dehydrogenase. About 50 deaths occur each year from the ingestion of ethylene glycol, a component in automobile antifreeze. Ethylene glycol itself is not lethal. However, oxalic acid, the oxidation product of ethylene glycol, crystallizes in the kidneys, leading to serious and sometimes fatal damage. The metabolism of ethylene glycol is given by equation (2). In the 1980s, a few Italian wine manufacturers added ethylene glycol to red wines to enhance their sweetness. This practice was discovered as a result of the increased number of cases of ethylene glycol poisoning reported by local hospitals.

$$
\begin{array}{ccc}
CH_2OH & & COOH \\
| & \xrightarrow[\text{dehydrogenase}]{\text{alcohol}} & | \\
CH_2OH & & COOH \\
\text{Ethylene glycol} & & \text{Oxalic acid}
\end{array}
\qquad (2)
$$

Ethylene glycol is broken down in the body by the enzyme alcohol dehydrogenase into toxic oxalic acid.

The unwanted oxidation of ethylene glycol can be inhibited by administering a nearly intoxicating dose of ethanol. This approach works because ethanol preferentially occupies the binding site in the enzyme, thereby giving the body more time to excrete ethylene glycol.

Serine Proteases

The serine proteases are a class of enzymes that have a reactive serine amino acid in the active site. Chymotrypsin (see Figure 8.2) is a serine protease that catalyzes the digestion of food in the small intestine. Another example is thrombin, an enzyme that plays a key role in the formation of blood clots. Hormone production and egg fertilization also require serine proteases. On the average, these enzymes lead to rate accelerations of 100,000 to 1 billion over that observed without the catalyst.

Blood clotting provides an excellent example of how enzymes and proteins regulate each other. It is obvious that the clotting process must be precisely controlled. There is a fine line between hemorrhage and premature coagulation of the blood. Clots must be confined to the site of injury, and they must grow at a rapid rate. In the human body there are proteins designed to turn the clotting process on *and off*. In particular, the enzyme thrombin, which initiates the growth of the clot, is regulated by another protein called antithrombin. The amounts of thrombin and antithrombin in the blood are precisely balanced so that rapid clotting occurs only at the site of an injury. This mechanism is triggered whenever blood leaks from the vascular system and encounters collagen (when blood in a surface cut contacts skin collagen). The blood platelets sense the collagen and then stimulate the production of thrombin at the site of the cut. Without the rate enhancement caused by these clotting enzymes, all living beings would bleed to death the first time their skin was cut!

Enzyme Inhibition

Many drugs are designed to block the active sites of enzymes. One example is penicillin, an antibiotic. Penicillin kills bacteria by blocking the last step of bacterial cell-wall synthesis. An important enzyme in bacterial cell-wall synthesis, glucopeptide transpeptidase, connects an alanine group of one polypeptide chain to a glycine of another. Forming this bond completes the cell wall. The active site of glucopeptide transpeptidase contains a serine group. Penicillin fools the protein and gains access to the active site. Once there, the drug irreversibly binds to the serine group, thereby inactivating the enzyme. When this happens, cell walls cannot be completed, and the bacterium dies. Figure 8.16 depicts this reaction.

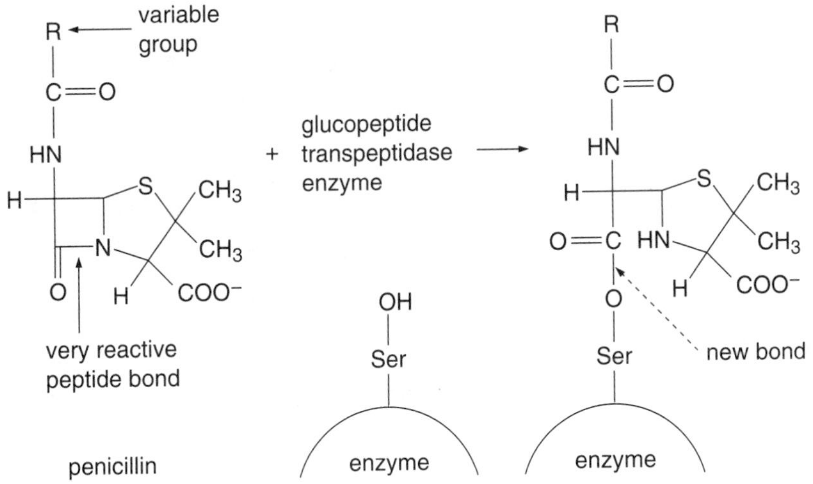

Figure 8.16. *Reaction of penicillin with the serine group at the active site of glucopeptide trans-peptidase. This reaction destroys the catalytic properties of the enzyme, thereby killing bacteria.*

Enzymes as Cleaning Agents

Soft contact lenses are polymeric materials made by cross-linking hydroxyethylmethacrylate (HEMA) with ethylene glycol dimethacrylate (EDMA, Figure 8.17). The abundance of OH groups makes this material very hydrophilic (water-loving). Mucus, proteins, and fats deposit on the surface of the lenses when they are worn, so they must be cleaned periodically. Detergents mobilize, emulsify, and dissolve fat and protein deposits. Antibacterial agents prevent growth of unwanted organisms, keeping the surfaces of the contacts sterile.

After a while, as most contact wearers know, normal cleaning solutions don't work very well. The protein and fat deposits stick strongly to the polymer substrate. When this occurs, it is necessary to break down these deposits chemically. This requires a cleaning solution that will break peptide bonds but *not the chemical bonds* that hold the contact lens together. The chemistry also needs to proceed at a reasonable rate at room temperature, so that cleaning is quick and easy. Enzymes are ideally suited for this purpose. Enzymatic lens cleaners usually contain several digestive enzymes. These molecules target proteins,

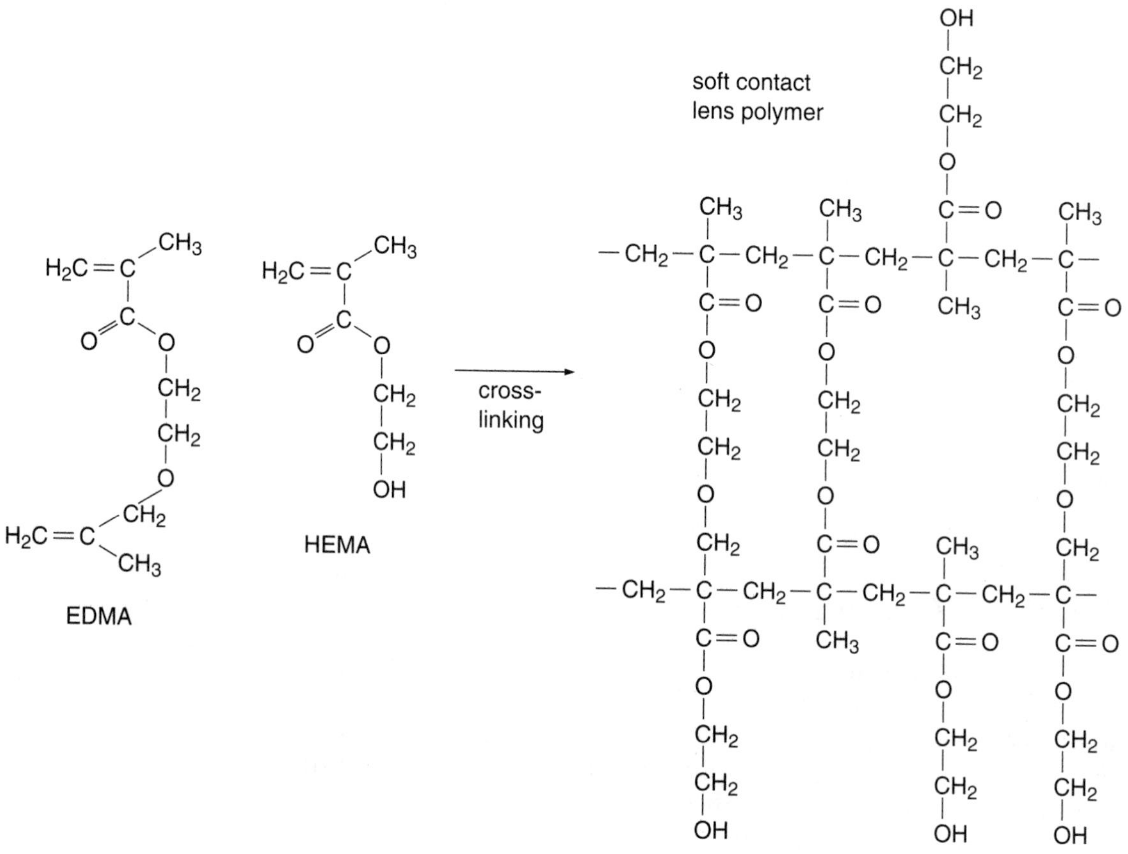

Figure 8.17.   *The chemical structure of soft contact lenses.*

fats, and other biological materials, which adhere to the surface of the lens. Many products use the enzyme papain to cleave peptide bonds in proteins. Papain does not degrade lipids, mucus, or other oily deposits. These molecules are removed with the enzymes lipase and mucinase, which decompose lipids and mucus, respectively.

Enzymes are also added to some laundry detergent formulations. They enhance the cleaning process by breaking down protein deposits in blood and food stains on clothes.

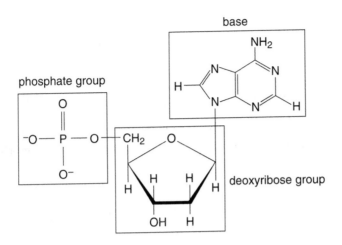

Figure 8.18. *Chemical structure of dAMP (the A nucleotide).*

DNA, RNA, and the Genetic Code

The Structure of Nucleic Acids

Long, chain-like molecules of deoxyribonucleic acid (DNA) contained in cells store all the information necessary to construct and operate the cell. DNA is a polymer, as are proteins, but its molecular building blocks are not amino acids. A DNA polymer is built from a set of four building blocks called *nucleotides.* A nucleotide contains three parts: a base molecule, a sugar group (deoxyribose), and a phosphate group. Figure 8.18 shows the structure of one of the four nucleotides. The name of the nucleotide comes from the base molecule. The molecule in Figure 8.18 is abbreviated dAMP (**d**eoxy**a**denosine **m**ono**p**hosphate). It is also called the A nucleotide, indicating the base molecule **a**denine. The base molecules are derivatives of aromatic compounds known as purine and pyrimidine. The purine derivatives are adenine (A) and cytosine (C). The pyrimidines are guanine (G) and thymine (T). Figure 8.19 gives the structures, names, and abbreviations of the four bases.

Figure 8.20 shows how a DNA polymer strand is formed. Chain growth occurs by bond formation between the sugar group on one nucleotide and the phosphate group on another. Examination of the molecular structure reveals that the two ends of the

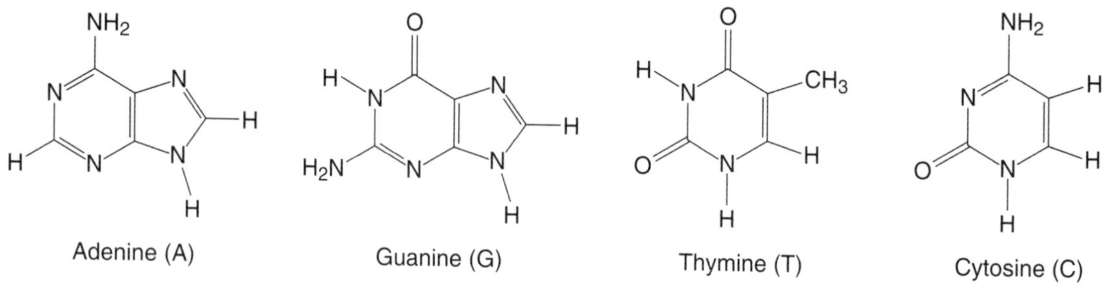

Figure 8.19. *The four bases found in nucleotide molecules.*

DNA strand differ. Like polypeptides, nucleic acids have a sense of directionality.

Normally two strands of DNA, running in opposite directions, associate to form a double helix. In this structure, two polynucleotide chains twist together around a common axis. Figure 8.21 shows the double helix; the structure has beautiful symmetry. The nature of the double helix plays a key role in DNA replication. Each strand of DNA can serve as a template to form a complementary strand, making possible the formation of two identical copies of the molecule.

James Watson and Francis Crick deduced the three-dimensional structure of DNA. Their initial report of the DNA double helix was published as a brief, two-page note in *Nature* in 1953. This discovery earned them the 1962 Nobel Prize for Chemistry. The double helix structure has bases pointing toward the center, and the sugar/phosphate groups lie on the outside. Thus the forces that hold the two helices together occur between base molecules on the individual strands.

The interactions between the two chains are the key to many properties of DNA. Looking perpendicular to the direction of the helix, the bases always come in the pairs A–T (adenine and thymine) and C–G (cytosine and guanine). Given the sequence of nucleotides in one chain, the sequence in the second chain is uniquely determined. This construct has been called the "base-pairing" model of DNA structure.

Specific pairing comes from two chemical properties of the DNA molecule. First, the space between the sugar backbone of two helices constrains the volume available for the bases. The

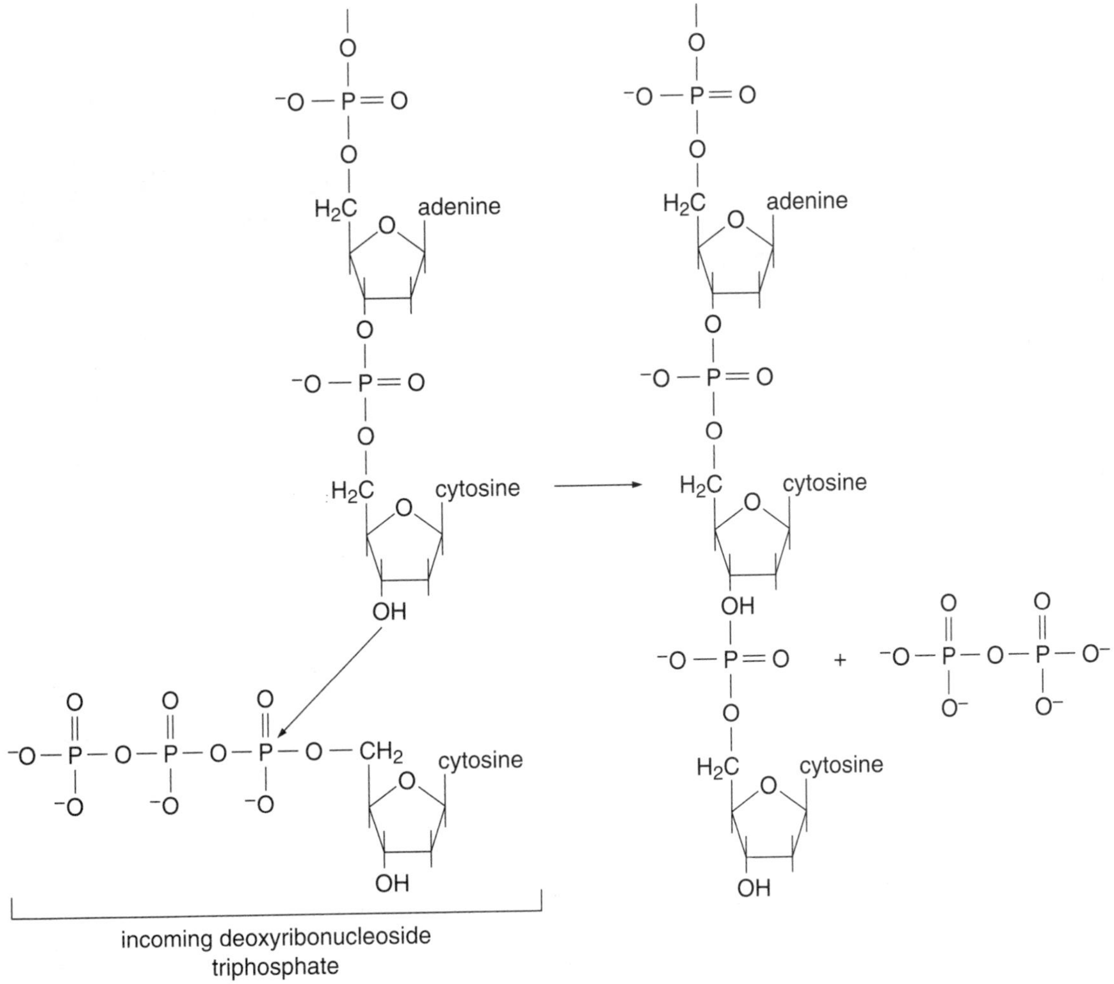

Figure 8.20. *The growth of a single strand of DNA. Addition of a nucleotide to the DNA backbone releases a diphosphate group. (Redrawn from* Genes and Genomes: A Changing Perspective *by Maxine Singer and Paul Berg. Copyright 1991. Used with permission of University Science Books.)*

distance is slightly larger than the sum of the dimensions of one purine (A or C) base molecule and one pyrimidine (T or G) base molecule. The space is too small to be occupied by a pair of T and G bases. The base pairing also maximizes favorable inter-molecular hydrogen bonds. Figure 8.22 shows the hydrogen

Figure 8.21. *Two views of DNA. (A) Space-filling model of the double helix; a complete revolution of the helix corresponds to ten nucleotides in the chain. (From Paul Berg and Maxine Singer,* Dealing with Genes: The Language of Heredity. *Copyright 1992. Used with permission of University Science Books. (B) Schematic representation indicating the pairing between the bases on the two strands. (From* Genes and Genomes: A Changing Perspective *by Maxine Singer and Paul Berg. Copyright 1991. Used with permission of University Science Books.)*

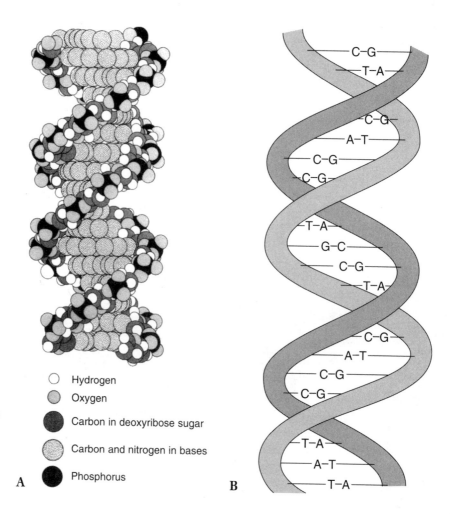

○ Hydrogen

● Oxygen

● Carbon in deoxyribose sugar

● Carbon and nitrogen in bases

● Phosphorus

A

B

bonds as dashed lines. The hydrogen bonding interactions, along with the spatial constraints imposed by the helical backbone, dictate the relative structures of the two helices.

Ribonucleic acid (RNA) is a second nucleic acid found in living systems. RNA resembles DNA in many respects, but two differences are important. First, RNA uses the base uracil (U) in place of thymine (T). Uracil, like thymine, still forms base pairs with adenine. Figure 8.23 shows the structure of uracil. Second, a different sugar group makes up the backbone of the helical structure. Ribose (also shown in Figure 8.23) is used in RNA.

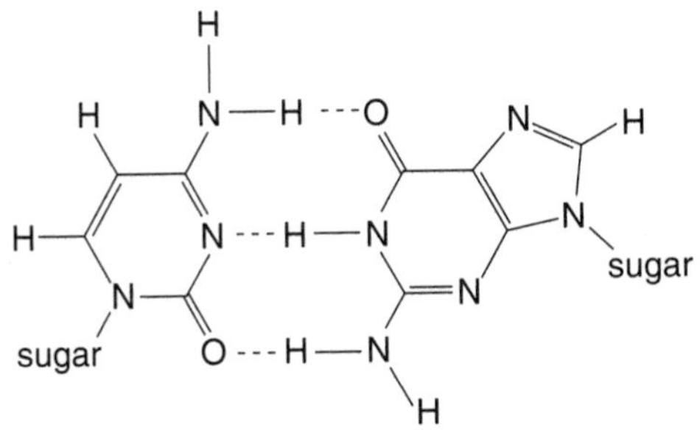

Figure 8.22. *The hydrogen bonding interactions between cytosine and guanine that occur when the two bases are across from each other in the double helix.*

This sugar contains one more hydroxyl group than DNA's deoxyribose. These molecular changes cause small changes in the geometry of the helices. RNA plays an important role in protein synthesis.

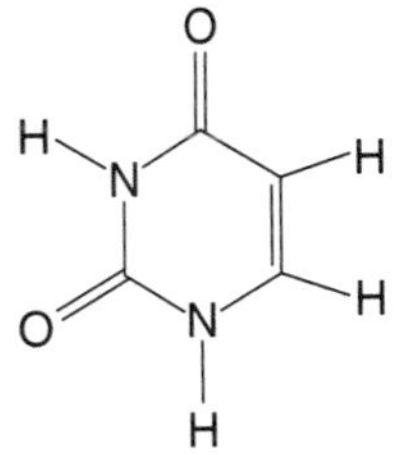

Deciphering the Genetic Code

How does DNA store the information needed for protein synthesis? The deciphering of the information contained in DNA is called *translation*, because the 4-letter alphabet of DNA is translated into the 20-letter alphabet of proteins. Translation is a complicated chemical process that involves many proteins, enzymes, and RNA molecules. Most of these reactions occur on a molecular framework called a *ribosome*. This molecular machine in the cell coordinates the actions of the various molecules involved in protein synthesis.

Many experiments performed in the late 1950s and early 1960s showed that the order of bases in a DNA strand dictates the order of amino acids in the resulting protein. There are only 4 different nucleotides in a DNA molecule. If a single nucleotide dictated a specific amino acid, only 4 amino acids could be uniquely specified, and that certainly wouldn't be enough. There are 20 different amino acids that must be uniquely specified by the DNA molecule. Considering pairs of adjacent bases, 4 nucleotides would yield 16 unique combinations (Figure 8.24), but even that would not be sufficient to specify all the amino acids. If the information were carried in a segment of 3 nucleotides, however, the DNA bases could be arranged in 64 unique

Uracil (U)

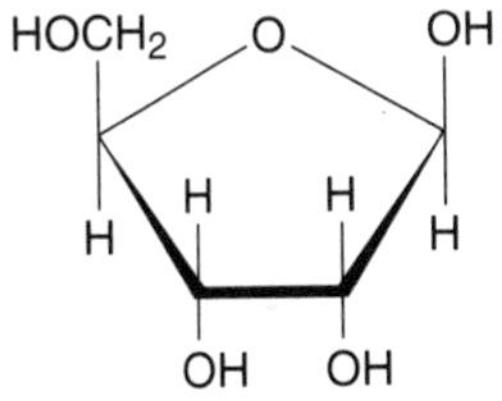

Ribose

Figure 8.23. *Two building blocks of RNA, uracil and ribose, that are not found in DNA.*

272

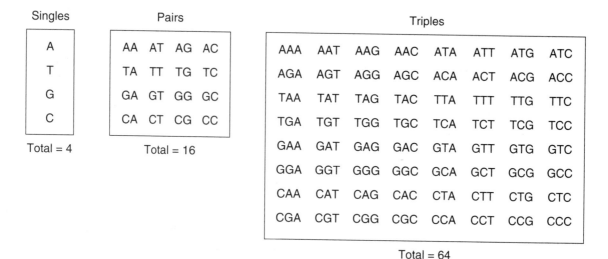

Singles
A
T
G
C
Total = 4

Pairs

AA	AT	AG	AC
TA	TT	TG	TC
GA	GT	GG	GC
CA	CT	CG	CC

Total = 16

Triples

AAA	AAT	AAG	AAC	ATA	ATT	ATG	ATC
AGA	AGT	AGG	AGC	ACA	ACT	ACG	ACC
TAA	TAT	TAG	TAC	TTA	TTT	TTG	TTC
TGA	TGT	TGG	TGC	TCA	TCT	TCG	TCC
GAA	GAT	GAG	GAC	GTA	GTT	GTG	GTC
GGA	GGT	GGG	GGC	GCA	GCT	GCG	GCC
CAA	CAT	CAG	CAC	CTA	CTT	CTG	CTC
CGA	CGT	CGG	CGC	CCA	CCT	CCG	CCC

Total = 64

Figure 8.24. *Unique combinations of nucleotide sequences of different lengths.*

ways—a number more than ample for determining protein sequences.

Clever research soon proved that amino acids are indeed coded in terms of "words" made up of three nucleotide bases. The amino acids dictated by a contiguous sequence of three nucleotide bases are now completely known, which makes it possible to tell what protein sequence that a given piece of DNA directs. Figure 8.25 shows a schematic for a piece of DNA that

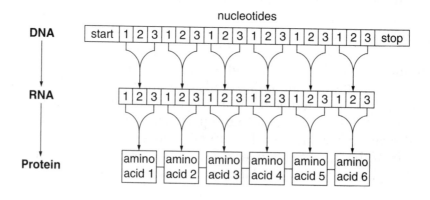

Figure 8.25. *A hexa-peptide encoded by a strand of DNA.*

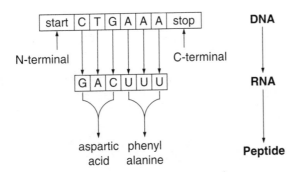

Figure 8.26. *The strand of DNA needed to synthesize NutraSweet®.*

encodes a polypeptide built of six amino acids. Special sequences of three nucleotides specify the start and stop positions of the amino acid chain. Between the stop and start signals, groups of three nucleotides specify the amino acids in order of their appearance in the polypeptide chain. The piece of DNA that directs synthesis of a protein constitutes a *gene*. A complete strand of DNA consists of many genes in sequence. The first step in protein synthesis is the translation of a single gene into a strand of RNA. The RNA molecule is then used to determine the sequence of amino acids. The cellular synthesis of the proteins from RNA molecules is carried out on the ribosomes.

Because NutraSweet® is a dipeptide, it can be made with the machinery of DNA. In this case, a DNA strand is needed that contains the start sequence, the sequences for the two amino acids—aspartic acid (Asp) and phenylalanine (Phe)—and the stop sequence. The three-base nucleotide sequence for aspartic acid is cytosine (C), thymine (T), guanine (G), and for phenylalanine it is adenine (A), adenine (A), adenine (A). If the strand of DNA shown in Figure 8.26 is inserted into the DNA of a cell (recombinant DNA), then the protein synthesis reactions of a cell (usually a bacterium) will include NutraSweet® as a product. This is genetic engineering!

The Chemical Basis of Evolution and Disease

In living systems, DNA molecules are constantly being replicated, damaged, and repaired. The replication process can be likened to the record industry. A single DNA molecule can be

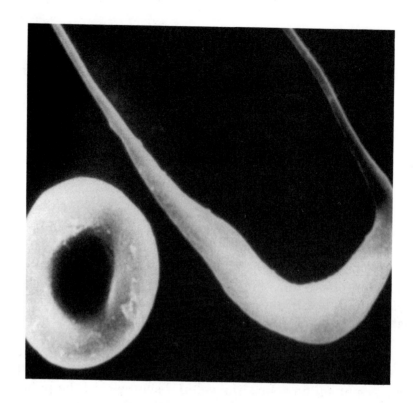

Figure 8.27.
*Micrograph showing
normal and sickle
cells. (From* The An-
nals of the New York
Academy of Sciences
Vol. 244, *copyright ©
1975 by the New York
Academy of Sciences.
Used by permission
of the New York Aca-
demy of Sciences and
Dr. Bruce Cameron.
Courtesy of NIH.)*

regarded as the master recording. Many copies are required so that the cells in the human body can access the stored information. In any recording process, there is the possibility of mistakes. The replication of a strand of DNA is no exception. The human body has systems to guard against mistakes. For instance, a host of enzymes will seek out errors in the DNA, cut out the bad section, and make repairs. This chemical quality control-system "tries" to ensure that all copies have the same fidelity as the original DNA. Considering that there are millions of bases in the DNA molecule that encode information for thousands of proteins, it is unsettling to think that a single mistake (one error in one gene) may be deadly. The dangers are real, as illustrated by sickle-cell anemia.

The red blood cells of a healthy person are round. The name for the genetic disease sickle-cell anemia derives from the crescent shape of the red blood cells found in people suffering from the disease (Figure 8.27). Discovered in 1904 by Dr. James

Herrick, this disease arises from a mutant form of hemoglobin. Hemoglobin has four interacting polypeptide chains. Two are called α-chains, and the other two β-chains. These four chains pack together, along with four oxygen-binding heme groups, to form a molecule of hemoglobin. Each chain contains about 150 amino acids. Herrick showed that diseased hemoglobin readily forms long fibers. Fifty years later, the molecular difference between diseased and normal hemoglobin was determined. It was shown that a *single* glutamic acid in one of the β-chains is replaced by valine. One mutation in the base sequence of DNA in the gene that directs hemoglobin synthesis causes sickle-cell anemia.

Mutations are not always bad. Over 150 different mutants of hemoglobin have been found in humans, and not all are fatal. It is even possible that some genetic mutations improve the properties of biological molecules. The occurrence of random mutations in DNA can be viewed as the molecular mechanism for evolution. In the Darwinian theory of evolution by natural selection ("survival of the fittest"), mutations that improve the ability of a molecule to carry out a biological function produce species with a better survival rate. Certain proteins, such as cytochrome *c* (which is involved in respiration), are found in a great many organisms. Minor changes occur from one species to another in the amino acid sequence of this protein. This is one piece of evidence that has been cited to support the theory that all life on Earth has a common ancestral background. Advanced organisms, such as mammals, also contain large regions of DNA called introns. These are inactive regions that may even occur in the middle of a gene. One hypothesis is that these inactive regions of DNA, which appear to serve no useful purpose, represent failed mutations over the course of evolutionary history.

Earlier in this chapter, we discussed the chemistry of penicillin. Many bacterial infections are now penicillin-resistant. This has spurred efforts to develop antibiotics such as the chemically modified penicillin known as amoxicillin. How did bacteria become resistant to penicillin? Mutations occurred in a hydrolytic bacterial enzyme that enabled it to break penicillin apart. These few mutated bacteria survived, flourished (most of

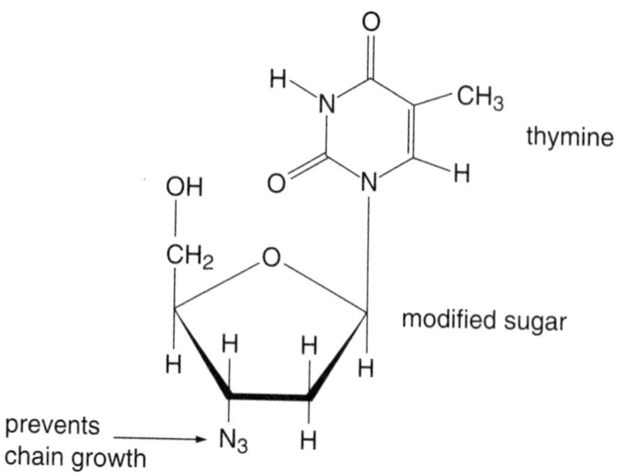

Figure 8.28. *The molecular structure of AZT, an antiviral drug used in the treatment of AIDS.*

the nonresistant competition having been killed), reproduced, and spawned a new penicillin-resistant strain.

The genetic code can also be disrupted to treat diseases. This is the logic behind administering AZT (Retrovir®) to people inflicted with AIDS. The structure of AZT is shown in Figure 8.28. AIDS is a viral disease. The genetic information is carried by an RNA molecule. [Several viruses, called retroviruses, store their genetic information in ribonucleic acid (RNA) polymer.] In the host cell, the RNA molecule creates a piece of DNA through base pairing. The DNA is then translated, producing more virus molecules. This is how the invading viruses use the normal operations of a cell to multiply. One approach to treating people with AIDS is to prevent the RNA molecule from synthesizing a strand of DNA. If this could be accomplished, the replication of the virus and the synthesis of the proteins encoded by the RNA strand would stop.

AZT works in this fashion. It is a nucleoside analog structurally similar to deoxythymosine, a precursor to thymine nucleotide (dTTP), one of the four building blocks of DNA. There is one very important difference, however. The sugar group has been modified. In place of the OH group, AZT has an azido group, N_3. This group is where the next nucleotide in a strand of DNA must attach itself. The chemistry of N_3 and that of OH are very different. The OH group undergoes a chemical reaction to link to another nucleotide; the azido group does not. If this base is incorporated into a DNA strand, it blocks the addition of further nucleotides. In order for the correct strand of DNA to be synthesized, an incorporated AZT molecule would need to be removed. The use of AZT is not without complications. Because the body has difficulty distinguishing AZT from dTTP, normal replication of human DNA also suffers. But the human body's ability to repair mistakes in DNA is vastly superior to that of a virus, so AZT is more toxic to the AIDS virus than to the human host.

In 1991 the FDA approved a second drug to treat AIDS. Like AZT, this drug, dideoxyinosine (DDI or Videx®), blocks the replication of the virus that causes AIDS. The long-term effects of DDI are not known yet, but the need for new AIDS therapies hastened its approval. Two other nucleoside analogs, Zerit® and HIVID®, won FDA approval for the treatment of advanced AIDS in 1992 and 1994.

Hormones

The human body contains trillions of cells. Each cell carries out important functions. Clearly, the actions of cells are orchestrated so that they work together. Chemicals that coordinate the activities of different cells are called *hormones*. In contrast to the proteins and nucleic acids we have considered so far, hormones are a family of chemically diverse compounds. These chemical messengers come in several forms, including small molecules derived from amino acids, small polypeptide chains, proteins, and steroids. Table 8.1 lists a few examples of hormone molecules. Hormones are synthesized by various glands in the body. Once produced, they travel in the blood stream to the sites where they carry out their chemical action. The molecules in Table 8.1 exemplify the diversity of hormones. Epinephrine (or adrenalin) is a derivative of an amino acid. Vasopressin is a small polypeptide. Insulin is a medium-sized protein. Estradiol is a steroid. The table also serves to point out that many regulatory systems are controlled by hormones. For example, epinephrine and insulin work in opposite ways to control the level of blood sugar.

Table 8.1. *Examples of Hormones and Their Functions*

Organ of Origin	Hormone	Structure	Function
Pituitary	Vasopressin	Polypeptide (9 amino acids)	Regulates the blood pressure.
Alimentary tract	Gastrin	Polypeptide (17 amino acids)	Causes secretion of stomach acid.
Adrenal medulla	Epinephrine	Amino acid derivative	Stimulates release of glucose into the blood.
Adenohypophysis	Prolactin	Polypeptide (197 amino acids)	Stimulates the synthesis of milk.
Pancreas	Insulin	2 Polypeptide chains (21 and 30 amino acids long, linked by disulfide bonds)	Stimulates the storage of glucose; stimulates protein synthesis.
Adrenal cortex	Glucocorticoids	Steroids (cortisol)	Have many effects on protein synthesis.
Ovary	Estrogens	Steroids (estradiol)	Govern the maturation and function of sexual organs.
	Progestins	Steroids (progesterone)	Function in the ovum implantation and in the maintenance of pregnancy.
Testis	Androgens	Steroids (testosterone)	Govern the maturation and function of sexual organs.

Many steroid hormones are synthesized from cholesterol (itself a steroid), and cholesterol is an essential component of cell membranes. The body makes cholesterol in the liver and intestines. People also eat a lot of cholesterol. Figure 8.29 shows four steroid hormones that the body synthesizes from cholesterol. Structurally similar steroid molecules may carry out very different biological functions. Estradiol, one of the female hormones, is responsible for the development of sexual characteris-

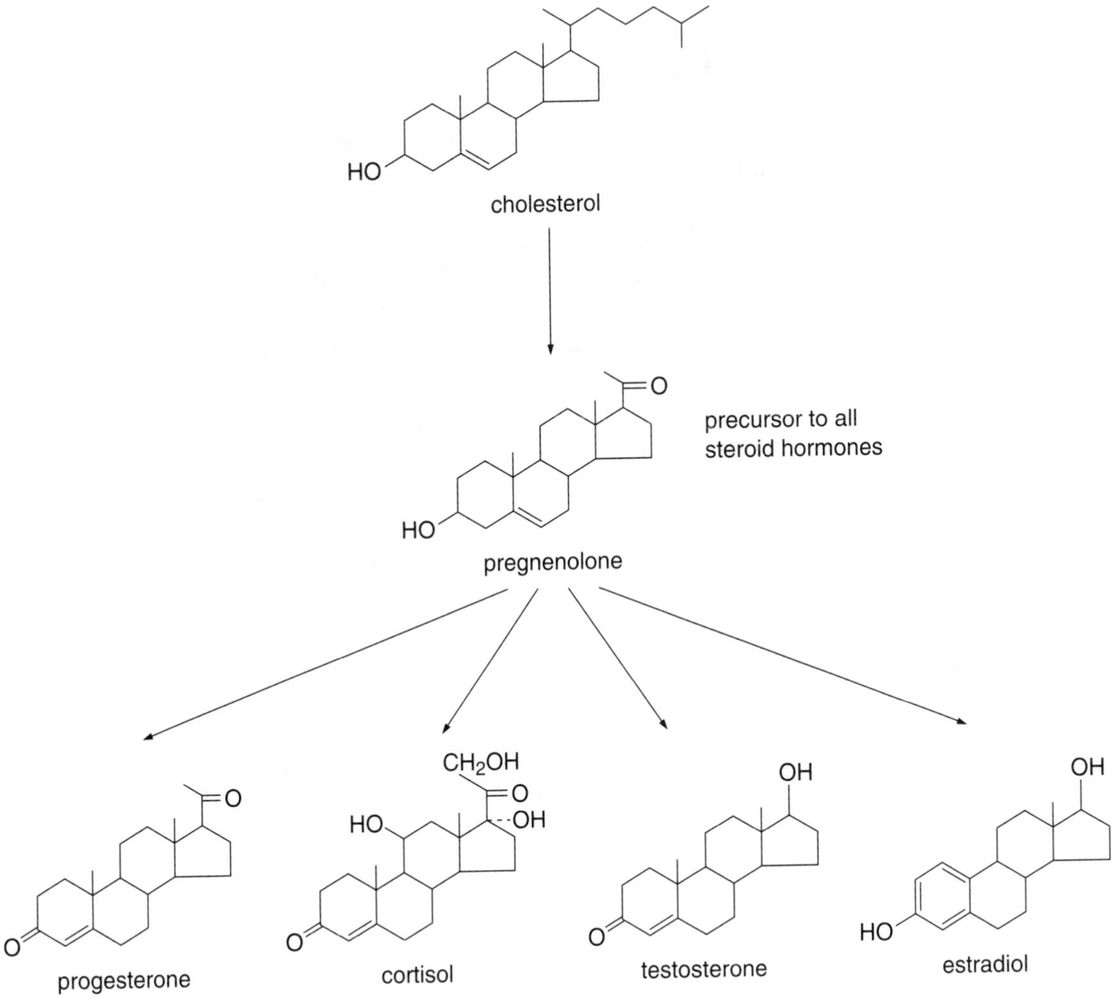

Figure 8.29. *Steroid hormones derived from cholesterol.*

tics that occurs at puberty. Progesterone prepares the uterine area for implantation of the fertilized egg. Testosterone is a male sex hormone that initiates the development of masculine sexual characteristics, such as deepening of the voice and the growth of body hair at puberty. Cortisol influences the metabolism of carbohydrates and proteins in the body.

Vitamins

A vast array of proteins, fats, carbohydrates, and minerals is included in the human diet. In addition to these chemicals, selected organic molecules are required in small amounts to prevent specific diseases. These compounds are called vitamins. Vitamins are not synthesized by human cells. They—or their immediate chemical precursors—must be eaten. Unlike the amino acids and the nucleic acid nucleotides, different vitamins do not share a common chemical structure. Figure 8.30 shows many of the vitamins required in the human diet.

Vitamins can be subdivided into two major groups. The *fat-soluble vitamins* include vitamins A, D, E, and K. The vitamin B complex (which consists of several molecules) and vitamin C are *water-soluble vitamins.* Fat-soluble (nonpolar) vitamins dissolve in fatty tissue. These molecules can be stored in fat for long periods of time. In a well-balanced diet, an adult human stores a long-term supply of vitamin A. If the diet becomes deficient in this vitamin, the reserves can be tapped to maintain good health. It is common knowledge that deficiencies in various vitamins cause diseases, but an overdose of some fat-soluble vitamins can also be harmful. Large excesses of vitamin A cause dry skin and a sensation of pressure in the head. In addition to ingesting vitamin A, the human body may synthesize this nutrient from the precursor molecule β-carotene. β-carotene is found in many vegetables, including carrots. The conversion of β-carotene to vitamin A is regulated by enzymes. Thus a safe way of building vitamin A reserves in the body is simply to eat foods high in β-carotene. Excess vitamin D can cause pain in the bones, nausea, diarrhea, and weight loss. The chemistry associated with excess vitamin D can also lead to the deposit of bone-like materials in joints, kidneys, and blood vessels, as well as in heart, stomach, and lung tissue. Vitamins E and K are also fat-soluble but are not stored so effectively as vitamins A and D. Vitamins E and K, therefore, seldom cause problems.

Water-soluble vitamins, on the other hand, are not stored effectively by humans. Excesses are excreted. Thus these vitamins need to be taken at frequent intervals (daily in some cases),

Fat-soluble vitamins

vitamin A

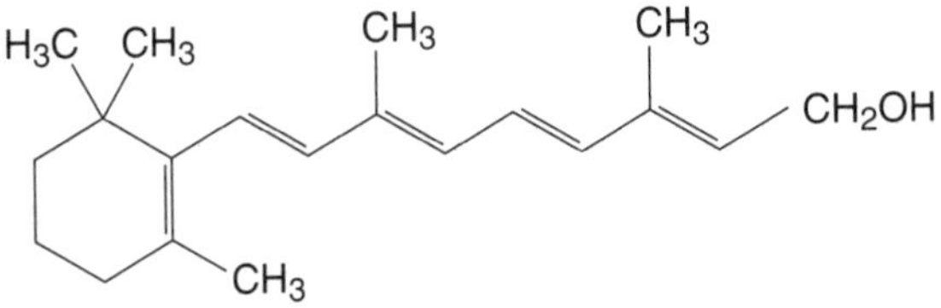

vitamin E

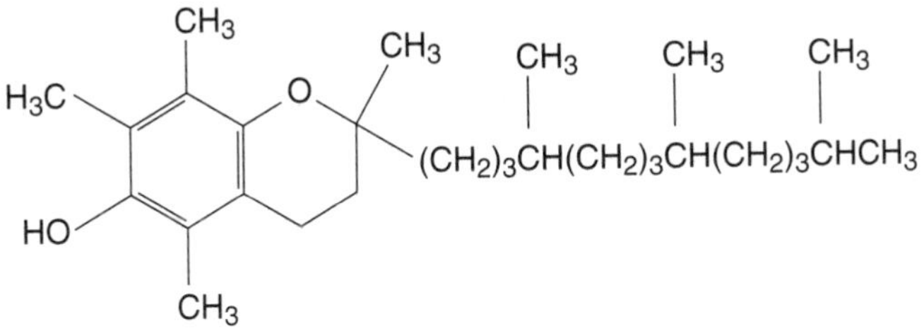

Water-soluble vitamins

vitamin B$_6$

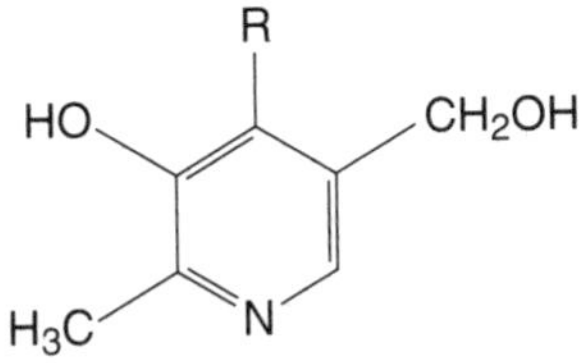

R	Chemical
HCO	pyridoxal
H$_2$COH	pyridoxol
H$_2$CNH$_2$	pyridoxamine

vitamin C

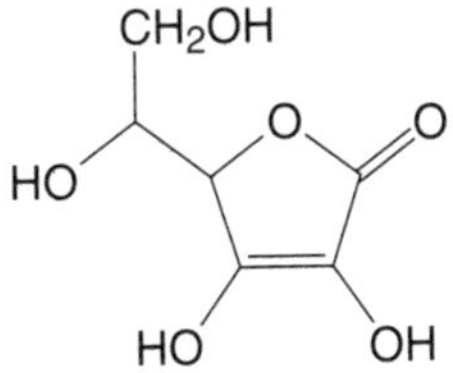

Figure 8.30. *Various vitamins needed in the human diet.*

whereas a single dose of a fat-soluble vitamin can last for weeks. A significant portion of the water-soluble vitamin content of foods can be lost when the food is cooked in water and then drained. The water-soluble vitamins dissolve and end up going down the drain. Water-soluble vitamins are generally not considered toxic. However, massive overdoses can cause some problems. Nerve damage has been associated with the ingestion of huge excesses of vitamin B$_6$.

Table 8.2. *Important Information on Vitamins*

Vitamin	RDA	Deficiency	Food Sources
Fat-Soluble			
Retinol (A)	0.12 mg	Night blindness	Liver, fruit, eggs, vegetables
Calciferol (D)	0.01 μg	Rickets	Cod liver oil, milk supplements
Tocopherol (E)	0.8 μg	Lack of hemoglobin	Plant oils, egg yolks
Linoleic acid (F)	—	Lesions, scales	Fatty foods
Phylloquinone (K)	» 0.1 mg	Blood loss	Leafy vegetables
Water-Soluble			
Thiamine (B$_1$)	1.5 mg	Beriberi	Pork, whole-wheat bread, nuts, milk, brewer's yeast
Riboflavin (B$_2$)	1.7 mg	Cheilosis	Milk, red meat, egg white, liver
Niacin (B$_3$)	2 mg	Pellagra	Liver, turnip greens, tomato juice
Pantothenic acid (B$_5$)	10 mg	Nervous disorders	Liver, eggs
Pyridoxal, Pyridoxol, Pyridoamine (B$_6$)	2 mg	Anemia, skin lesions	Liver, yeast, peas, beans
Biotin (B$_7$)	0.3 mg	Dermatitis	Liver, grains, peanuts, chocolate
Folic acid (B$_9$)	0.4 mg	Anemia	Leafy vegetables, liver, kidney, mushrooms
Cobalamin (B$_{12}$)	6 μg	Pernicious anemia	Liver, meat, eggs, fish (not found in plants)
Ascorbic acid (C)	60 mg	Scurvy	Fruits, vegetables

Table 8.2 lists the recommended daily allowance (RDA) published by the FDA for each vitamin. In addition, it lists diseases associated with deficiencies and some dietary sources of the vitamins.

Unlike enzymes, a single vitamin can be important in many different biochemical reactions. Vitamin B$_6$ is involved in over 60 reactions. Most of these reactions are important in the metabolism and synthesis of proteins. Thus it is not surprising that a deficiency of vitamin B causes anemia. Vitamin A is converted to retinal, the light-absorbing molecule in rhodopsin, the protein involved in vision. This explains why night blindness results from a deficiency in this nutrient.

Vitamin C also serves many functions. It is involved in the destruction of invading bacteria. It helps the body reduce the adverse effects of a variety of toxic substances. Vitamin C also participates in the synthesis and activity of interferon, which prevents the entry of many viruses into cells. The ability of vitamin C to "cure the common cold" has been hotly debated for many years. Scientific studies show that ingesting vitamin C supplements results in a decrease of 30 percent in various illnesses, especially upper respiratory infections. Less widely publicized is evidence that vitamin A is also effective in preventing or "curing" colds. Some people respond better to vitamin A than to vitamin C; others respond better to vitamin C than to vitamin A. Not surprisingly, some people don't respond to either. This wide range of results continues to fuel the debate on vitamin cures for diseases. In any case, for a vitamin to be effective, it needs to be taken before one gets a cold (preventively) or in the very early stages of a cold. Although both vitamins C and A can individually play a role in relieving the symptoms of the common cold, doctors do not recommend taking both together, because doing so seems to prolong cold symptoms.

Vitamin E is an important antioxidant. It is particularly effective in preventing the oxidation of polyunsaturated fats. If left unchecked, the products from the oxidation of polyunsaturated fats lead to many unwanted reactions in cells. The antioxidant role of vitamin E helps maintain the integrity of cell membranes and helps maintain the circulatory and nervous systems. Vitamin E also helps the body detoxify poisonous substances and aids in maintaining proper kidney, lung, and liver function.

Energy and Metabolism

The energy needs of the human body are met by carrying out oxidation reactions on energy-rich compounds. Oxygen (O_2) is the primary oxidant. Humans breathe in oxygen and exhale carbon dioxide (CO_2) and water in a process called respiration.

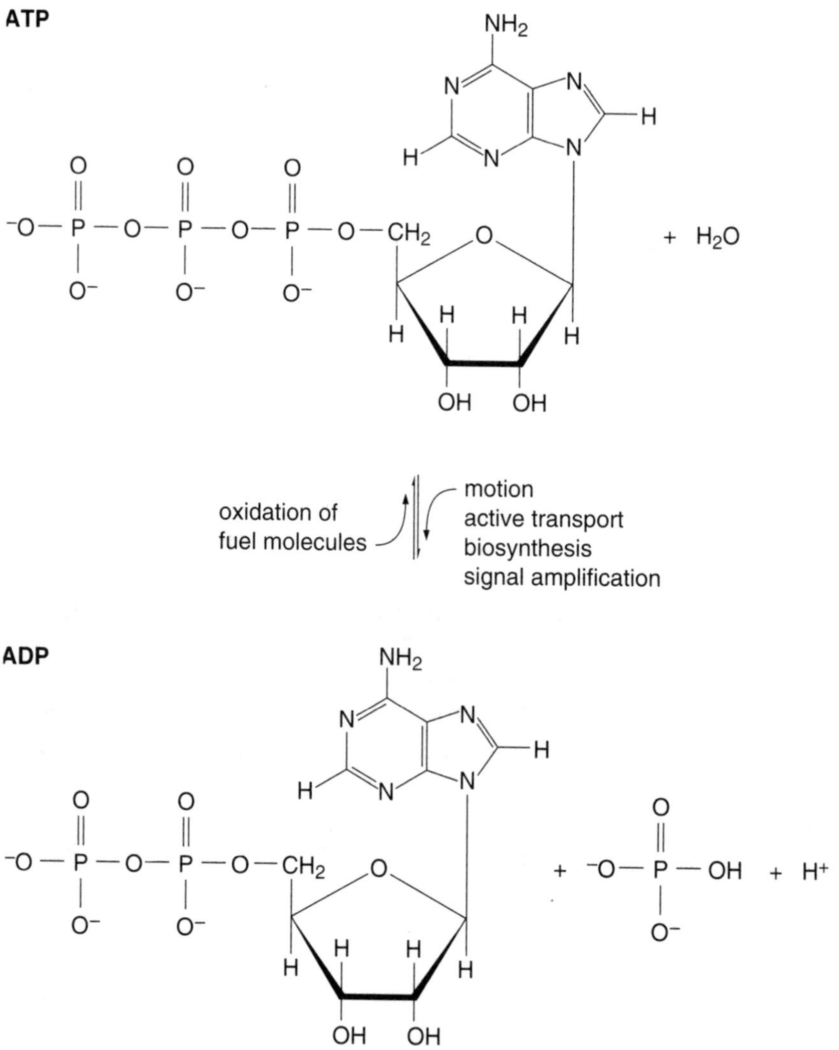

Figure 8.31. *The chemical equilibrium between ATP and ADP. The biological functions associated with the two reactions are listed next to the arrows.*

If energy is obtained in one form (from breakage of chemical bonds or light from the Sun) and used in another form (muscle contraction or protein synthesis), there must be some means of *efficiently* coupling the energy-yielding process to the energy-requiring process. The energy source common to a large number of biochemical reactions is the molecule adenosine triphosphate (ATP). This energy-rich molecule is made by respiration and

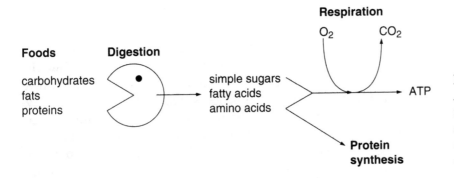

Foods   **Digestion**

carbohydrates
fats
proteins

simple sugars
fatty acids
amino acids

Respiration

O_2 CO_2

ATP

Protein synthesis

Figure 8.32. *How ATP and other important ingredients for biosynthesis are produced from digestion and respiration.*

then used to promote energy-requiring reactions throughout the body. ATP is similar to the DNA building block dAMP.

One of the molecular mechanisms responsible for the generation and storage of chemical energy is the reversible reaction between ATP and adenosine diphosphate, ADP (Figure 8.31). This reaction involves breaking an oxygen–phosphorus bond. The role of the ATP–ADP system is to serve as an acceptor and a donor of energy-rich phosphate groups. ATP is the immediate energy source for most energy-requiring reactions in cells. It follows that living organisms have a chemical mechanism for generating ATP, thereby keeping a constant amount of this high-energy molecule around. The balance between the generation and use of ATP is well controlled. If something caused a living cell to stop generating ATP, the supply of this molecule would be exhausted in only a few minutes. Survival depends on manufacturing the amounts of ATP necessary to carry out biological reactions.

Ultimately, the source of energy and raw materials (amino acids) for carrying out life-sustaining processes comes from breathing, eating foods, and drinking water. Food provides carbohydrates, lipids, proteins, and nucleic acids. However, these substances cannot be used directly in the form in which they are ingested. Before food can be utilized by the body, it must be broken down into small molecules. This process, called digestion, involves converting carbohydrates into simple sugars, large fat molecules into fatty acids and glycerol, and proteins into amino acids. All these processes involve the action of enzymes. The transformation of food into useful molecules follows the general scheme shown in Figure 8.32. The metabolism of the sugar sub-units of carbohydrates produces the high-energy ATP molecule. This is why water solutions of glucose can be used for intravenous feeding. The chemical process (respiration) requires oxygen, which is supplied by hemoglobin and myoglobin. The inhalation of oxygen from the atmosphere enables oxidation reactions of the sugars to occur, producing ATP and liberating CO_2. The CO_2 by-product is removed from the body by exhaling.

CHAPTER 9

Photochemistry

An Evening at Christies

"Seventy-three million," said the nervous bidder.

The auctioneer was elated. Everyone in the room was standing, and each bid evoked a long round of applause. As the auctioneer dimmed the spotlights, the yellow and orange pigments in the portrait became more pronounced and the blue background mellowed. The exhibitors had worked for days to get the lighting perfect. This particular painting, the *Portrait of Dr. Gachet,* was completed six weeks before the artist's suicide in 1890. Paul-Ferdinand Gachet was not royalty; he had been van Gogh's friend and physician. But the value of the painting had nothing to do with its subject. It was sought after because of the unique way van Gogh used color and light to evoke emotion. The liveliness of the brush strokes around the man's form and

Vincent van Gogh's Portrait of Dr. Gachet *sold at Christies. (Reuters/Bettmann.)*

on his clothes gave a shimmer to the painting that contradicted the melancholy expressed on Gachet's face. In a letter to Paul Gauguin, van Gogh wrote that Gachet had been painted "with the heartbroken expression of our time." There were two van Gogh portraits of Gachet. One hung in the Musée d'Orsay; the other has just been bid up to seventy-three million dollars!

"I have seventy-three million, do I hear seventy-four?"

"Seventy-four million" echoed from the back of the room. The room filled with applause.

Hideto Kobayashi was determined to buy this work. He gave the hand signal for seventy-five million dollars. The applause resumed for a minute and then silence filled the room.

"Seventy-five, do I hear seventy-six?" No bids were made. The auctioneer raised the mallet, "I have a bid of seventy-five million dollars for Vincent van Gogh's *Portrait of Dr. Gachet.* Going once . . . , twice . . . , sold!" The mallet struck the top of the podium, and the entire crowd cheered.

A new record had been set for art purchased at an auction. Including the buyer's fee, Kobayashi had spent eighty-two and a half million dollars for a collection of colorful compounds slapped onto a cheap piece of canvas.

Absorption and Emission of Light: What Is Color?

Matter absorbs, reflects, and transmits light. The color of an object depends on the light that is reflected. If an object reflects all colors in sunlight, it appears white. Conversely, if all the colors are absorbed, no light is reflected toward the eyes and the object is black. A red shirt reflects red light while absorbing green and blue. Many common items transmit various colors of light and absorb others. Colored glass bottles and colored sodas contain molecules that absorb certain colors of sunlight. By adding these chemicals, manufacturers hope to make their products more appealing to the consumer. But colored bottles are not used only as a marketing strategy. Some liquids (especially medicines) contain molecules that undergo bond breakage when

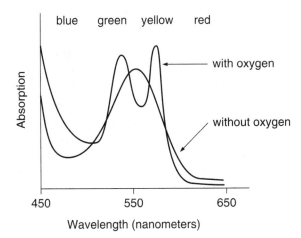

Figure 9.1. *The absorption spectrum of hemoglobin before and after it has taken up oxygen in the lungs. (Redrawn from Biochemistry 3rd ed. by Lubert Stryer. Copyright © 1988 by Lubert Stryer. Used by permission of W. H. Freeman and Company.)*

exposed to certain colors of light. Dark-colored bottles protect the contents from exposure to the wavelengths of light that cause chemical decomposition.

It is the wavelength of the light wave that determines its color. The wavelength of a visible light wave is very small—on the order of 0.00001 cm (10^{-5} cm). For example, red light has a wavelength of 6.4×10^{-5} cm. Thus in a distance of 1 meter, the light wave oscillates over 1.5 million times! Light's wavelength also determines its energy content. The larger the wavelength, the less energy contained. The complete range of wavelengths of light is called the electromagnetic spectrum.

A plot of the light absorbed as a function of energy is called an absorption spectrum. Figure 9.1 shows the absorption spectrum of hemoglobin, the oxygen-transporting protein in red blood cells before and after it has bound oxygen. The wavelength of light is given in nanometer (nm) units (1 nm = 10^{-9} cm). The spectrum indicates that green, yellow, and blue light is absorbed and that red light is transmitted. The absorption properties of hemoglobin give red blood cells their color. Hemoglobin with bound oxygen is found in the arteries, which supply blood throughout the body. When oxygen is released from hemoglobin, the absorption spectrum changes. This is why blood in veins has a slightly different color from that in arteries.

Following absorption of light, a molecule is said to be in an excited state. After a short time (usually less than a millionth of a second), the electrons change back to their original spatial arrangement and release energy. Energy is usually liberated as light and heat. Colorful fireworks displays result from the light emitted from molecules in an excited state. Fireworks are composed of several chlorine-containing compounds such as chlorinated rubber, polyvinyl chloride, and perchlorate or chlorate oxidizers. The metals strontium (Sr), barium (Ba), and copper (Cu) are added to this mixture. The high temperatures in a fireworks explosion decompose the organic materials and form free chlorine atoms. The chlorine atoms react with metal additives to yield excited diatomic molecules (excited molecules are designated by placing an asterisk (*) after the chemical formula). These unstable molecules return to the ground state by the emission of light, as indicated in the following three equations.

$$Cl + Sr(metal) \rightarrow SrCl^* \rightarrow \text{red light} \qquad (1)$$

Chlorine atoms react with strontium metal to make an excited strontium chloride molecule that relaxes by emitting red light.

$$Cl + Ba(metal) \rightarrow BaCl^* \rightarrow \text{green light} \qquad (2)$$

Chlorine atoms react with barium metal to make an excited barium chloride molecule that relaxes by emitting green light.

$$Cl + Cu(metal) \rightarrow CuCl^* \rightarrow \text{blue light} \qquad (3)$$

Chlorine atoms react with copper metal to make an excited copper chloride molecule that relaxes by emitting blue light.

In addition to emission, excited molecules can undergo chemical change. Light energy can break bonds and change molecular geometries. Because the energy for these reactions comes from an absorbed photon, the chemical processes are referred to as photochemical reactions or photoreactions, and this area of chemistry is called photochemistry.

Vision

Photochemistry makes it possible for people to see. In the human eye, there are two kinds of photoreceptors: rods and cones. Rods make possible sight in low light but offer no color perception. Cones function in bright lights and enable us to differentiate colors. The perception of color is possible because the eye contains three different types of cones, each sensitive to a different region of the visible spectrum. A human retina contains close to 3 million cones and 100 million rods.

In 1938 Selig Hecht discovered that the human rod cell could be stimulated by a single photon. What is the molecule that absorbs the light, and how is the excitation energy used to start the visual process? The answer to this question was found by studying the rod cells. A schematic of a rod cell is shown in Figure 9.2. Rod cells are about 1 millionth of a meter in diameter and 40 millionths of a meter long. The outer segment (that closest to the back surface of the eye) contains a stack of about 1000 disk-like structures, each of which is about 16 nm thick. The inner segment contains the chemical machinery necessary for generating ATP and proteins. The disks in the outer segment have a lifetime of about a month and are continuously replaced by the chemistry of the rod cell. Early studies showed that light is absorbed in the disk structures. Careful examination of the contents of the disks revealed that a protein—rhodopsin—absorbs the light. Figure 9.3 shows the absorption spectrum of rhodopsin.

Rhodopsin in rod cells absorbs strongly at 500 nm, the green part of the visible spectrum. The absorption is greatly reduced in the blue and yellow regions. Essentially no absorption is observed for the colors orange and red. Rhodopsin must absorb light for us to perceive an image. The broad absorption spectrum of this protein reveals why rod cells are not able to distinguish colors. Investigations into the chemical composition of rhodopsin reveal that the amino acid sequence of this protein is bonded to the organic molecule 11-*cis*-retinal. In 1958 George Wald and his coworkers showed that when 11-*cis*-retinal absorbs a photon, the molecule twists around a carbon–carbon bond, a process known

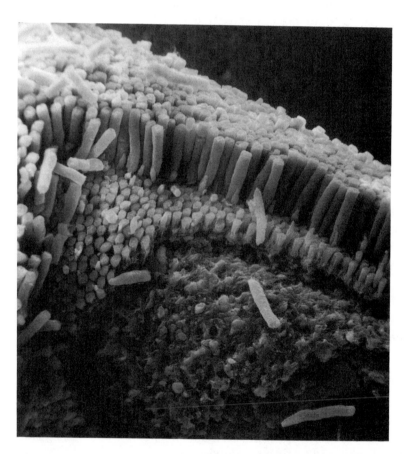

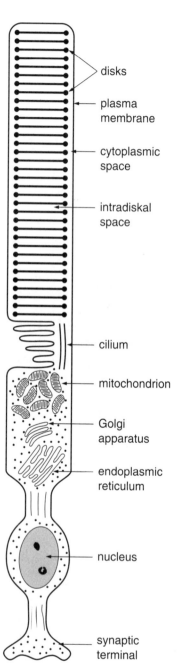

disks

plasma
membrane

cytoplasmic
space

intradiskal
space

cilium

mitochondrion

Golgi
apparatus

endoplasmic
reticulum

nucleus

synaptic
terminal

Figure 9.2.   *(left) An electron micrograph of rod cells. (Courtesy of M. Deric Bownds and Stan Carlson.) (right) The schematic of an individual rod cell found in the retina of the eye. (From Bio-chemistry 3rd ed. by Lubert Stryer. Copyright © 1988 by Lubert Stryer. Used by permission of W. H. Freeman and Company.)*

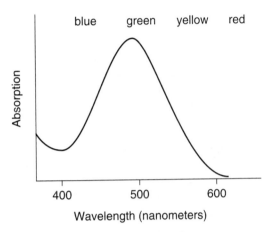

Figure 9.3.   *Absorption spectrum of rhodopsin in rod cells. The maximal absorption occurs in the green region of the spectrum, which explains why the eye is most sensitive to green light.*

as isomerization, forming 11-*trans*-retinal (equation 4). The prefixes *cis* and *trans* describe the three-dimensional molecular geometry around the double-bonded carbons. This change in the molecular geometry of the retinal molecule occurs upon photoexcitation of rhodopsin and serves as the trigger for the visual process. The transformation of 11-*cis*-retinal into 11-*trans*-retinal moves the bonding site between the retinal molecule and the surrounding protein by 0.5 nm. This is a significant distance, given the geometrical constraints imposed by the protein structure. The isomerization reaction stores light energy. This energy drives further chemical reactions involved in the visual process.

11-*cis*-Retinal $\xrightarrow{h\nu}$ 11-*trans*-Retinal   (4)

11-cis-Retinal in the rod cells of the eye absorbs light and changes to 11-trans-retinal.

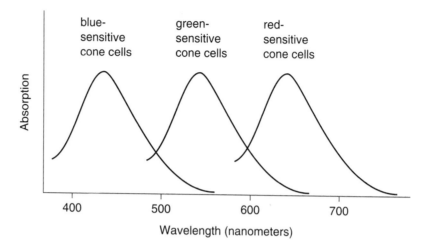

Figure 9.4.
Absorption spectra of the different rhodopsins found in cone cells. These cells are responsible for color vision.

Isomerization of retinal takes place in less than a picosecond (1 picosecond is 1 trillionth of a second). Following this initial photochemical event, rhodopsin undergoes a series of chemical reactions. These reactions transform rhodopsin into an activated state that then interacts with other enzymes. This complicated series of biochemical reactions, initiated by the original absorption of a photon, stimulates the optic nerve, giving rise to the sense of sight.

Rod cells enable us to see black and white, but they play no part in the perception of color. Rod cells are very sensitive and are essential for night vision. This explains why some people are color blind at low light levels. Color vision involves the other visual receptors, the cones. Specific cone cells detect blue, green, and red. Again, the molecule responsible for the absorption is rhodopsin. The differences in the absorption properties (Figure 9.4), and thus the color of light detected, arise from alterations in the structure of the protein in the vicinity of the 11-*cis*-retinal. Color blindness results when one of the color-sensitive forms of rhodopsin is missing. Cone cells are not as sensitive to light as rod cells. Nocturnal animals tend to rely primarily on rod cells for vision and have far fewer cone cells than diurnal species.

Photography

In 1727 Johann Heinrich Schulze reported that solid silver chloride, $AgCl_{(s)}$, darkens when exposed to light. Toward the end of the same century, Thomas Wedgwood used a lens to form an image on a piece of paper treated with $AgNO_3$. Wedgwood produced a picture that faded quickly, but it was a picture. Photography was born.

Before 1935 all photography was black and white, though color photography is taken for granted today. Films that record wavelengths of light not sensed by the human eye are also available. For example, infrared-sensitive film can be used to detect heat. It is very useful for identifying insulation leaks in homes in cold climates and for military reconnaissance photography.

Photography can be understood by examining the molecules that make up film and the chemistry that turns exposed film into photographic negatives. Both color and black-and-white photography are based on the photochemistry of silver bromide crystals. On the molecular level, excitation of a silver bromide crystal moves an electron from a Br^- anion to a neighboring Ag^+ cation, as shown in equation (5). This photochemical reaction produces atomic silver and bromine.

$$Ag^+Br^- + light \rightarrow Ag + Br \qquad (5)$$

Light excitation of silver bromide transfers an electron from Br^- to Ag^+.

Color photographic film should record all colors visible to humans, but film manufactured from pure silver bromide is sensitive only to the blue part of the visible spectrum. Taking pictures with a photographic film that detects only blue is not very appealing. To solve this problem, chemists developed compounds known as sensitizers. These species alter the energy levels of electrons in the silver halide crystals so they absorb light in the green and red parts of the visible spectrum. In this manner, the silver becomes "sensitized" to all parts of the visible spectrum.

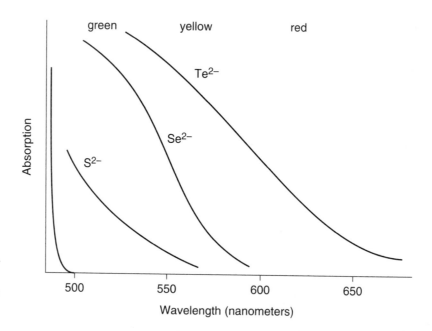

Figure 9.5. *Change in the ability of AgBr crystals to absorb visible light brought about by the addition of sensitizing agents.*

Silver bromide is an ionic crystal that consists of Ag^+ cations and Br^- anions. A simple approach to sensitizing the crystal is to replace some of the Br^- anions with impurity anions. This substitution changes the absorption properties of the crystal. As an example, Figure 9.5 shows the absorption curves for pure AgBr and AgBr containing about $10^{19}/cm^3$ atoms of S^{2-}, Se^{2-}, or Te^{2-} impurity ions. A pure AgBr crystal contains over 10^{22} Br^- ions/cm^3. In the manufacture of these modified solids, less than 1 out of every 1000 bromines is replaced by an impurity ion. This low level of impurity doping greatly extends the absorption spectrum of the silver halide crystal across the visible region.

Figure 9.6 shows the structure of black-and-white film. The film consists of two main layers. A plastic (or paper) substrate is used to support a gelatin that contains uniformly dispersed silver halide crystals doped with sensitizers. The plastic substrate makes the film flexible so that it can be rolled up. Photochemical action occurs in the gelatin. In manufacturing film, the concentration of AgBr crystals in the gelatin is made uniform. This guarantees a constant sensitivity to light over the whole piece of film. Gelatin is made from the protein collagen. Collagen is one of the most abundant proteins in animals. It forms

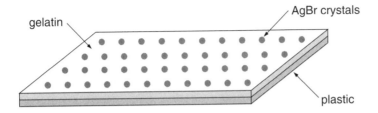

Figure 9.6. *A schematic of a piece of film. The light-sensitive AgBr crystals are spread uniformly in a gelatin that is mounted on a plastic backing.*

skin, bone, cartilage, and ligaments. Flexible celluloid film was introduced by George Eastman in 1898 with his invention of the Kodak camera.

In photography, a lens images onto film the object to be photographed. The number of photons that strikes a certain area of the film depends on the brightness of the object being imaged. The exposure time can be adjusted so that only a fraction of the silver bromide crystals absorbs photons. The variation in the number of photons absorbed per area of film produces different shades of gray after development.

The presence of a Ag atom makes the crystal easier to reduce chemically to a solid piece of silver than a pure ionic AgBr crystal. This difference in the rates of reduction of excited and nonexcited crystals is the chemical foundation of film development. Developers are solutions of mild reducing agents. The chemistry of hydroquinone, a developer, is shown in equation (6).

$$\underset{\text{Hydroquinone}}{\text{OH}\bigcirc\text{OH}} + 2AgBr^* \xrightarrow{H_2O} 2Ag_{(s)} + 2HBr_{(aq)} + \underset{\text{Quinone}}{\text{O}\bigcirc\text{O}} \qquad (6)$$

Hydroquinone reduces the light-exposed silver bromide crystals in film; metallic silver is deposited, and the HBr and quinone byproducts are washed away.

Silver bromide crystals that have been sensitized by exposure to light are reduced to silver metal, while hydroquinone is oxidized to quinone. Hydroquinone has the ability to reduce both excited and nonexcited AgBr crystals to metallic crystals. However, because the time allowed for the reaction to take place is controlled, only those crystals that have absorbed photons are actually reduced to metallic silver. The finely divided silver metal deposit appears black or a shade of gray, depending on the amount produced.

There is a tradeoff between film speed and the resolution of the final picture. With increasing film speed, the time that the shutter is open decreases. This means that fewer photons are collected and imaged onto the film. As a result, fast films use large AgBr crystals to capture the small numbers of photons that pass through the open shutter. Slower films use smaller AgBr crystals, because more photons can be collected with the slower shutter speeds. The development process converts to metallic silver only those crystals exposed to light. In the fast films, these crystals are large and therefore give less resolution. Because of the large imaging crystals, photos taken with fast film may look grainy. The initial image in the developed film is a negative image. This is because bright areas in the initial image correspond to photoexcited areas of the film, which become dark deposits of silver after development. When light shines through the negative image onto print paper, the image is reversed again. This double-reversing process results in a normal image on the printing paper.

To prevent the reduction of all the AgBr crystals in the film, a "stop" solution is used to arrest the reduction chemistry after the light-sensitized crystals have been developed. The stop bath consists of a strong acetic acid solution. Unfortunately, this two-step chemical process does not lead to the production of a usable negative. It took over 30 years after the discovery of developers and stop solutions to devise methods for producing permanent pictures. The problem is as follows: Once the excited AgBr crystals are reduced to metallic silver, any additional exposure of the film to light outside the darkroom photoexcites the unreacted AgBr crystals. As a result of this reaction and subsequent exposure to air, the entire film turns black over time. Because the printing step requires shining light on the negative, this unwanted chemistry needs to be prevented. Unreacted AgBr crystals must be removed from the film after the photoexcited crystals are reduced to silver. The chemicals used to remove the unreacted AgBr crystals must be inert toward gelatin and metallic silver. Otherwise, the photographic image would be destroyed in the process. Furthermore, the chemicals used must completely dissolve the AgBr crystals, leaving no precipitates on the film. A water solution of sodium thiosulfate ($Na_2S_2O_3$) was found to preserve, or fix, the photographic image. The thiosulfate anion dissolves the silver bromide by the reaction shown in equation (7).

$$AgBr_{(s)} + 2S_2O_3^{2-}{}_{(aq)} \rightarrow Ag(S_2O_3)_2^{3-}{}_{(aq)} + Br^-{}_{(aq)} \qquad (7)$$

Thiosulfate ion ($S_2O_3^{2-}$) binds silver ions and dissolves unexposed AgBr in the photographic negative.

After this step, the film is washed with large quantities of water to remove all the fixer solution, $Na_2S_2O_{3(aq)}$. Otherwise, further chemical reactions that affect the image might occur and ruin a good picture. This three-step process of developing, stopping, and fixing produces a photographic negative.

Color film also relies on the photochemistry of a AgBr crystal. Figure 9.7 shows a basic design of color film. Colors are differentiated by modifying black-and-white film in two major ways. First, filters are used to control the wavelengths of light (or color) that strike particular areas of the film. Second, organic dye molecules are dissolved in the gelatin layer.

All visible colors can be made by appropriate mixtures of the three primary colors red, blue, and yellow. Color film contains closely spaced regions that are sensitive to the three primary colors. Each region is much smaller than the eye can detect. A green object imaged onto the film is stored by exciting AgBr crystals in patches sensitive to yellow and blue. In developing color film, it is necessary to preserve the color separation caused by the filters. Each color-sensitive region of the film contains an organic dye molecule. These molecules, called couplers, are transparent to visible light but form permanent color deposits on the negative during development.

The colors in a negative are complementary to the colors of the photographed objects. In the developing of color film,

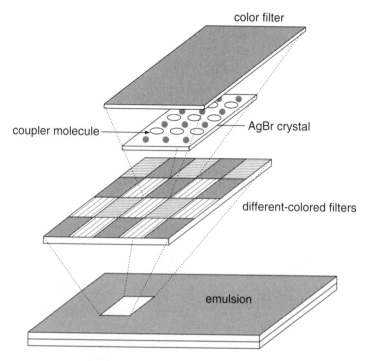

Figure 9.7.   *Color film consists of an emulsion that contains many small patches that are sensitive to different colors. The presence of coupler molecules that generate different colors upon development creates the color image.*

regions that are sensitized to a certain color by filters need to be stored as that color's complement. This entails linking the reactions of the coupler molecules to the oxidation of the excited AgBr crystals. One example of this chemistry is shown in equation (8) for the coupler molecule 1-naphthol. In this reaction, the molecule DEPD is the developer. Neither DEPD nor 1-naphthol alone absorbs visible light, but the coupled molecule, indophenol blue, is highly colored. The coupling of the dye molecule and the developer requires simultaneous reduction of AgBr crystals. During development, the blue dye forms in regions of the film where AgBr crystals had absorbed orange light (the complement of blue). By proper choice of the coupler molecule, it is possible to obtain any desired color.

$$\text{DEPD} + 4AgBr + \text{1-naphthol} \longrightarrow \text{Indophenol blue} + 4Ag + 4HBr \qquad (8)$$

Coupling reaction between the developer molecule (DEPD) and 1-naphthol, which forms a blue dye in the area of color film where the AgBr had been exposed to light.

The first commercially successful three-color dye-coupling process was introduced by Kodak in 1942. The film consisted of three separate emulsion layers. Each layer was sensitized to respond to a different part of the visible spectrum. Developing produced the three colors yellow, magenta, and cyan. Each layer had to be developed individually, using different couplers and developers.

Modern color film consists of an emulsion layer that contains all three dye couplers. A single developer reacts with all three couplers, making the processing of color film much easier. Figure 9.8 shows several dyes that are used. The size of the dye molecule helps determine the resolution of color film. The photographic emulsion is not a crystalline solid. As a result, coupler molecules can diffuse. With increasing size, the coupler molecule is less able to wander in the emulsion, resulting in a higher-resolution film. The dyes generated by the development process also undergo slow photochemical reactions. Over time, this photochemistry can cause the colors to fade in both color negatives and prints.

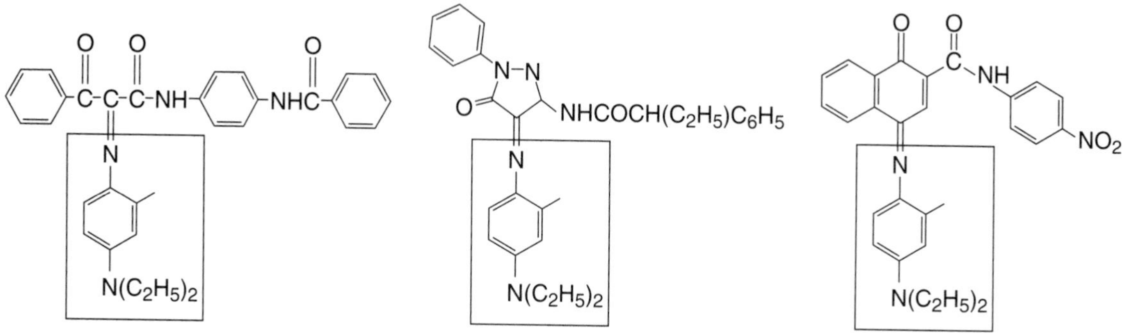

Yellow dye
λ_{max} = 435 nm

Magenta dye
λ_{max} = 526 nm

Cyan dye
λ_{max} = 713 nm

Figure 9.8. *Common dyes used in color photography. The dye molecule is a combination of the developer (DEPD, boxed) and coupler molecules that are placed in the film during manufacturing.*

Sun Tans, Sunblocks, and Light-Sensitive Sunglasses

Molecules in skin undergo photochemical rearrangements when exposed to ultraviolet light. These reactions not only cause tans but also play a role in skin cancer. Skin consists chiefly of the protein collagen, but it also contains many smaller molecules that are required for a variety of biochemical processes. One of these molecules is 7-dehydrocholesterol (7-DHC). When 7-DHC absorbs an ultraviolet photon, the energy that is gained leads to the formation of vitamin D, as shown in equation (9). Vitamin D is a very important molecule. A hormone derived from Vitamin D plays an essential role in the uptake of calcium and phosphorus for building teeth and bones. Lack of vitamin D causes many ailments, including rickets, a disease that deforms children's bones. But too much vitamin D can be harmful to health. It can cause the body to accumulate excess calcium and phosphorus, creating bone-like deposits in the skin; it also can lead to deadly problems such as kidney failure.

In addition to generating vitamin D, ultraviolet light causes photochemical changes in the DNA of skin cells. Upon exposure to UV light, adjacent pyrimidine bases (cytosine or thymine) in a DNA strand undergo a photochemical reaction that creates new covalent bonds between the bases. As a result, the damaged DNA cannot be replicated or translated to produce proteins. To reactivate the DNA, this dimer must be removed and replaced by the correct nucleotides. The human body has enzymes that cut the DNA, remove the damaged segment, and repair the DNA strand. If the DNA damage exceeds the repair capacity of the cell, then cell death (sunburn peeling) or cancerous mutations may occur.

(9)

7-Dehydrocholesterol is converted to vitamin D in the presence of sunlight.

Fortunately, the body has a mechanism for reducing the amount of UV light that penetrates the skin. Sunlight promotes biochemical reactions in the skin's outer layer; these reactions darken the skin. The more tanned the skin, the harder it is for UV radiation to penetrate and damage DNA.

Tanning of the skin is a photochemical process. It involves the formation of melanins, which are large polymer molecules that absorb light throughout the visible and ultraviolet regions of the spectrum. (Polymeric red and black melanins control the skin color in different races of people.) The photochemical synthesis of black melanins occurs in two stages. In the outer layer of skin, the body stores melanin molecules in a reduced state. In the first hour of exposure to sunlight, these colorless molecules absorb ultraviolet light and are oxidized. This causes an immediate buildup of the melanin concentration in the skin.

Prolonged exposure to UV light stimulates the body to synthesize more melanin. The chemical processes involved in the synthesis of black melanins are shown in equation (10). Melanin synthesis is initiated by tyrosinase, an enzyme that reacts with the amino acid tyrosine in the skin and produces molecules that polymerize to form melanins. (A genetic deficiency of tyrosine occurs in albinos.) Too much sunlight can overwhelm the body's protective tanning mechanisms and damage the skin. Excess exposure to sunlight is the leading cause of skin cancer. Many photochemical reactions that occur in skin are irreversible and permanent. Thus precautions to prevent excessive exposure to UV radiation should be taken. *Sunblocks* and *sunscreens* were invented for this purpose.

Chemical reactions used in the skin to make melanins, the dark pigment of a sun tan.

Most sunblocks are pigments of zinc oxide (ZnO) or titanium dioxide (TiO_2). These substances are white and absorb ultraviolet light, while reflecting all colors in the visible spectrum. Most sunscreens contain organic molecules and are applied as a colorless oil or cream. Figure 9.9 shows the structure and absorption spectrum of one commonly used molecule, amyl dimethylaminobenzoate (DMAB). DMAB strongly absorbs ultraviolet radiation. In this particular case, absorption does not lead to any photochemistry. The excess energy is released as heat. DMAB is very soluble in water. It washes off when the individual swims or sweats. This is why many sunscreens need to be applied to the skin frequently. Commercial sunscreens have rating numbers ranging from 2 to greater than 50. These numbers are a measure

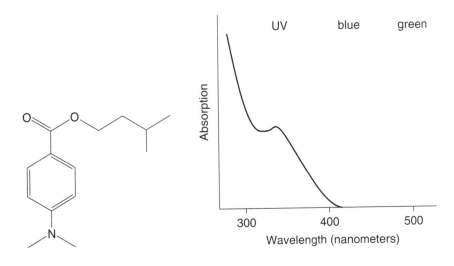

Figure 9.9. *Structure and absorption spectrum of amyl dimethylaminobenzoate, a common ingredient in sunscreens.*

of the absorbing power of the active sun-blocking molecules (the larger the number, the greater the blocking power for UV light).

Photochromic sunglasses have the property of adjusting their darkness in response to changes in brightness. Inside buildings, the glasses are essentially clear. However, as soon as they are exposed to sunlight, the glasses turn dark. And once back indoors, they return to being nearly transparent. These color changes involve a photochemical reaction. The photochromic process utilizes silver photochemistry and the availability of ultraviolet light from the Sun. During the manufacturing process, small crystallites of copper-doped silver halides are randomly distributed in the normal glass mixture. These crystallites absorb weakly in the visible region but absorb UV light strongly. In particular, the doped copper ions absorb UV light from the Sun. The photochemical energy causes an electron to transfer from an excited copper ion to a nearby silver ion, producing atomic silver. After several UV photons are absorbed by a single small crystal, the silver atoms begin to aggregate and form a small, dark spot of solid silver. Because this happens for all the small crystallites randomly positioned in the glass, the glass darkens. The degree of darkening depends on the amount of UV light. Once the glass is no longer exposed to UV radiation, electrons transfer from silver atoms back to the copper cations, and the glass clears. Indoor lighting contains little UV radiation, so the lenses remain transparent indoors. The difference between this process and the chemistry of photographic film is due to the copper ions, which supply electrons when excited but then take them back in the absence of UV light. Basing the photochemistry on the properties of copper ions makes the photochemical reaction reversible. If pure silver halide crystals were used, as in photographic film, the glass would simply turn black when exposed to sunlight and would not clear when the illumination stopped.

Everything Under the Sun

The asteroid sped through the solar system from a far region of the Milky Way galaxy. It contained significant amounts of iridium, a metal rare in this particular part of the universe. There were no telescopes tracking its path, because humans would not inhabit this space for 65 million years. Although the asteroid had been deflected from its path by many planets, it had always managed to escape their gravitational pull and head for the Sun. Now it veered toward a beautiful blue planet, a place humans would later call Earth.

Dawn had come and gone on the Yucatan peninsula. The Sun burned high in the sky; its golden rays nourished the lush plant life. On the edge of a marsh, a *Tyrannosaurus rex* was imbibing some tepid water. Safe across the lake, several hadrosaurs dined on some reeds. Two placeriases along a hillside methodically munched giant palm fronds. Nature provided water, an ample supply of plants, and a temperate climate. The enormous reptiles flourished in the warm sunlight of the Cretaceous period.

When the asteroid entered the upper atmosphere of the planet, at a speed of 55,000 miles per hour, it began to glow red hot. Collisions with gaseous molecules heated the surface of the alien projectile to several thousand degrees. With a diameter of 10 kilometers, its shadow blocked sunlight from an area covering hundreds of square kilometers. The sudden nightfall startled the creatures below. Then they were blinded by a white-hot asteroid, which burrowed relentlessly into the lower atmosphere and ignited the sky. With an explosive force exceeding that in all present nuclear arsenals com-

THE FAR SIDE By GARY LARSON

Chronicle Features, 1982 *Larson*

The real reason dinosaurs became extinct

(1982 The Far Side cartoon by Gary Larson is reprinted by permission of Chronicle Features, San Francisco, CA. All rights reserved.)

bined, the large mass pierced the Earth's surface. Instantly, the asteroid and surrounding rock melted, gushing lava plumes into the sky. The blast mercifully vaporized all living things within several hundred kilometers beneath a sky glowing of fire and ash. Violent earthquakes shook the ground for days, and volcanoes spouted magma from within the planet, as if to answer the galactic challenge.

The entire surface of the Earth was affected by the impact. Forest fires continued to spread and spewed black clouds of smoke into the perpetual darkness. A dense cloud of ash spread throughout the stratosphere, blocking all sunlight. Instead of bringing relief, rain washed soot, ash, nitric acid, and sulfuric acid from the sky, destroying plant life. The loss of plants caused a decline in the amount of oxygen in the atmosphere. Atmospheric carbon dioxide levels increased dramatically from the fires, creating a runaway greenhouse effect. Surface temperatures soared worldwide. Without sunlight to regulate life processes, fungi multiplied on the dark planet's surface. Lakes became festering pools of algae and bacteria. Airborne organisms blossomed and infected the few remaining dinosaurs. The giant creatures were ill equipped to survive these harsh conditions. Their 200-million-year reign on the planet abruptly ended.

With time the atmosphere healed. Rain cleansed the debris from the skies. Deposits of the rare element iridium would remain and provide a geologic clue to the graves beneath it. Centuries after the impact, a brief instant in the geologic record, the Sun shone again on the blue planet. Many small, adaptable reptiles and mammals had survived. Plants and other life returned, but the Earth's former rulers had been silenced. New life would emerge under the Sun.

Photosynthesis

Photosynthesis is the process by which green plants and certain bacteria transform light energy into chemical energy. The energy to drive photosynthesis comes from sunlight. Equation (11) indicates the overall chemistry of plant photosynthesis.

$$\text{Carbon dioxide} + \text{water} + \text{light} \rightarrow$$
$$\text{carbohydrates} + \text{oxygen} \tag{11}$$

Plants assimilate carbon dioxide from the air and combine it with water, in the presence of light, to make oxygen and carbohydrates.

In making carbohydrates from CO_2, plants require a source of hydrogen atoms. Water molecules provide the hydrogen atoms. The chemistry of photosynthesis produces oxygen molecules, which are released from the plants into the atmosphere.

The energy content of sunlight that strikes the Earth in 1 year is 25 times greater than all proven reserves of fossil fuel and uranium on the planet. In earlier times, green plants and the small organisms that fed on plants increased faster than they were consumed. The remains from these living systems were buried in the Earth's crust. There, protected from oxidation, these remains were slowly compressed and converted into the fossil fuels (coal, gas, oil) that provide most of the energy used by modern society.

In many plants, the primary chemical reactions of photosynthesis take place inside cells called chloroplasts. Protein complexes embedded in membranes of these cells contribute to two light-driven reactions involved in photosynthesis. The smallest protein complexes capable of performing these photochemical reactions have been isolated from several plants and subjected to extensive study. These subassemblies are referred to as Photosystem I and Photosystem II.

Each photosystem consists of a light-harvesting complex and a core complex. The light-harvesting (or antenna) complex contains between 40 and 60 chlorophyll molecules. These chromophores absorb sunlight and transfer the energy they acquire to

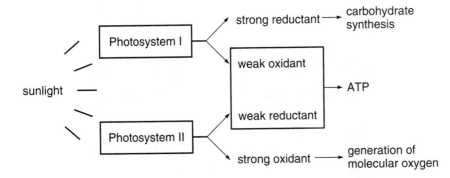

Figure 9.10. *How Photosystems I and II work together to convert light energy into chemical energy.*

the molecules inside the core complex. Once this energy reaches the core, the chemical reactions of photosynthesis take place.

The two photosystems work together to drive the life-sustaining chemical processes of cells. Photosystem I can be excited by light of wavelengths shorter than 700 nm. This subassembly photochemically generates a strong reductant that runs the carbohydrate synthesis biochemistry of the plant. Photosystem II requires light of wavelengths shorter than 680 nm. This unit produces a strong oxidant that leads to the splitting of water molecules and the formation of molecular oxygen. In addition, the two photosystems produce a weak oxidant and a weak reductant. The chemical interactions between these two products provide the cell with the chemical energy to manufacture its molecular energy currency, ATP. This entire process is shown schematically in Figure 9.10. The molecular structure of the photosystems is not known. Experimental studies reveal that Photosystem II spans the membrane in the chloroplasts.

Many blue-green algae and bacteria also survive by photosynthesis. These microscopic organisms, which appeared on the Earth between 2 and 3 billion years ago, are believed to have greatly increased the oxygen content of the atmosphere and made possible the evolution of oxygen-breathing species. The chemistry of bacterial photosynthesis is less complicated than that of plants. Unlike plants, bacteria exhibit a single photoactive system called the reaction center. The reaction center contains all

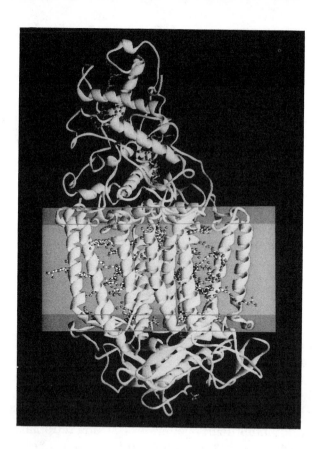

Figure 9.11.
Structure of the bacterial photosynthetic reaction center. (Courtesy of Johann Deisenhofer.)

the molecules that are involved in the primary steps of bacterial photosynthesis. The structure of the reaction center of *Rhodopseudomonas viridis,* a purple sulfur bacterium, was reported by Johann Deisenhofer, Hartmut Michel, and Robert Huber in 1985. For this work, they received the Nobel Prize for Chemistry in 1988.

The structure of the reaction center is shown in Figure 9.11. It consists of approximately 5000 atoms, which are arranged in 4 interacting polypeptide chains. Embedded in this protein structure are 12 organic molecules and several metal ions. The entire complex spans the bacterium's inner membrane. The chemistry begins when one of the embedded chlorophyll molecules absorbs light. Following this absorption, several reactions take place. Their net result is to create different concentrations of protons on the two sides of the membrane. This charge separation is similar to that in a battery. The chemical energy stored by this miniature battery drives the biochemical reactions that sustain the cell.

Solar Cells

Solar cells are devices that transform sunlight into electrical energy. Most commercial solar cells use semiconductors made of silicon. Other important semiconducting materials are cadmium selenide (CdSe), cadmium telluride (CdTe), and gallium arsenide (GaAs).

Silicon is the one of the most abundant elements on Earth. It makes up almost 20% of the Earth's crust. The most common form of silicon dioxide, sand, is useless for solar cell fabrication because of the impurities in it. Industrial manufacturers refine silicon for solar cells from silicon dioxide (SiO_2, silica) taken from the ore quartzite. For the fabrication of electronic materials, the number of impurities in the silicon must be reduced to the level of parts per billion (ppb).

Silicon, a shiny, gray-black solid, is in the same column of the periodic table as carbon. Like carbon, silicon forms a hard network solid with bonding similar to that in diamond. Each silicon atom in the solid bonds to four other silicon atoms, forming a tetrahedron. The network of these tetrahedra gives the three-dimensional array shown in Figure 9.12. Pure silicon reflects more than 35% of the sunlight that strikes its surface. If left uncorrected, this would spell a dramatic loss in the efficiency of the device. Fortunately, there are ways of coating a silicon surface to reduce reflection to about 5%.

Silicon absorbs visible light. The absorption process excites electrons in the crystal. On an atomic level, excitation removes an electron from a silicon–silicon bond, generating a free electron that can travel through the crystal. The remaining half of a chemical bond with the missing electron is referred to as a hole. The electron that is freely moving in the crystal is in the crystal's conduction band. Holes also move in the crystal by the transfer of electrons between neighboring chemical bonds. The generation and movement of the electrons and holes are what solar cells are all about. In pure silicon, however, there is no preferential direction for the electrons and holes to migrate. They meander randomly in the crystal until they recombine. To

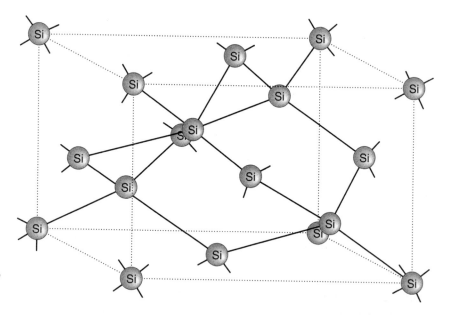

Figure 9.12. *Three-dimensional structure of solid silicon.*

exploit the charge separation between electron and hole and use the free electrons to generate electricity, a bias is created so that the electrons and holes travel in opposite directions.

To separate the holes and electrons, a solar cell is fabricated from silicon crystals doped with specific impurity atoms. Two types of dopants are used: negative carriers and positive carriers. The resulting materials are called n-type (negative carrier) and p-type (positive carrier) silicon. The n-type silicon crystal has small amounts of an added element that provides the crystal with excess electrons. In many commercial applications, phosphorus is used. The phosphorus atom occupies a crystal site that normally contains silicon. Because phosphorus has one more electron in its outermost shell than silicon (it is one column to the right of silicon in the periodic table), it can provide one electron for each of the four bonds that silicon forms and have one electron left over. Compared with an electron involved in a chemical bond, this extra electron is relatively free. The energy available from the heat in a comfortable room can excite this electron into the conduction band. Thus, in a phosphorus-doped n-type silicon crystal, there are always many electrons moving freely in the solid. The whole crystal remains neutral because

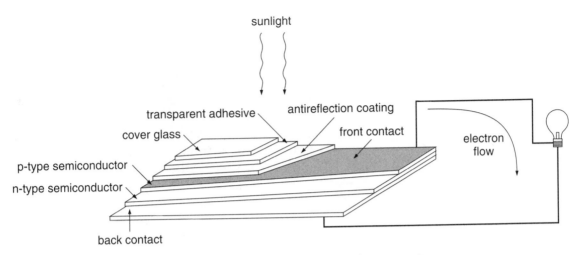

Figure 9.13. *The components of a solar cell, by which light energy is converted into electrical energy.*

the phosphorus atoms that lose electrons become positively charged.

The p-type semiconductors work the opposite way. In this case, impurity atoms to the *left* of silicon in the periodic table are substituted into the crystal. (Boron, which contains one less electron in its outermost shell than silicon, is commonly used.) As a result, the impurity atom does not have enough electrons to bond to the four neighboring silicon atoms. Doping a silicon crystal with boron creates holes.

A solar cell consists of a layer of p-type silicon on top of a layer of n-type silicon (Figure 9.13). Light absorbed by the hole-rich p-type layer generates electrons in the conduction band. Using very thin layers of silicon causes these electrons to be accelerated into the n-type layer. Once in the n-type layer, the electrons are in no danger of recombining with a hole. A similar situation occurs when light is absorbed by the electron-rich n-type layer. In this case, holes from the n-type layer are accelerated into the p-type layer, leaving the electrons behind. This photochemical action creates a charge separation across the junction. Electrons (negative charges) are pumped into the n-type layer, and holes (positive charges) are pumped into the p-type layer.

The charge separation provides an energy source similar to a battery. Connecting the two semiconducting layers with a wire causes electrons to flow from the electron-rich n-type layer to the electron-poor p-type layer.

Solar cells are still not widely used. One problem is that these devices cost about 5 to 10 times as much as conventional energy sources. In the early 1980s, silicon for solar cell applications cost over $1000 for 1 kilogram. The cost of silicon-based technology is decreasing dramatically, however, and there have been many improvements in purification techniques. Today, the cost of materials is about $20 for 1 kilogram. With further improvements in the manufacturing of solar devices, solar power could compete effectively with fossil fuel and nuclear energy technologies.

CHAPTER 10

Atmospheric Chemistry

In a 1984 interview in *The New Yorker,* F. Sherwood Rowland (a professor at the University of California, Irvine) was quoted as saying, "From what I've seen over the past 10 years, nothing will be done about this problem until there is further evidence that a significant loss of ozone has occurred. Unfortunately, this means that if there is a disaster in the making in the stratosphere we are probably not going to avoid it." Ten years of bantering among politicians, lawyers, corporate executives, and scientists had discouraged him. Rowland and Mario Molina's experiments in 1974 had provided laboratory evidence that chlorofluorocarbons (CFCs) threatened the Earth's protective ozone layer. But convincing politicians and the leaders of industries that manufactured CFCs was a tough task.

Unknown to most of the world, including Rowland and Molina, a small band of British researchers under the direction of Joe Farman had been making measurements of the levels of ozone and various trace gases from a ground-based station at Halley Bay, Antarctica, since 1957. Until the early 1980s, this work had been a routine exercise and showed only minor changes in the atmospheric gases. The team's measurements were taken with a variety of ground-based Dobson spectrometers. These instruments, developed by Sir G. M. B. Dobson at Oxford University, measured the amount of solar ultraviolet radiation that reaches the surface of the Earth. Because ozone absorbs ultraviolet light, the intensity of the detected surface radiation could be used to calculate the concentration of atmospheric ozone. The oldest spectrometer in operation made its first measurement in 1931. Keeping the old equipment running was a full-time job, and any measurement far from the norm was suspect. Every year the House of Commons appropriated a paltry amount for stratospheric research; it just didn't seem very important. Farman's funding was never enough to enable him to replace the antiquated machines.

After analyzing the data from 1982, Farman and his group detected a decrease in the amount of stratospheric ozone as compared to previous years. Nervous about their findings, they decided to review earlier data sets closely. Perhaps their 25 years of quantifying the ozone levels in the sky would not be wasted. Careful inspection of their data suggested that a gradual drop in stratospheric ozone levels started as early as 1977. Was this decline real or simply an artifact of aging equipment? American *Nimbus* satellites, with more sensitive equipment, reported no ozone depletion in the area. And furthermore, natural fluctuations in ozone values were known to occur. Unbeknownst to Farman, the satellite equipment was actually recording a similar seasonal decrease in Antarctic ozone—but the satellite data processing software threw out the low ozone measurements as outside the expected limits!

Winter of 1983 arrived with similar results, but Farman, still skeptical about the findings, did not publish his work. In 1984 the research team decided to extend their studies to double-check their data. By establishing a second measuring station to look at the skies above Argentine Island, about 1000 miles to the northwest of Halley Bay, they could check the consistency of the observations at an independent site. At the end of the year, the British scientists found themselves staring at data that showed a 40% loss in ozone concentration during the period from September to October at both monitoring sites.

There was no question now. At the end of the austral winters, ozone levels in the skies over Antarctica were dropping dramatically to create an ozone hole. When the NASA satellite records were reexamined, a similar pattern in their observations emerged as far back as 1977. Farman's additional measurements of trace gases showed the presence of high levels of CFCs in the ozone hole. Though he was familiar with Rowland's work, it had taken Farman years to make the connection between his observations and the chemistry of CFCs. Instead of a gradual decline in ozone, huge, rapid, seasonal changes were occurring. There was a correlation between the drop in ozone levels and the concentration of CFCs, but it was not so simple as the predictions that were based on the results of laboratory experiments.

The Atmosphere

The atmosphere consists of the gases that surround the Earth. The chemistry of the atmosphere varies significantly with altitude because of the different molecules present and the incoming radiation from the Sun. Scientists divide this large expanse into shell-like regions. Figure 10.1 shows some of the important chemical species present in each layer. Some molecules in the atmosphere (O_2 and O_3) help filter out harmful ultraviolet radiation emitted by the Sun.

The part of the atmosphere that extends from sea level to an altitude of about 10–16 km is called the troposphere. The troposphere is in contact with plants and animals. All gaseous molecules needed for sustaining life are present. The two major components (by volume) of dry air at sea level are nitrogen (N_2, 78.08%) and oxygen (O_2, 20.9%). Next come argon (0.94%) and carbon dioxide (CO_2, 0.035%). Water normally constitutes between 1% and 3% of air (the full range is 0.1–5% by volume). Percentages of the noble gases vary significantly: Ar, 0.94%; Ne, 0.0018%; He, 0.00052%; Kr, 0.00011%; Xe, 0.0000087%; and Rn, ≤ 0.00000001%.

The chemistry of the troposphere balances many biological and photochemical reactions that generate and destroy molecules. Molecules cycle in and out of the atmosphere on dramatically different timescales. The average "residence times" for nitrogen, carbon dioxide, methane, and sulfur dioxide in the troposphere are 10^6 years (N_2), 5–6 years (CO_2), 4–6 years (CH_4), and 2–3 days (SO_2).

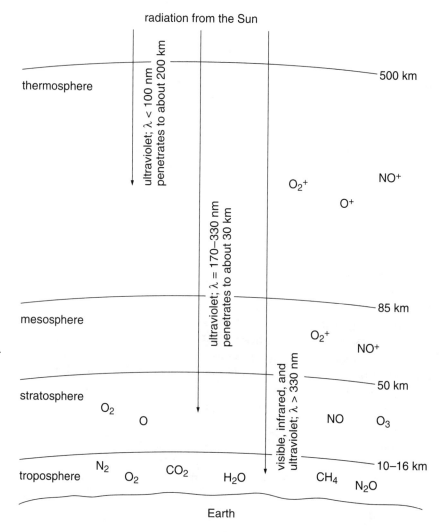

Figure 10.1. *An illustration of the layers that make up the Earth's atmosphere. Varying wavelengths of sunlight penetrate to different depths. Some of the more common molecules and ions found in each layer are indicated.*

The stratosphere, which extends from the top of the troposphere up to an altitude of about 50 km, contains the ozone layer. Stratospheric ozone absorbs ultraviolet light emitted from the Sun. If this radiation impinged on the surface of the Earth, photochemical reactions caused by the high-energy ultraviolet radiation could destroy the proteins and nucleic acids in living systems. According to one theory of evolution, life did not emerge from the sea until a protective ozone layer developed in the atmosphere.

When a molecule of ozone absorbs ultraviolet light, a photodissociation reaction occurs to form molecular oxygen (O_2) and atomic oxygen, as shown in equation (1).

$$O_{3(g)} + \text{light } (\lambda < 320 \text{ nm}) \rightarrow O_{2(g)} + O_{(g)} \qquad (1)$$

Ozone absorbs ultraviolet light in the stratosphere to form an oxygen molecule and an oxygen atom.

To maintain a constant concentration of ozone in the stratosphere, there must be an efficient mechanism for ozone formation. Ozone is made by a photochemical reaction. The first step is the photochemical breaking of the oxygen–oxygen bond of molecular oxygen by high-energy ultraviolet light (equation 2). The dissociated atoms are very energetic and react with oxygen molecules to form ozone (equation 3).

$$O_{2(g)} + \text{light } (\lambda < 240 \text{ nm}) \rightarrow 2O_{(g)} \qquad (2)$$

Oxygen molecules absorb ultraviolet light in the stratosphere to form oxygen atoms.

$$O_{(g)} + O_{2(g)} \rightarrow O_{3(g)} \qquad (3)$$

Oxygen atoms and oxygen molecules collide in the stratosphere to form ozone.

Whereas the natural process of ozone formation is well understood, the various chemical reactions that lead to its destruction are still under extensive study. The earliest known reaction of ozone involves the formation of molecular oxygen by reaction with oxygen atoms, as shown in equation (4).

$$O_{3(g)} + O_{(g)} \rightarrow 2O_{2(g)} \qquad (4)$$

Collisions between ozone and an oxygen atom can produce two molecules of oxygen.

This reaction, along with UV photochemical decomposition (equation 1), accounts for about 20% of the total ozone removal. Research has shown that another 10% is removed by reaction with the reactive hydroxyl radical (OH). The mechanism of the reaction of ozone and the hydroxyl radical depends on altitude. The reactions shown below are the mechanism for depletion of ozone by OH for altitudes below 30 km. Above 30 km, the chemical species that reacts with HOO in equation (6) is the oxygen atom (O) instead of ozone. This difference in mechanism is due to the relative concentrations of O and O_3 found in the atmosphere as a function of altitude.

$$O_{3(g)} + OH_{(g)} \rightarrow O_{2(g)} + HOO_{(g)} \qquad (5)$$

Ozone combines with a hydroxyl radical to form molecular oxygen and the hydroperoxyl radical.

$$HOO_{(g)} + O_{3(g)} \rightarrow 2O_{2(g)} + OH_{(g)} \qquad (6)$$

A hydroperoxyl radical combines with ozone to make two molecules of oxygen and a hydroxyl radical.

In this two-step process (equations 5 and 6), two ozone molecules are converted into three molecules of molecular oxygen; the hydroxyl radical is a catalyst. After destruction of one ozone molecule, the reactive OH radical is ready to continue on its quest to destroy many more ozone molecules. Thus a very small amount of OH can lead to the destruction of a large number of ozone molecules. On the basis of much scientific research, it is now believed that there are several catalytic mechanisms that destroy ozone. The steady concentration of ozone in the stratosphere results from the dynamic balance between the reactions that constantly form and destroy it. Using data from field measurements and laboratory studies, investigators are developing models to account for the origins of the ozone hole and the thinning of the ozone layer.

Nitrogen oxides and chlorine oxides also lead to ozone destruction. Equations (7) and (8) show the reaction steps for nitric oxide, NO. Similar to OH, NO catalytically decomposes ozone into molecular oxygen.

$$O_{3(g)} + NO_{(g)} \rightarrow O_{2(g)} + NO_{2(g)} \qquad (7)$$

Ozone reacts with nitric oxide to form oxygen and nitrogen dioxide.

$$NO_{2(g)} + O_{(g)} \rightarrow NO_{(g)} + O_{2(g)} \qquad (8)$$

Nitrogen dioxide reacts with oxygen atoms to form nitric oxide and oxygen.

Where do OH and NO come from? Both are made by naturally occurring chemical processes. The OH comes mainly from the photochemical reactions of O_3, O_2 and H_2O. Most NO arises from the reaction of electronically excited oxygen atoms (O^*) with stratospheric nitrous oxide (N_2O), as shown in equation (9).

$$N_2O_{(g)} + O^*_{(g)} \rightarrow 2NO_{(g)} \qquad (9)$$

Nitrous oxide reacts with electronically excited oxygen atoms to make nitric oxide.

Nitrous oxide (laughing gas) is a major byproduct of bacterial nitrification and denitrification processes in soils and oceans. Figure 10.2 shows the fraction of ozone loss attributed to each catalyst as a function of altitude from 30 to 70 km. The relative importance of oxygen atoms, nitrogen oxides, chlorine oxides, and hydroxyl radicals at different altitudes reflects changes in the relative concentrations of the molecules present. In the lower stratosphere, recent experiments (1994) suggest an increased role for hydroxyl and a decreased role for nitrogen oxides than are suggested by this figure.

Above the stratosphere lie the mesosphere (50–85 km) and the thermosphere (85–500 km). In the mesosphere, gas-phase ionization processes are important. Above altitudes of about 70 km, a portion of the sunlight is energetic enough to kick electrons off molecules and create ions. Ionization also results from collisions between molecules in the upper mesosphere and high-energy particles from the Sun. These particles, carried by the solar wind, enter the upper atmosphere and travel along the magnetic field lines of the Earth. As these field lines converge at the poles, the amount of ionization peaks. Recombination of the ionized products can produce neutral molecules and atoms that are in excited states. These excited molecules can then release their excess energy by emitting light. The northern lights (*aurora borealis*) are an example of this type of light emission, which is called chemiluminescence.

Ions in the upper atmosphere are useful in communication. The concentration of ions in the mesosphere is sufficient for reflecting radio waves from one region on the surface of the

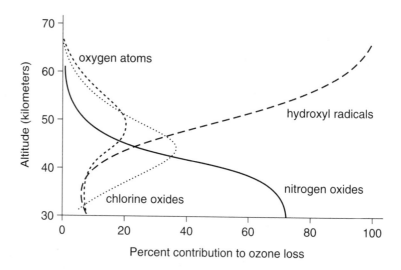

Figure 10.2. *The relative contributions of hydroxyl radicals, nitrogen oxides, chlorine oxides, and oxygen atoms to ozone destruction as a function of elevation from 30 to 70 km. (Data from* Atmospheric Ozone 1985, World Meteorological Organization, Geneva, *1986.)*

Earth to another. That is why short-wave radio transmissions can travel around the world. Some radio frequencies, such as FM, are not reflected efficiently, which limits their transmission path to a line of sight.

Photochemistry of Chlorofluorocarbons and the Ozone Hole

The development of chlorofluorocarbons (CFCs) was a major technological breakthrough. These gases (also called Freons) liquefy readily under pressure; as liquids, they are used in the compressors of air conditioners and refrigerators. Before the 1920s, refrigerator compressors required ammonia as the working fluid. CFCs are much less toxic, easier to handle, and safer for consumer applications. Many commercial solvents for cleaning electronics components (computer chips) are also made of CFCs. And until the 1970s, CFCs were widely used as propellants in spray cans.

The worldwide production of CFCs is nearly half a million metric tons annually. These molecules are now a homogeneous component of the global atmosphere; there are several hundred CFC molecules for every trillion molecules in the air. With each breath you are inhaling several trillion CFC molecules.

CFCs are extremely inert molecules; they do not undergo chemical reactions in the troposphere. Because of their chemical inertness, it was thought that these molecules would not damage the environment. But CFCs are not photochemically inert: Excitation with short-wavelength ultraviolet light breaks a carbon–chlorine bond (equation 10).

$$CCl_2F_{2(g)} + light\ (\lambda < 240\ nm) \rightarrow Cl_{(g)} + CClF_{2(g)} \quad (10)$$

A molecule of CFC-12 absorbs ultraviolet light in the stratosphere to form a chlorine atom and a CClF₂ radical.

Because the solar UV light required to break this bond does not penetrate below the stratosphere, this reaction does not occur in the troposphere. However, air currents carry CFC molecules into the upper atmosphere, where short-wavelength UV light is available. Here the photogenerated chlorine radical, Cl, efficiently reacts with ozone molecules, producing molecules of oxygen and chlorine oxide (equation 11).

$$Cl_{(g)} + O_{3(g)} \rightarrow ClO_{(g)} + O_{2(g)} \quad (11)$$

A chlorine atom reacts with ozone to form chlorine oxide and oxygen.

Chlorine oxide reacts further with atomic oxygen to regenerate the Cl radical (equation 12).

$$ClO_{(g)} + O_{(g)} \rightarrow Cl_{(g)} + O_{2(g)} \quad (12)$$

Chlorine oxide reacts with an oxygen atom to form a chlorine atom and molecular oxygen.

Taken together, equations (11) and (12) show how chlorine atoms catalytically decompose ozone. Laboratory studies suggest that a single chlorine atom destroys up to 2 million molecules of ozone before it is consumed by other trace chemical species in the stratosphere.

During the past 15 years, the concentration of stratospheric ozone has decreased steadily. Between 1976 and 1984, the National Academy of Sciences released four major reports on the impact of CFCs on the ozone layer. The estimates of worldwide

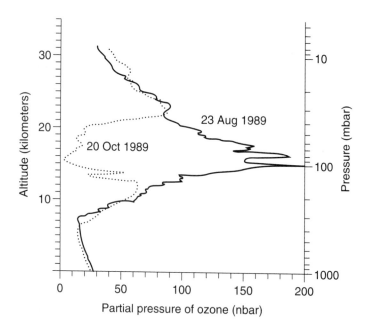

Figure 10.3. *Ozone concentrations over Antarctica in August and October of 1989. (Source: J. Deshler, D. J. Hoffman, J. V. Hereford, C. B. Sulter,* Geophys. Res. Lett. *17, 151 (1990).)*

ozone depletion by CFC chemistry range from 2 to 20% of the natural concentration. Some estimates suggest that the average amount of ozone will decrease by an additional 5 to 10% by the year 2050. Rowland and Molina's predictions about the effect of CFCs on the atmosphere have come to pass.

How does the worldwide picture fit with Joe Farman's findings that a dramatic decrease in ozone occurs over the South Pole at the end of each austral winter? Unlike most areas inhabited by humans, the poles do not experience daily light and darkness. Instead, extended periods of many months of darkness follow similar periods of sunlight. An interesting phenomenon in the Antarctic is that a vortex of air forms over the pole during winter. The polar vortex is a mass of very cold, stagnant air contained by the surrounding strong westerly winds. It is an extreme weather condition spawned by the unique geographic features of the polar ice masses. The stability of the vortex is so great that air at polar latitudes is almost sealed off from that at lower latitudes. When the Sun returns in September, the air in the stratosphere warms, the winds weaken, and the vortex generally breaks up in November. Vortices appear at both poles, but the one in Antarctica is better defined. In 1989 the ozone hole returned with a vengeance, covering an area of 26 million square kilometers. At the worst part of the hole, ozone losses were about 45%. And at some altitudes, all of the ozone was destroyed (Figure 10.3). In 1991 ozone losses approached 60%!

Farman's observations showed that the greatest depletion occurs just before the annual breakup of the polar vortex. This clear signal that the Earth's UV radiation shield is vulnerable to chemical attack motivated policy makers into action by early 1986. In that year, an Antarctic ozone expedition led by Susan Solomon of the National Oceanic and Atmospheric Administration provided the first evidence that CFCs play a major role in the destruction of stratospheric ozone. When the vortex breaks up, the ozone-depleted air mixes with the surroundings and migrates across the globe. In 1987 ozone levels dropped about 10% over New Zealand and Southern Australia for nearly three weeks as a result of such mixing. And bad news keeps coming: In 1992 ozone loss was detected over the Arctic ice cap. Ozone depletion over Northern Europe, Asia, and Canada is now a possible cause for concern.

Research on the mechanism of ozone depletion in the Antarctic polar vortex reveals the importance of polar stratospheric clouds. In fact, it is not possible to account for the huge ozone losses without including reactions that occur on the surfaces of cloud particles. In particular, in addition to its catalytic reaction with ozone, chlorine reacts with other trace gases in the stratosphere to form HCl and chlorine nitrate, $ClONO_2$. These molecules do not undergo reaction in the gas phase. However, on the surface of a polar stratospheric cloud, they react with each other as indicated in equation (13).

$$HCl + ClONO_2 \rightarrow Cl_2 + HNO_3 \qquad (13)$$

HCl reacts with $ClONO_2$ on the surface of a polar
stratospheric cloud to form Cl_2 and nitric acid.

The surface converts inert molecules that contain chlorine into Cl_2. The Cl_2 is photodissociated by the UV light in the stratosphere, thereby generating Cl atoms. The temperature in the Arctic vortex is not cold enough to cause the formation of polar stratospheric clouds. This is why the ozone depletion at this pole is significantly less than that found in Antarctica.

Prompted by a National Academy of Sciences recommendation in December of 1986, the United States proposed a worldwide 95% reduction of CFCs by 1990. In 1987 many countries signed the Montréal Protocol to control substances that deplete the ozone layer. Major worldwide reductions in the use of CFCs by the year 2000 were mandated; however, as a political compromise, developing countries were allowed to increase CFC use to 0.3 kg per person. Without the development of CFC substitutes, the worldwide population growth anticipated by the year 2000 suggests an estimated 50% increase in CFC use over current levels. In February of 1988, three U.S. senators asked du Pont, the major manufacturer of CFCs, to halt production. Although du Pont denied this particular request in March, three weeks later company officials agreed to stop making CFCs as soon as substitutes become available. (The technical problems are not trivial, because any substitute must be compatible with all existing refrigeration and air-conditioning systems; and the economic consequences of a CFC ban extend well beyond the chemical industry.) In September of 1988, the EPA reported that it had underestimated the effects of ozone depletion and called for an immediate 85% cutback in the manufacture and use of CFCs. In March of 1989, European countries and the United States agreed to implement faster CFC reductions. Developing countries opposed the new timetable, citing the costs and the difficulty in developing substitutes. Figure 10.4 displays predictions of future concentrations of CFC-12 in the atmosphere for two scenarios. These calculations clearly show that even under the stipulations of the Montréal Protocol, CFC-12 levels will increase. A global reduction of 77% is needed to result in a constant concentration for the next half-century. A complete halt in CFC-12 emission would only halve the amount of this molecule in the atmosphere by the middle of the next century. In the spring of 1991, more accurate measurements of ozone depletion revealed even greater losses than had been predicted before. Many countries are currently considering a complete ban on CFCs by the year 2000.

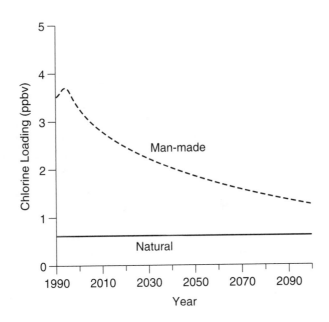

Figure 10.4. *Predictions of the average concentration of total chlorine in the atmosphere resulting from the atmospheric concentrations in 1989 projected forward at the maximum production rate allowed by the 1992 adjustments to the Montréal Protocol. (Adapted from Fig. 4.3, page 22,* Stratospheric Ozone 1993 *by the United Kingdom Stratospheric Ozone Review Group. Prepared at the request of the Department of the Environment and the Meteorological Office. Crown copyright is reproduced with the permission of the Controller of HMSO.)*

If production stopped tomorrow, the CFC molecules now in the atmosphere would cause further ozone loss until at least the first decade of the next century. This is because of the long atmospheric lifetime (hundreds of years) of the CFCs currently present in the troposphere. These molecules will continue to diffuse up into the stratosphere for centuries. The depletion of stratospheric ozone allows increased amounts of UV radiation to impinge on the Earth's surface. Increased exposure to UV light will affect both plant and animal life. About 90% of all skin cancers result from exposure to UV light. It is estimated that each 1% decrease in ozone causes a 2–5% increase in treatable basal-cell skin cancers. In the United States, several hundred thousand cases occur annually. And the more serious squamous-cell skin cancers are expected to increase by 4–10% for each 1% decrease in ozone. At present, about 100,000 cases are reported annually in the country. The projected increases are only rough estimates; the real extent of the problem won't be known for several years. Perhaps more serious will be the stress to plant life (the food supply) from radiation damage.

Currently, many corporations are switching from CFCs to hydrochlorofluorocarbons (HCFCs). HCFCs differ from their CFC counterparts by replacing one or more halogen atoms (F or

Cl) with a hydrogen atom. As a result, these molecules are more reactive and to a large extent are destroyed in the troposphere. These CFC substitutes are not so damaging as their CFC counterparts, but they do carry some ozone-depleting chlorine into the stratosphere. HCFCs are a short-term solution to the problem; they are scheduled to be phased out by 2030. The preferred substitutes are hydrofluorocarbons, HFCs, which are compounds that contain only hydrogen, fluorine, and carbon. The U.S. appliance industry has chosen HFC-134a (CH_2FCF_3) to replace CFC-12 in the compressors of home refrigerators. Polyurethane foam insulation in refrigerator walls will be blown with HCFC-141b (CH_3CCl_2F) instead of CFC-11. Most of the new cars being sold in the United States are now equipped with air-conditioning units that use HFC-134a. The EPA currently plans to phase out HCFC use in stages. In 1994 all HCFCs will be banned from use in aerosols (except medical devices) and plastic foams (except foam insulation). Then, in 2003, HCFC-141b production and imports will be banned. In 2010 production and imports of HCFC-22 ($CHClF_2$, used in refrigeration and air conditioning) and HCFC-142b (CH_3CClF_2, used in refrigeration and foams) will be frozen at their 1989 levels. In 2015 all remaining HCFCs will be frozen at their 1989 levels. In 2020 there will be a ban on production and imports of HCFC-22 and HCFC-142b. Finally, in 2030, production and imports of all HCFCs will be banned.

The photochemistry of CFCs is not the only man-made cause of ozone depletion. Solid propellants, such as those used by the space shuttle, introduce small amounts of chlorine into the atmosphere. And nitric oxide (NO), a component of supersonic and subsonic jet exhaust, catalyzes ozone destruction. The potential for ozone depletion by NO led Congress to abandon plans to help finance the construction of 300–400 supersonic transports (SSTs). These planes were to fly at supersonic speeds through the lower stratosphere. There are fewer than 20 European Concorde SSTs in use today.

Nitrous oxide, N_2O, is also active in the catalytic destruction of ozone, because it generates NO in the stratosphere by equation (9). Among its various uses, N_2O is an anesthetic gas employed in surgical and dental procedures. Most N_2O in the

The Need for Scientific Communication with the Public

Remarks by F. Sherwood Rowland

... From my own experience, I see that the most serious problems are related to faulty communication about science among the various segments of society, including the scientific segment itself. Each of us is bombarded daily by messages from television, radio, magazines, newspapers, and so on. We live in the midst of massive information flow, but those items connected with science itself are often badly garbled, sometimes with potentially serious negative consequences. The remedy must lie in greater emphasis by all of us on increasing both the base level of knowledge of science and communication about science with all levels of society.

... During the last 2 or 3 years, a new generic question appears regularly which runs more or less as follows: "Don't volcanoes cause the Antarctic ozone hole?"

... A simple answer of course is that the very large losses of ozone over Antarctica during October—now exceeding 60% each Southern Hemisphere spring—first became noticeable around 1980 in measurements conducted continually since 1956 and became public knowledge in 1985. Because volcanoes have been around for geologic time, with no unusually violent activity during the past decade, they can't be the direct cause of these Antarctic ozone losses.

This answer, however, is usually not satisfactory by itself because some of the doubters are not prepared to accept that there has been any springtime loss of ozone over Antarctica and put their dis-

F. Sherwood Rowland. (Courtesy of F. Sherwood Rowland.)

sent as an assertion that volcanoes are the primary source of stratospheric chlorine, totally overshadowing any possible effect from man-made compounds. Asserting this to be so, they then conclude that there cannot be an Antarctic ozone hole and therefore the whole ozone depletion story is a hoax.

In discussing this scientific situation in some detail, I am ultimately raising a cause for great concern over the role of science in a democracy in which the general population has not enough understanding of science itself; does not entirely trust "science experts" and does not want to; and is left with no way to distinguish between the competing claims of apparent experts on both sides of any question.

... The volcanic chlorine question is part of a background which has been confronted over the past 20 years in atmospheric chemistry and which asks the extent to which "natural" processes have contributed chlorine-containing chemical species to the stratosphere. ... "Natural" is thus loosely defined as not being influenced by the activities of mankind, although obviously human beings and their activities are very much part of the natural world. Sunspots, volcanoes, most oceanic biology, primeval forests—these are among the processes which are clearly designated as natural.

The general consensus of the active atmospheric scientists on a worldwide basis, after 20 years of study, is that the major chlorine sources to the stratosphere in 1993 are almost entirely compounds that contain both carbon and chlorine and are released at the Earth's surface. Furthermore, the great majority of these are man-made chlorocarbon compounds such as carbon tetrachloride and the chlorofluorocarbons, or CFCs, together representing now about five times as much chlorine as carried by the only significant natural source, methyl chloride. The possible inorganic sources such as hydrogen chloride or sodium chloride from the evaporation of ocean spray dissolve in cloud droplets and are removed by rainfall, with negligible quantities reaching the stratosphere.

... Obviously, confirmation of the hypothesis of volcanic delivery of hydrogen chloride into the stratosphere, or a decisive choice between these two hypotheses, is not possible without further experimental information.... As you all well know, the progress of science depends on the formulation and testing of hypotheses, discarding those which are not consistent with the accumulating mass of observations.

In this case, the appropriate stratospheric experimentation has been carried out over the past 15 years through observations of the global accumulations of stratospheric hydrogen chloride with infrared spectrometers operated from aircraft, from the ground, and from space. These observations showed only a small increase (less than 10% in total stratospheric hydrogen chloride) accompanying the major eruption of the Mexican volcano El Chichon in April 1982 and have shown even less increase from the other recent large volcanic eruption, that of Pinatubo in the Philippines in June 1991. Furthermore, the measured amount of stratospheric hydrogen chloride has increased regularly over this time period. If stratospheric hydrogen chloride is increasing with chlorofluorocarbons as the major source, an obvious corollary question is whether hydrogen fluoride is also on the increase. A severalfold increase in stratospheric hydrogen fluoride since 1977 has been observed by infrared techniques, and the amounts of both hydrogen chloride and hydrogen fluoride are consistent with expectations from the stratospheric decomposition of chlorocarbon and chlorofluorocarbon compounds—in other words, the stratospheric evidence does not support any other significant input to the stratosphere of either chlorine or fluorine. The working atmospheric science community has therefore rejected volcanoes as an important source of chlorine (and fluorine) for the stratosphere, at least for the past few decades and specifically for the past 15 years during which significant ozone depletion has been observed.

(Excerpted with permission from "President's Lecture: The Need for Scientific Communication with the Public," by F. Sherwood Rowland in *Science*, Vol. 260, 11 June 1993, pages 1571–1576. Copyright 1993 American Association for Advancement of Science.)

atmosphere is generated by naturally occurring biological processes. However, concentrations of this gas have been increasing in recent years, which suggests that industrial sources are no longer negligible. As is the case for many CFCs, the decomposition of N_2O in the troposphere requires hundreds of years. In 1991 N_2O was identified as a byproduct of the large-scale industrial synthesis of nylon. All the major chemical producers (du Pont, ICI, Monsanto, Rhône-Poulenc, and BASF) have voluntarily adopted emission controls to prevent the escape of chemically generated N_2O into the atmosphere.

Global Warming

The Earth is heated by radiation from the Sun. About half the total solar radiation that enters the Earth's outer atmosphere reaches the surface. The rest is either absorbed by molecules in the upper atmosphere or reflected back into space. The Sun delivers about 1340 watts to each square meter of the Earth's surface, most of which is absorbed. If all this energy were retained, the planet would have vaporized long ago. Fortunately, the energy is reemitted in the form of infrared light, and some is stored as chemical energy by photosynthesis. Some of the infrared light is converted into heat by molecules in the atmosphere; the rest escapes into outer space. The temperature at the surface of the Earth reflects a balance between the absorption and reemission of radiation, and the amount of infrared light converted to heat by molecules in the troposphere. Tropospheric gases that absorb infrared radiation are collectively referred to as greenhouse gases. The major players are water vapor (1–3% by volume) and carbon dioxide (CO_2, 0.04% by volume). Without the greenhouse-gas molecules in the troposphere, the Earth would be an icy planet (about 33°C colder).

The insulating effect of water is well known. In humid climates, the ambient temperature does not change much between day and night, because the high concentration of water vapor in the air efficiently traps the infrared emission from the Earth and thereby inhibits cooling after dark. Desert climates, however, exhibit large swings between daytime and nighttime temperatures. Once the Sun sets, temperatures drop quickly. Unlike water, the concentration of atmospheric CO_2 (the second most important greenhouse gas) is fairly uniform around the globe. This really isn't surprising, because CO_2 is a gas at all temperatures and pressures found near the Earth's surface.

Human activities produce many greenhouse gases. Scientific studies show a significant increase in the concentrations of carbon dioxide, methane, ozone, nitrous oxide, and CFCs in the troposphere. In discussing global warming, it is important to separate fact from hypothesis. It is a fact that infrared-absorbing gases in the troposphere warm the Earth. And it is a fact that the concentrations of infrared-absorbing gases in the troposphere are increasing at unprecedentedly rapid rates. It is a hypothesis, based on these observations, that continued greenhouse emissions will lead to problematic global warming.

There is much disagreement on how to model the temperature increases associated with global warming. However, available evidence seems to suggest that man-made sources of greenhouse gases have already caused an increase of about 0.5°C. The present concern is exacerbated by projections that CO_2 levels will double in the next 25 years. Computer simulations indicate that this increased CO_2 concentration will cause the global temperature to rise between 1.5 and 4.5°C. The wide range of these predictions reflects the considerable uncertainty in assessing several potentially large effects on the projected

The Ice Record of Atmospheric Gases

Gas molecules trapped in polar ice provide a record of the amounts of several greenhouse gases present in the atmosphere over the last 160,000 years. These records reflect the interplay of ecosystems, oceans, and climate, providing a detailed picture of how the global atmosphere has responded to change.

The accompanying figure plots the concentrations of tropospheric carbon dioxide from 1750 to 1989 and methane from 1850 to 1989. The majority of the data was derived from the analysis of ice-core samples taken in Greenland and Antarctica. After 1958 for CO_2, and after 1978 for CH_4, direct atmospheric measurement was used. During the last 200 years, the concentrations of both gases have increased steadily. Until the 1950s, agricultural activities (deforestation) dominated the growth in CO_2 levels. Since then, the largest source has been the combustion of fossil fuels.

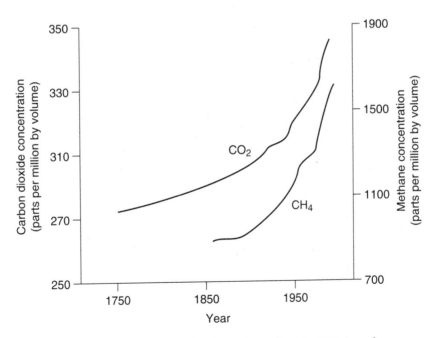

The concentrations of atmospheric carbon dioxide (CO_2) and methane (CH_4) since the Industrial Revolution.

temperature increases, such as the scattering of light by clouds, the absorption of CO_2 by the oceans, and the utilization of CO_2 by plants. In any case, it is believed that a 4°C change would dramatically alter the global climate. (The resulting melting of the polar ice caps could cause a rise in sea level.)

In 1987 human activity generated about 8.5 billion metric tons of carbon as CO_2, 255 million metric tons of methane, and more than 770,000 metric tons of CFCs. These three molecules represent about 86% of the greenhouse-gas emissions attributed to industrial sources. The heating ability of each molecule is different: A molecule of methane traps 20–30 times, and a CFC molecule 20,000 times as much heat as a molecule of CO_2. Thus in addition to causing stratospheric ozone depletion, CFC molecules are extremely potent greenhouse gases.

The responsibility for the greenhouse effect is worldwide, as can be seen by examining the origins of CO_2. It is a common view that the only significant source is the burning of fossil fuels. However, activities such as cement manufacture and the burning of trees associated with large-scale deforestation contribute significantly to increased CO_2 levels. Estimates of the amount of tropical forest being destroyed vary widely. In areas such as Brazil, deforestation rates of over 1.5 million hectares per year are being reported. In 1987 the National Space Research Institute of Brazil used satellite imaging to determine that 8 million hectares were destroyed that year. Although the figures are controversial, current estimates indicate that between 5 and 7% of the Amazon has been deforested. Roughly 2.8 billion metric tons of carbon are released into the atmosphere annually as a result of the deforestation, an amount comparable to that generated worldwide by the manufacture of cement (1.4 million metric tons) and by combustion of various forms of fuel (5.7 billion metric tons).

Only a small percentage of the carbon on Earth exists as atmospheric CO_2 (the major forms found on the planet are given in Table 10.1). Carbon changes among various forms through a global process called the carbon cycle. Oceans and soils readily absorb atmospheric CO_2. The exchange between these reservoirs occurs so rapidly that only about half of the gas generated by

Table 10.1. *The Distribution of Carbon on Earth*

Storage Form	Billions of Metric Tons
Atmospheric CO_2	700
Living organisms	800
Soil organic matter (mainly humus and peat)	1,500
Dissolved organic carbon in the oceans	3,000
Dissolved inorganic carbon near the ocean surface	600
Dissolved inorganic carbon below 100 m in the ocean	40,000
Carbonate minerals in the Earth's crust	15,000,000,000

human activities stays in the atmosphere. Scientists have yet to determine where the remainder goes. A complete picture of the effects of increased atmospheric levels of CO_2 will require a much deeper understanding of the dynamics of the carbon cycle.

Increased atmospheric CO_2 levels affect the equilibrium with dissolved CO_2 in oceans, lakes, and rivers. In water, dissolved CO_2 reacts to form bicarbonate and a proton (equation 14).

$$CO_{2(g)} + H_2O_{(l)} \rightleftharpoons HCO_3^-{}_{(aq)} + H^+{}_{(aq)} \qquad (14)$$

Carbon dioxide dissolves slightly in water
to form the bicarbonate ion and a proton.

The bicarbonate ion (a weak acid) dissociates into the carbonate ion and a proton. (equation 15).

$$HCO_3^-{}_{(aq)} \rightleftharpoons CO_3^{2-}{}_{(aq)} + H^+{}_{(aq)} \qquad (15)$$

The bicarbonate ion partially dissociates to the carbonate ion and a proton.

Reactions (14) and (15) are coupled; a change in either one causes a change in the other. An increase in atmospheric CO_2 leads to an increase in the concentration of dissolved bicarbonate. This helps remove much of the excess CO_2 from the atmosphere.

Photochemical Smog

Photochemical smog is a major problem in many industrialized cities. One source of the brown haze observed is the molecule nitrogen dioxide, NO_2. Nitrogen dioxide forms as a result of oxidation reactions of nitric oxide, NO. Several oxidants can cause this reaction, including O_2 and O_3. In addition, oxidation can be carried out by peroxyradicals, as exemplified by equation (16).

$$NO_{(g)} + CH_3O_{2(g)} \rightarrow NO_{2(g)} + CH_3O_{(g)} \qquad (16)$$

Nitric oxide and a peroxymethyl radical react
to form nitrogen dioxide and a methoxy radical.

Nitric oxide is generated by the partial combustion of N_2 (equation 17).

$$N_{2(g)} + O_{2(g)} \rightleftharpoons 2NO_{(g)} \qquad (17)$$

At high temperatures, nitrogen begins
to react with oxygen to form nitric oxide.

At ambient temperatures, this reaction between molecular nitrogen and oxygen is extremely slow and unfavorable, so very little NO_2 forms in the atmosphere. However, inside a running automobile engine, the high temperatures accelerate the rate of nitrogen combustion, and small amounts of NO appear in the exhaust. Atmospheric oxidants then readily oxidize NO to the toxic, red-brown gas nitrogen dioxide (equation 16).

The absorption spectrum of NO_2 is shown in Figure 10.5; its extensive absorption of light over a large portion of the visible region contributes to the orange-brown hue seen over most large cities (compare Figures 10.6 and 10.7). In the atmosphere, NO_2 undergoes photochemical decomposition—hence the name *photochemical smog*. When NO_2 absorbs sunlight, the molecule dissociates into two fragments, nitric oxide and atomic oxygen, as shown in equation (18).

$$NO_{2(g)} + light \rightarrow NO_{(g)} + O_{(g)} \qquad (18)$$

Brown nitrogen dioxide in smog absorbs visible
light and forms nitric oxide and an oxygen atom.

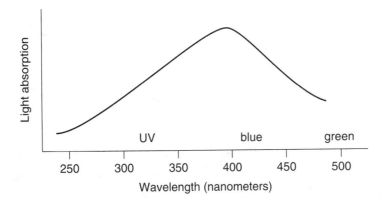

Figure 10.5. *The absorption spectrum of NO_2, a common component of smog.*

The resulting oxygen atom is highly reactive and initiates many chemical reactions. For example, reaction of oxygen atoms with molecular oxygen forms ozone. Although ozone plays an important role in the stratosphere, it is toxic to most forms of plant and animal life. Ozone is only one of the harmful products that result from the photochemistry of NO_2 in the troposphere.

Scientific studies show that in a major metropolitan area, the relative concentrations of NO_2, NO, and O_3 change over the course of a day (Figure 10.8). In the early morning hours, all three molecules are present in low concentrations. In most industrial cities, automobile activity starts picking up around 6 A.M., causing an immediate increase in the atmospheric concentration of NO. Over the next several hours, NO becomes oxidized to NO_2. The NO_2 concentration reaches its maximum in late morning. Nitrogen dioxide levels then decrease through photochemical reactions that give rise to increased ozone levels. Ozone concentrations peak in the early afternoon. By combining results like those shown in Figure 10.8 with data from laboratory experiments, investigators have developed models to explain the chemistry of photochemical smog.

The Clean Air Act of 1970 set allowable emission levels of various pollutants. During the two decades since this law was passed, there have been periodic modifications to require reductions in the emissions of specific molecules from automobile engines. Although NO and NO_2 are the major contributors to photochemical smog, automobiles also release other chemicals

Figure 10.6 *Manhattan skyline on a clear day. (Charles Merkle FPG.)*

into the air. For example, toxic organic molecules are generated as byproducts in the combustion of gasoline. Several of these undergo reactions with nitrogen oxide molecules, producing noxious organic smog components. Three commonly found molecules are peroxyacetyl nitrate (PAN), peracetic acid, and peroxybenzoyl nitrate (PBN); see Figure 10.9. PAN is a very potent oxidizing agent and lacrymator. It causes most of the eye irritation on smoggy days. Photochemical smog degrades many materials. Oxidation by ozone causes rubber to age and crack. Ozone and other strongly oxidizing molecules such as PAN and PBN are toxic to plant life. Exposure of plants to PAN at levels of only about 0.05 parts per million causes damage. Ozone-induced crop damage in the San Joaquin valley reduces crop yields by

Figure 10.7   *Manhattan skyline on a smoggy day. (Charles Merkle FPG.)*

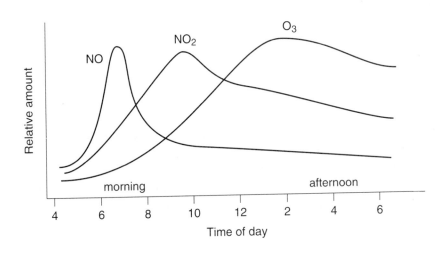

Figure 10.8.   *Daily changes in the levels of various gases in the atmosphere over an industrial city.*

Cleansing the Atmosphere—Naturally

Even in the absence of photochemical smog, ozone is naturally found in the troposphere. Ozone constitutes about one billionth of the molecules in the troposphere. Here, it is of great importance because of its role in the generation of the hydroxyl radical, OH, by the following reactions.

$$O_3 + light \rightarrow O^* + O_2$$

Ozone absorbs light and generates an excited oxygen atom and an oxygen molecule.

$$O^* + H_2O \rightarrow 2OH$$

Excited oxygen atoms react with water to produce hydroxyl radicals.

The hydroxyl radical serves as nature's atmospheric detergent. Many molecules emitted into the atmosphere by natural and industrial sources contain one or more hydrogen atoms. Some of these pollutants are toxic; others contribute to the greenhouse effect. These molecules react slowly with oxygen and are not easily precipitated. Fortunately, chemistry occurs that enables these molecules to be removed from the troposphere. The first step in the process involves reaction of a pollutant with hydroxyl radicals. Hydroxyl radicals abstract hydrogen atoms to form water and create reactive sites on the pollutant molecule. The reactive sites combine with oxygen and water in a series of reactions that either form particulates that settle out of the atmosphere or form water-soluble gases that are removed by precipitation. Without the hydroxyl radicals to start this cleansing process, the atmosphere would be extremely susceptible to air pollution. There is concern that rising concentrations of carbon monoxide and methane, which react with hydroxyl radicals, are reducing the self-cleansing power of the atmosphere.

20–25%, according to University of California estimates. The yearly damage to crops amounts to about $200 million for growers and $184 million for consumers in all of California.

Catalytic converters were invented to reduce harmful automotive emissions. These devices promote the oxidation of exhaust

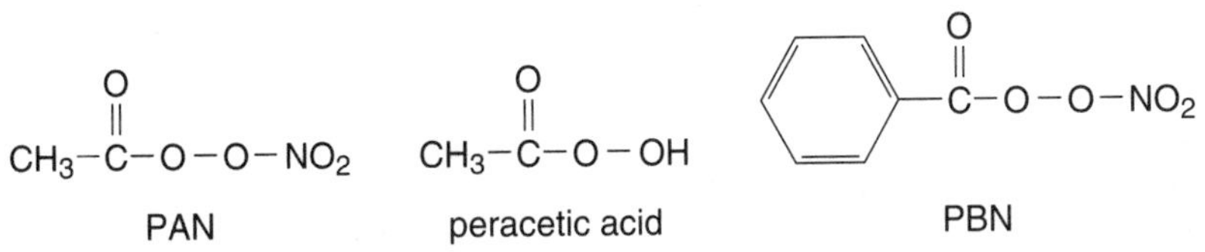

Figure 10.9. *The molecular structures of PAN, peracetic acid, and PBN.*

carbon monoxide and unburned hydrocarbon fuel to CO_2. Conventional converter catalysts use palladium (Pd), platinum (Pt), and rhodium (Rh) metals dispersed on porous alumina (Al_2O_3) pellets. Although catalytic converters are not 100% efficient, they reduce CO and hydrocarbon emissions from automobiles by as much as 90%. They also reduce emissions of NO by promoting its reaction with CO to form CO_2 and N_2. However, NO emission is still a problem. Catalytic converters have made a significant contribution to reducing air pollution. Today, half of the automotive air pollutants are emitted by the 10% of the vehicles on the road that lack emission-control devices or whose systems have been damaged.

Acid Rain

During the late 1960s, concern arose about the increased acidity of rain and snow in many areas of the world. Studies of acid precipitation and lake acidity in the Scandinavian countries first brought the issue to public attention. Acid precipitation affects lakes, rivers, and soils. The lakes in the Adirondack mountains, located in the northeastern part of New York State, exemplify the adverse effects of acid rain. Figure 10.10 plots the pH of lakes surveyed in this region during the 1930s and the 1970s. Over this 40-year period, the average pH dropped by 1.7 pH units, signaling a 50-fold increase in the acid concentration! The change

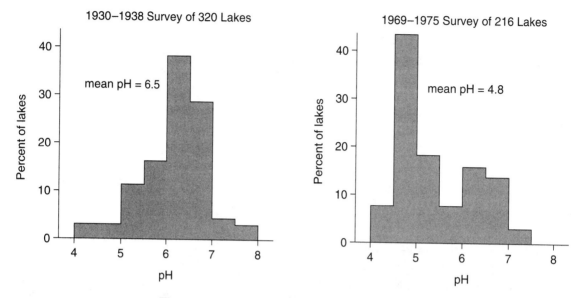

Figure 10.10. *The pH in the Adirondack lake region in upper New York State in the 1930s and in the 1970s.*

in the average pH was a direct consequence of the increased acidity of the rain and snow in this area. The problem is especially severe in this region because the soil is rocky and naturally acidic. The soil and rock formations of the northeastern United States lack natural basic components, such as limestone, which help neutralize acid rain in many other regions of the world.

What is the chemistry that leads to acid rain and snow? In an unpolluted world, rainwater is slightly acidic, pH = 5.7 (a neutral solution has a pH of 7.0). This slight acidity results from the reaction between water and CO_2 molecules in the clouds, which is similar to equations (14) and (15).

The major component of acid pollution is sulfuric acid (H_2SO_4). Sulfuric acid forms from the reaction between sulfur dioxide (SO_2), molecular oxygen, and water, as shown in equation (19).

$$2SO_{2(g)} + O_{2(g)} + 2H_2O_{(l)} \rightarrow 2H_2SO_{4(aq)} \qquad (19)$$

Sulfur dioxide, oxygen, and water aerosol (clouds) combine to form sulfuric acid as a liquid aerosol in the atmosphere, which precipitates as acid rain.

To a lesser extent, nitrogen dioxide (NO_2) also contributes to the acidity of rain. Atmospheric NO_2 reacts with hydroxyl radicals to form nitric acid (HNO_3), as shown in equation (20). The dominant source of NO_2 is the oxidation of nitrogen gas in automobile engines, so it is not surprising that acid rain occurs in most metropolitan areas.

$$NO_{2(g)} + OH_{(g)} + M_{(g)} \rightarrow HNO_{3(g)} + M_{(g)} \qquad (20)$$

Nitrogen dioxide and hydroxyl radicals combine with the help of collisions with gas molecules (M) in the atmosphere to form nitric acid. The gaseous acid is taken up by water droplets, generating acid rain. The species $M_{(g)}$ is needed to carry away the excess energy when the $NO_{2(g)}$ and $OH_{(g)}$ combine.

Understanding the increase in concentration of atmospheric sulfuric acid requires identifying the sources of SO_2. Fossil fuels contain small amounts of organic sulfur compounds that form SO_2 when burned. Coal-burning power plants generate SO_2 as a byproduct of the combustion process, because coal contains 0.2 to 1.3% sulfur by weight; this source alone accounts for nearly 60% of all industrial sulfur emissions. Other major sources include the combustion and refining of petroleum fuels (about 20%), ore smelting (about 10%), and various industrial (coke production, sulfuric acid manufacture) and waste disposal (incineration) processes.

Table 10.2 gives estimates of the annual emissions of SO_2 for several countries for periods from the mid-1970s to the mid-1980s. The numbers given are in units of thousands of metric tons. These data show that many countries have decreased the amounts of SO_2 that are emitted into the atmosphere. However, developing countries such as Poland and Bulgaria show increased outputs.

Since the early 1980s, sulfur emissions from coal power plants have been reduced by using scrubbers on smokestacks. These devices remove particulates and gases such as SO_2. Inside

Table 10.2. *Sulfur Dioxide Emissions in Selected Countries. The Numbers Are in Units of Thousands of Metric Tons.*

Country	1973–1975	1979–1981	1982–1984
United States	25,600	23,300	21,200
Japan	2,620	1,640	1,610
France	3,760	3,410	2,305
United Kingdom	5,430	4,740	3,750
Finland	540	570	360
Sweden	690	480	285
Poland	2,080	2,600	3,700
Bulgaria	—	1,030	1,140
Portugal	180	260	305

the scrubbers, the gases generated from the combustion of coal pass through a fine mist of water and lime (CaO). The mist traps more than 99% of the particulates and more than 80% of the total SO_2 emissions. Although the removal of such a large amount of sulfur dioxide from the exhaust has decreased the acid pollution caused by coal power plants, the scrubbers themselves have raised new environmental concerns. The devices produce a sludge that is rich in sulfur compounds and mineral matter. The sludge, which contains some toxic trace elements, must be disposed of with care; otherwise, it can adversely affect nearby ecosystems.

Unlike that of sulfur oxides, the output of nitrogen oxides (NO_x) has climbed in most countries. The problem is the lack of an inexpensive technology for eliminating this gas from the output of large-scale industrial combustion processes. Current technology for the removal of NO_x from the exhaust of power plants relies on the selective catalytic reduction (SCR) of nitrogen oxides (forming nitrogen gas) with ammonia over a heterogeneous metal oxide catalyst. The process is costly, and some of the added ammonia remains in the exhaust gas as a new pollutant. Current SCR technology reduces NO_x emissions below 300 ppm for coal-fired power plants, below 150 ppm for oil-fired plants, and below 10 ppm for natural gas-fired facilities. Given the grand scale of electric power generation, there are great economic opportunities for the development of new technologies.

CHAPTER 11

Chemistry and Cancer

Cancer is the second leading cause of death in the United States. Causes of cancer include diet, smoking, certain viral diseases, exposure to mutagenic chemicals, and radiation. Experts disagree about the health risks from traces of industrial and natural chemicals present in food, drinking water, and air. Carcinogenic chemicals usually behave as mutagens in experiments with cell cultures; however, not all mutagenic chemicals are carcinogenic. Many drugs used to treat cancers are potent mutagens! Nonetheless, a test for mutagenicity in cell culture is the first step in assessing the potential carcinogenicity of a chemical. The only sure way to test a chemical for carcinogenicity is by testing with human subjects, which can be accomplished in retrospective surveys of factory workers, smokers, asbestos miners, and other groups exposed to certain agents before the hazards were suspected. In most cases this is not possible, however, so animals are used to test the toxicity and carcinogenicity of chemicals. Of course, using animals introduces uncertainty into the results, because the physiology of animals differs considerably from that of humans. In addition, chemical carcinogenesis is a low-probability event. Only a small fraction of animals exposed to a given chemical dose develops cancer. The expense and "political problems" of using animals in research require two more compromises. Inexpensive animals whose use generates relatively little public objection (mice and rats) are selected for most studies. Optimists argue that humans have a more advanced defense system against chemical carcinogens. Pessimists

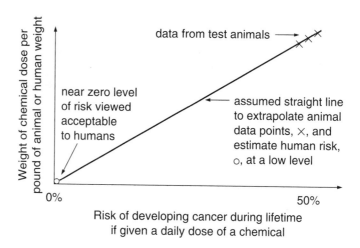

Figure 11.1. *Graph used to estimate the cancer-causing potential of a chemical. The vertical axis is the dose (normalized to the weight of the test animal). The horizontal axis is a measure of the risk of developing cancer.*

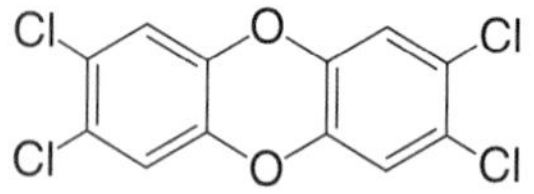

Figure 11.2. *The molecular structure of dioxin, a controversial carcinogen.*

argue that the animal's short life span doesn't allow a test of the long-term exposure a human would receive. The second compromise brings up the contentious issue of dose versus response.

Society demands that the risk to humans from chemical carcinogens be very low: less than one in a million (0.0001%). In an animal study, it is impractical to determine the dose of chemical that causes cancer at such low probabilities. Instead, chemical carcinogens are administered daily to the test population at the maximally tolerated dose (MTD). This greatly increases the cancer risk and reduces the number of animals needed. Skin tests for the tumor-inducing power of a chemical are another part of the test. The MTD differs for each chemical. It corresponds to the highest daily dose an animal can receive without manifest toxic effects (such as loss of weight or death). Results from the MTD testing procedure are controversial. The approach requires extrapolating results obtained with high doses in stressed animals to that expected for lower (realistic) exposures. In general, a linear extrapolation is used in assessing dose versus response, as shown in Figure 11.1. The validity of assuming a straight line, instead of some curve, is the subject of vigorous debate. Optimists contend that this procedure overestimates the risks to humans, whereas pessimists believe it underestimates those risks.

The chemical dioxin (Figure 11.2) illustrates the problems created by a lack of reliable information. An accident in 1976 at a chemical plant in Seveso, Italy, spread dioxin over the town.

Residents showed high levels of the chemical in their bodies. Doctors attributed skin rashes (chloracne) and various other illnesses to the effects of dioxin. Dioxin also appeared in the chemical sludge at Love Canal and as a contaminant in Agent Orange. Its toxicity became a focal point for Vietnam veterans, who claimed that using it as a defoliant during the war had caused them physical harm. In 1983 the United States government purchased and evacuated the entire town of Times Beach, Missouri, in response to dioxin contamination of the soil.

Initial research with guinea pigs showed that 6 micrograms of dioxin per kilogram of body weight killed 50% of the test animals. This led many to conclude that dioxin was the most toxic industrial chemical known! In 1978 Dow Chemical researchers showed that dioxin acts as a potent carcinogen in rats. Given that Dow was a defendant in an Agent Orange lawsuit, these findings, which hurt *its own case,* were taken seriously. On the basis of these alarming facts, the EPA set the maximum daily intake level for dioxin in humans as 0.0000000004 gram for a 68-kg person. This was the dose estimated to entail a cancer risk of one in a million. The EPA recommendation is about 200 to 1700 times smaller than the intake levels adopted by European nations, whose risk-factor estimates differ.

Recent testing of dioxin suggests that it might not be so potent a toxin in humans as originally feared. Dioxin is 1000 times less toxic if hamsters are used as test animals instead of guinea pigs. This discrepancy illustrates the difficulty of identifying human risks with tests in other species. An epidemiological study of chemical plant workers exposed to high levels of dioxin was conducted by the National Institute for Occupational Safety and Health. Workers exposed to dioxin for more than a year had a cancer rate that was 46% higher than that of the general population. Results obtained at the cellular level suggest that the dose-versus-response curve deviates from the linear extrapolation model assumed in normal cancer-risk analyses, but this has been disputed on the basis of the results of some recent studies. As this manuscript went to press, an EPA review of the available scientific evidence led them to rank dioxin as a "probable" human carcinogen.

Proven Carcinogens

Most scientists and health officials agree that chemicals on the list of proven human carcinogens (Appendix II) should be used with extreme caution. As of 1993, over 150 chemicals were listed. Epidemiological studies of workers exposed to these chemicals show an increased cancer rate. Industries and research laboratories now avoid using known carcinogenic chemicals because of these hazards and also because federal and state regulations make them inconvenient and expensive to use. Employees who might be exposed to known human carcinogens must be medically monitored on a regular basis.

Natural Toxins in Food

Diet is estimated to be an important factor in about 35% of all cancer cases. In addition to various pesticides and fungicides used in food processing, some foods contain natural chemicals suspected of causing cancer. Aflatoxin B_1 (Figure 11.3) is a mycotoxin product of various molds found in peanuts, corn, and many other grains. Molds that produce aflatoxin B_1 can grow in nearly any food. Some human and animal deaths have been attributed to moldy food contaminated with this agent. Scientific evidence links aflatoxin with an increased risk of liver cancer. The FDA has placed a limit of 20 parts per billion (ppb) as the maximal level acceptable for human consumption and a limit of 100 ppb for consumption by mature livestock. The 1988 midwest corn harvest was threatened by high concentrations of aflatoxin. In Texas alone, 60–70% of the fields in the northeast part of the state had aflatoxin levels exceeding the FDA limit. Dairies in Texas were forced to destroy 2 million pounds of milk from cows fed with contaminated grain. In Iowa, grain elevator operators had to test each delivery of corn, and many of the shipments tested in excess of 100 ppb.

Consumer groups have even reported that "natural" foods (fresh ground peanut butter) occasionally contain higher levels of this mycotoxin than processed brands. This has led some peo-

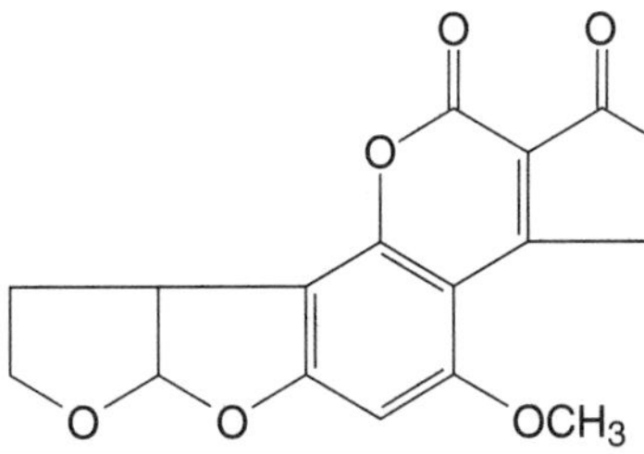

Figure 11.3. *The molecular structure of aflatoxin B₁, a chemical produced by molds.*

ple to suggest that reducing the use of pesticides and grain fungicides on food crops could actually boost the cancer risk from exposure to increased levels of mold toxins.

Plants also contain natural pesticides (Figure 11.4). A recent debate has focused on their toxic and carcinogenic effects on humans. Allyl isothiocyanate, a chemical-warfare agent and known carcinogen, occurs in mustard seed. A precursor to it occurs in cabbage and kale. Phorbol esters—potent tumor-promoting agents and co-carcinogens—occur naturally in some herbal teas. Sodium oxalate, a toxic salt used to finish textiles and leather, occurs in spinach and rhubarb.

Consider the potato. This tuber must persist underground for months without being eaten by insects and microorganisms. What makes it so resistant to attack by predators? Potatoes contain solanine (Figure 11.5), a chemical that can also be used as an agricultural insecticide. Disease-resistant potatoes, with increased levels of solanine, have been bred genetically to avoid the need for soil fumigants. Production of these potatoes can cause serious health problems, because solanine is very toxic. The lethal dose in mice is about 42 mg/kg of body weight. This is comparable to the lethal dose of parathion (Figure 11.6), a synthetic pesticide that was banned by the United States because of its high toxicity. These and other observations have renewed interest in studying the health effects of natural carcinogens in the food supply.

Fortunately, the human body has an elaborate chemical defense system against toxic molecules. For example, the enzyme cytochrome P450 performs two function in the body. First, it catalyzes reactions that are important in the synthesis of hormones. Second, it plays a key role in the liver by removing toxic organ-

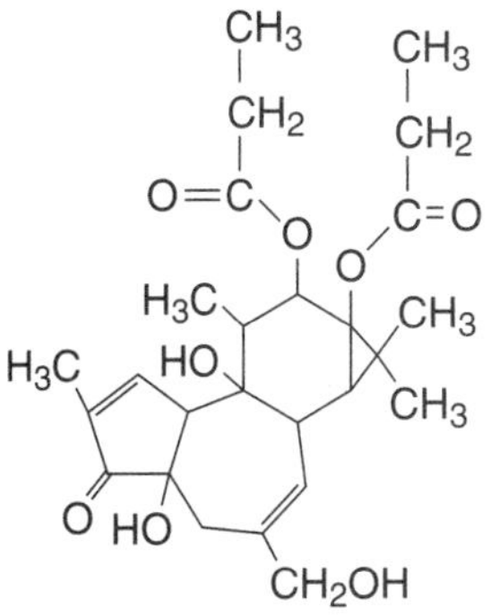

A phorbol ester

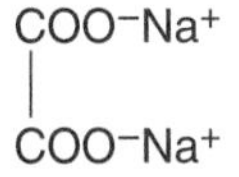

Sodium oxalate

$H_2C{=}CHCH_2NCS$

Allyl isothiocyanate

Figure 11.4. *Natural pesticides that occur in plants.*

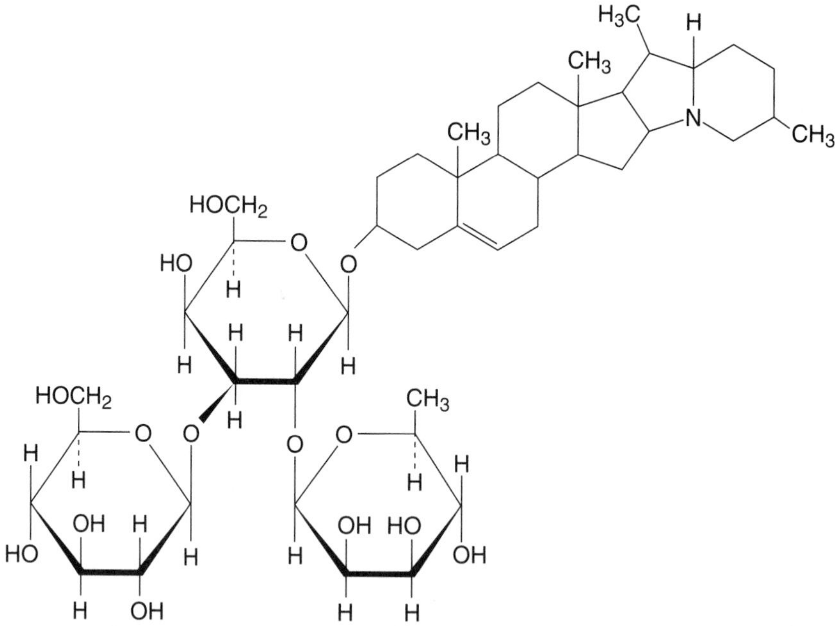

Braving the Elements

Figure 11.5. *The molecular structure of solanine, a natural pesticide found in the skin of the potato.*

ic molecules from the body. The enzyme catalyzes the oxidation of foreign chemicals, which makes them more readily excreted. Unfortunately, in some cases the oxidation reaction produces a more potent carcinogen. This is suspected to be the case in the metabolism of benzopyrene.

How much danger do the large numbers of chemicals that a person encounters each day pose? We don't really know. A few general studies of cancer rates help define the possible magnitude of the problem. A recent survey of the change in cancer rates in the industrialized nations suggests that cancer incidence rates for those 65 years of age or younger decreased between 1973 and 1992. For individuals above the age of 65, the cancer

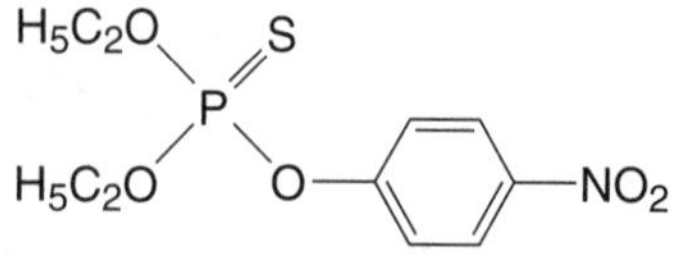

Figure 11.6. *The molecular structure of parathion, a synthetic pesticide.*

incidence rate increased slightly. Overall, the total incidence of cancer in the United States increased 16%, which can be partially credited to earlier detection. Except for an increase in lung cancer and a decrease in stomach cancer, the cancer *death* rate in the United States has been relatively constant for the past 30 years. The increase in the death rate for lung cancer is attributed to tobacco-related problems. The decrease in stomach cancer fatalities reflects improved food storage, dietary changes, and the use of less salt and nitrate preservatives.

The most recent health concerns about chemicals in the environment center on organic compounds that contain chlorine (chlorocarbons). An aryl hydrocarbon (Ah) receptor has been identified in cells. When chlorocarbons, such as dioxin, bind to the Ah receptor it issues a signal that turns on several genes in the cell's nucleus. This includes the increased production of a cytochrome P450. There are also concerns about the ability of chlorocarbons to bind with other cellular receptors for estogen hormones and cause disorders in the endocrine system. Some suggest this may explain the increased risk of breast cancer and endometriosis for females, and the decreased sperm counts observed for males during the past 30–50 years. Because there are potential alternative explanations, this is a controversial hypothesis, but one that will receive scientific scrutiny over the next decade.

Hazardous to Inhale

Asbestos

Asbestos is a generic name for a variety of fibrous silicate minerals. Asbestos was once widely used as insulation for furnaces, heating ducts, and steam pipes. In addition, automotive brake linings and protective clothing for firefighters have been fabricated out of asbestos. Despite its usefulness, some asbestos materials pose severe health hazards. Long-term exposure is estimated to increase the risk of lung cancer by a factor of 5. (Cigarette smoking increases the risk by a factor of 10.) The effects of low-level exposure are not known. Exposure to asbestos fibers is also associated with mesothelioma, a rare and incurable cancer of the linings of the body cavities.

Two natural asbestos minerals are mined: crocidolite and chrysotile. Crocidolite is carcinogenic. It has a low solubility in water and persists in tissues. Its fibers are long, thin, and straight and penetrate the lung passages. Chrysotile, a white mineral, is slightly soluble in water and tends to disappear from tissues. Its crystals are curly and do not penetrate lung tissue. Four epidemiological studies of the female population near Québec chrysotile mines showed no statistically excess disease. About 95% of the asbestos used in the United States is made with chrysotile. However, public fear has resulted in government regulations that require the removal of asbestos from public buildings, regardless of the mineral form used. Estimates of the total cost for this work range from $50 to $150 billion. One fact often neglected is that asbestos removal releases fibers into the air, which could result in the building's occupants running a greater risk of developing cancer than if the asbestos were left in place.

The EPA has fostered the idea that a single asbestos fiber causes cancer. However, this has not been proved. Each of us breathes in about a million fibers per year that arise from natural geologic processes. Inside most of the buildings in question, the content of asbestos fibers in the air is harmlessly small and essentially the same as that in outdoor air. A 1990 study of comparative risks reported in *Science* magazine found that children exposed to low levels of asbestos in the classroom for 5 years face 10 times as much risk of dying from a whooping cough vaccine as of developing a cancer caused by asbestos.

Radon

Radon is a colorless, odorless, tasteless, and unreactive gas. Radon is released as a result of the radioactive decay of uranium-238, a natural component of many rocks (granite, shale) and minerals (pitchblende, phosphate ores). Outside, the gas dissipates and poses a negligible health hazard. Inside homes, how-

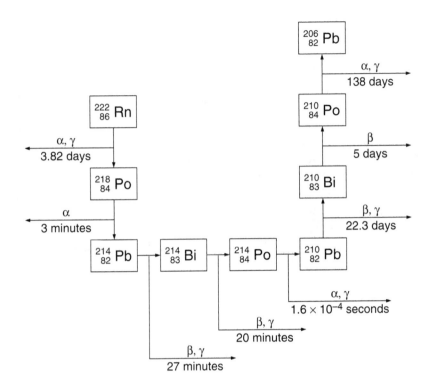

Figure 11.7. *The radioactive-decay pathway of radon gas.*

ever, the gas levels can build up and exceed the maximal safe level established by the EPA (0.15 disintegrations per second per liter of air). Radon itself is not carcinogenic. The dangers arise from the fact that its radioactive-decay products (Figure 11.7) are also radioactive and become suspended in air as fine particles. When inhaled, the various decay products (polonium, bismuth, and lead) are trapped in the lung tissue. The α- and β-particles generated by subsequent radioactive decay of these elements can cause cellular damage.

Although it is generally accepted that radon can cause lung cancer, how much risk results from radon exposure in the home is a controversial issue. In the late 1980s, the EPA estimated that 21,000 deaths occur each year from radon exposure. However, in February of 1991, the EPA announced that it may have overestimated those risks by between 20 and 30%, revising downward to 16,000 its estimate of annual deaths caused by radon exposure in the home.

Recently, the EPA proposed a standard to reduce the level of natural radon in drinking water. The Association of California Water Agencies (ACWA) estimates that it would cost California nearly $4 billion to meet the proposed standard. And this tremendous expense would result in reduced radon exposure for only 1% of the public. In an appeal to President Bush, the ACWA and 27 California state legislators questioned "regulations which place a considerable financial burden on our citizens (about $1,000 for a household of four) without providing appreciable public benefit." Good scientific evidence is needed to assess human-health risk and ecological risk, inasmuch as society benefits most by allocating resources to solve the most serious problems first.

Smoking

The health risks associated with smoking are widely known. Smokers face higher risks of cancer of the lung, bladder, cervix, esophagus, and larynx than nonsmokers. Almost a third of all cancer deaths are associated with tobacco. Research has identified over 40 carcinogens in the more than 4000 chemicals in cigarette smoke. In a study reported in 1993, a group of researchers developed a model of cancer risks based on data collected in a 6-year study started by the American Cancer Society in 1982, which cataloged the habits of 900,000 people. Smokers who had quit still faced higher risks of developing cancer than people who had never smoked. Although the risks remain consistently higher for former smokers than for nonsmokers, the study suggests that it is possible to reduce the cancer risk by giving up smoking at a young age.

One of the most controversial topics in the media is the proposed link between second-hand smoke and lung cancer. Researchers agree that exposure to second-hand smoke puts the nonsmoker at risk. However, the magnitude of the effect is still uncertain. In 1993 the EPA released a report concluding that second-hand smoke is a human carcinogen that each year causes about 3000 lung-cancer deaths in nonsmokers. It is likely that the debate on the effects of second-hand smoke will continue for several years.

Pesticides and Herbicides

Agricultural activities pour an enormous quantity of pesticides and herbicides into the soil. Pesticides are used to kill rats, insects, and other pests. Herbicides are chemical agents that control the growth of weeds. Some pesticides and herbicides are poisonous to microorganisms and animal life. Others generate toxic materials through chemical degradation. Current legislation requires that the environmental impact of potential pesticides be determined before they are used.

Public concern over the adverse effects of pesticides came to the forefront in 1962 when Rachel Carson published her book *Silent Spring*. At that time, the most commonly used pesticide was DDT (Figure 11.8). Despite an enormous research effort, the exact mode of biochemical action of DDT is still largely unknown. Like many pesticides, DDT affects the central nervous systems of insects. It dissolves preferentially in the fatty tissues, accumulating in regions that surround nerve cells. It is believed that DDT kills insects by interfering with the transmission of nerve impulses. The chemical side effects of this molecule were not known when its use as a pesticide began. It was employed widely to control mosquitoes, thereby dramatically decreasing the incidence of malaria. The agricultural and medical benefits of DDT earned its discoverer, Paul Müller, the Nobel Prize for Physiology or Medicine in 1948.

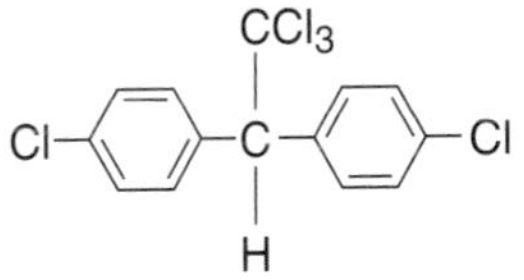

When they become concentrated in the food chain, DDT and its chief breakdown product, DDE (Figure 11.9), reach high levels in various animals. The most intensively studied such animals are brown pelicans and peregrine falcons. These poisons did not kill the birds outright but rather interfered with the deposition of calcium in their eggshells, resulting in a thinning of the shell. The fragile shells broke easily, increasing the mortality rate of embryos. Although it is now banned in the United States, DDT is still used in many other countries. The curtailed application of DDT has led to a rise in the worldwide incidence of malaria.

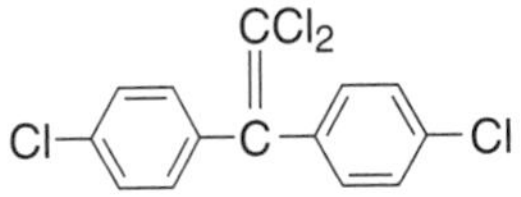

Does DDT cause cancer? A 1989 study reported in the *American Journal of Public Health* reviewed a decade in the lives of

Figure 11.8. *The molecular structure of the pesticide DDT (dichlorodiphenyl-trichloroethane).*

Figure 11.9. *The molecular structure of the breakdown product of DDT, DDE.*

nearly 1000 people who had been exposed to DDT for at least 25 years. The results revealed no statistically significant link between the amount of DDT in these people's bodies and the risk of death by cancer. Because of the small sample size, however, the researchers could not rule out small increases in cancer risks. In 1992 researchers reported that people involved in the manufacture of DDT had 4.8 times the normal incidence of pancreatic cancer, suggesting a link between long-term DDT exposure and pancreatic cancer. The risk to ordinary consumers of DDT, however, was estimated to be minimal.

One of the greatest environmental disasters ever to result from pesticide manufacturing involved the production of Kepone (Figure 11.10). Designed to control the tobacco wireworm, ants, and other small insects, Kepone exhibits acute, delayed, and cumulative toxicity in birds, rodents, and humans. In addition, this chemical causes cancer in rodents and appears on the EPA list of proven carcinogens.

Kepone was manufactured in Hopewell, Virginia, in the mid-1970s. During this time, many workers who were exposed to this toxin reported health problems. Their nervous system disorders became known as the "Kepone shakes." The processing plant was connected to the Hopewell sewage system. Frequent influxes of Kepone into the sewer system periodically caused the water treatment plant to be inoperative. It is estimated that 53,000 kilograms of Kepone were dumped into the river. The waste affected more than the operations of a sewage treatment plant. Effluent from this facility ended up in the James River and Chesapeake Bay, thereby dispersing the toxin into the surrounding ecosystem. Over 100 million cubic meters of river sediment would need to be dredged up and detoxified to decontaminate the river. The cost of such a cleanup would be several billion dollars. As of 1985 the $5.2 million fine paid by Allied Chemical was exhausted.

The flooding of Kepone into the ecosystems of the Hopewell area is only one example of the many problems that arise from the use of pesticides. Pesticides contaminate many of the world's water supplies, because the soil is unable to absorb all the pesticides applied. The runoff from precipitation dissolves these

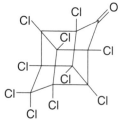

Figure 11.10. *The molecular structure of the highly toxic pesticide Kepone.*

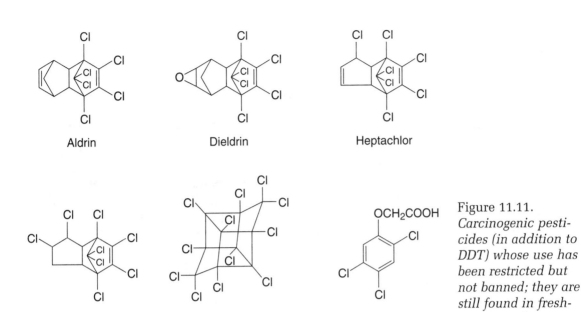

Aldrin

Dieldrin

Heptachlor

Chlordane

Mirex

2,4,5-T

Figure 11.11.
Carcinogenic pesticides (in addition to DDT) whose use has been restricted but not banned; they are still found in freshwater supplies.

molecules and carries them into the water table. Figure 11.11 shows some of the carcinogenic pesticides found in fresh water. The United States restricts the use of these molecules.

In many cases, pesticides categorized as restricted by the EPA are reported by the news media as having been banned. This has misled the public about the use of these agricultural chemicals. In canceling the major uses of a particular pesticide, the EPA often allows a slow phasing out. It is also common to permit the sale and distribution of existing supplies. Such sales have made it possible for some companies and individuals to use these compounds even after legislation has been enacted to ban them.

Agricultural chemicals are used for purposes other than crop cultivation. Herbicides were employed as defoliants in the Vietnam War. The most effective and controversial herbicide was Agent Orange, a defoliant that is a 50:50 mixture of 2,4,5-T (see Figure 11.11) and 2,4-D (Figure 11.12).

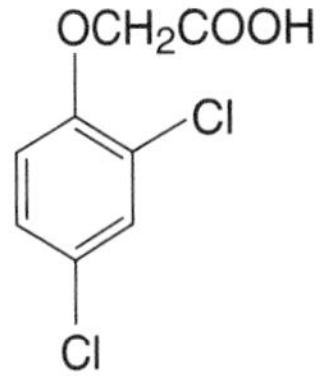

United States planes sprayed over 42 million kilograms of Agent Orange on Vietnamese soil, decimating 1.8 million hectares of countryside and at least 200,000 hectares of farmland.

OCH$_2$COOH

Figure 11.12. *The molecular structure of the defoliant 2,4-D. This compound is often contaminated with dioxin.*

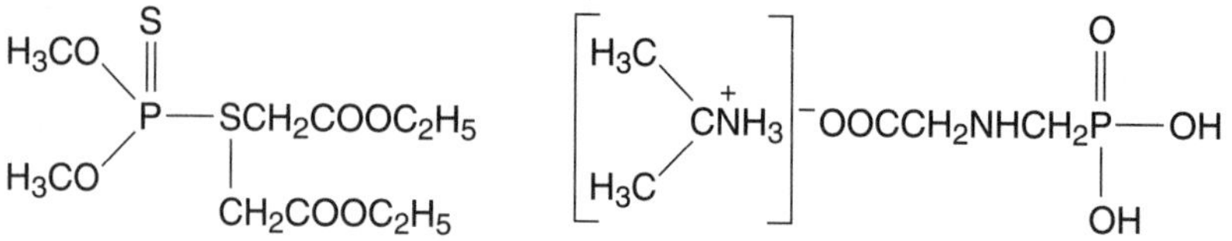

Figure 11.13. *The molecular structures of the low-toxicity pesticide malathion and the low-toxicity herbicide Roundup®.*

The National Academy of Sciences has estimated that the defoliated areas will require over a century to recover. In recent years, many news reports have appeared on the medical problems experienced by soldiers exposed to this herbicide. In 1970 the United States government stopped using Agent Orange in Vietnam. In 1985 the EPA terminated all registered uses for 2,4,5-T on rice fields, orchards, sugarcane fields, range lands, and other noncrop sites. However, 2,4-D and 2,4,5-T are still employed in the United States in evergreen forests that grow hardwood trees for commercial uses. 2,4-D is also a common additive to lawn fertilizers. Unlike such molecules as DDT, 2,4-D is broken down by bacteria in the soil within a few months.

Agent Orange, DDT, and Kepone belong to a family of compounds known as polychlorinated hydrocarbons. They consist primarily of chlorine, carbon, and hydrogen. Such compounds tend to be stable, so they persist for years—or even decades—in the environment. They tend to concentrate in the fatty tissues of higher mammals to cause liver and reproductive damage. Less toxic and less persistent nonchlorinated hydrocarbon pesticides (malathion) and herbicides (Roundup®) have been developed (Figure 11.13). Both represent a more sophisticated attack on the biochemistry of pests. Malathion is a nerve agent tailored to react specifically with insect acetylcholineesterase enzyme. Roundup® blocks the synthesis of an essential amino acid in plants by inactivating an enzyme not found in mammals. Both malathion and Roundup® degrade to nontoxic fragments after their use.

In 1990 the herbicide Alar (Figure 11.14) garnered much attention in the media. Alar prevents ripe apples from dropping to the ground. As a result of public outcry, the FDA planned to

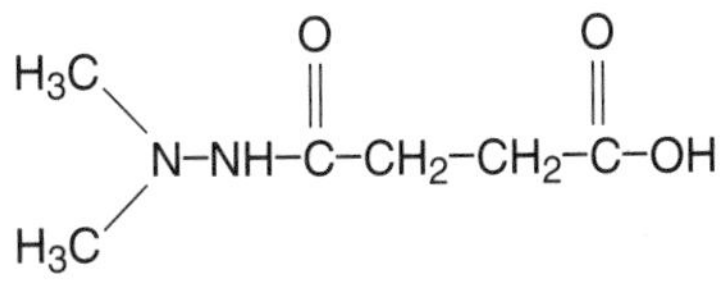

Figure 11.14. *The molecular structure of Alar, a chemical that prevents fruit drop from trees.*

phase out the chemical, but political pressures forced a voluntary withdrawal before the official date. The objection to the use of Alar was the presence of the trace metabolite known as dimethylhydrazine, a rodent carcinogen. But what was the risk? Bruce Ames, a biochemist at the University of California, Berkeley, pointed out that one mushroom contains more hydrazine than the amount present on an Alar-treated apple. Yet mushrooms are common ingredients in many foods. And what about the increase in natural mold toxins in the apples that, in the absence of Alar, are allowed to fall to the ground? How is that factored in? Much money and publicity were focused on an issue that may or may not have led to an outcome that is important for human health. Other problems (such as lead pollution in water, urban air pollution, mold toxins in food, and the levels of polychlorinated hydrocarbon pesticides in imported food) would appear to represent greater hazards to consumer health but have not received as much attention. When regulators are ill-informed about risks, money may be spent with little advancement of the public good. In 1991 the cost of U. S. environmental regulations amounted to $115 billion.

Attention has focused on the "Delaney clause," which requires the federal government to ban from processed foods any substance found to cause cancer. Though it has been written into law since 1958, there is broad agreement that the Delaney clause is too rigid a standard for assessing potential human health hazards. Since the enactment of this law, advances in analytical techniques have made it possible to detect minute amounts of material. In 1989 the EPA, responding to these technological changes, proposed a new standard of *negligible risk* for processed foods. The risk level adopted was set at a maximum of one additional case of cancer per million people resulting from

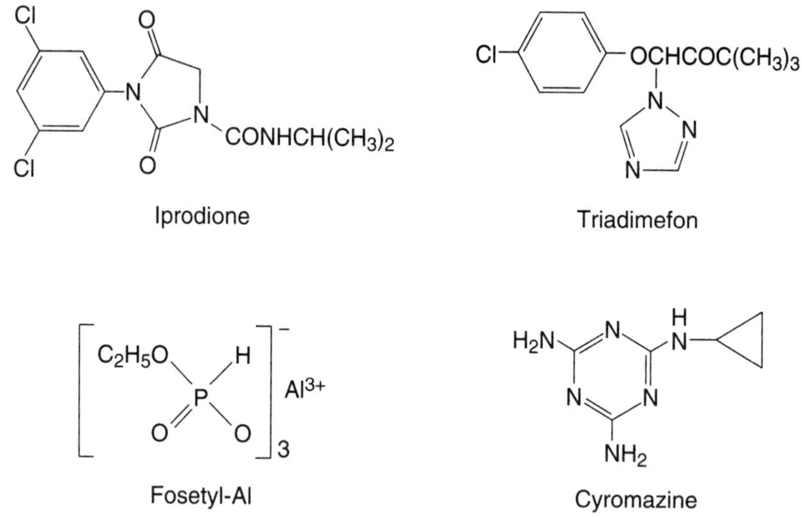

Figure 11.15.   *Four pesticides banned from use in processed foods by the EPA in 1993.*

daily exposure to the chemical over a lifetime. In a lawsuit brought by consumer advocates and environmental groups, an appeals court overruled the negligible-risk standard in July of 1992. In February of 1993, the EPA released a list of 35 agricultural chemicals whose use could be prohibited under the Delaney clause, emphasizing that the list was not definitive. Many of the chemicals on this list, as well as many still to be reviewed, could be restricted or banned unless the ruling on negligible risk is reversed by the Supreme Court.

In May of 1993, the EPA, in compliance with a 1992 federal court decision upholding the Delaney clause, banned the use of four fungicides in processed foods. Carol Browner, administrator of the EPA, released a statement that the substances banned "pose only a negligible risk to public health." The four substances are cyromazine (used on potatoes and tomatoes in Florida), fosetyl-Al (used on hops in Oregon), iprodione (used on apples in North Carolina), and triadimefon (used on tomatoes in California); see Figure 11.15.

Love Canal

Since 1978 the name "Love Canal" has symbolized the effect that unchecked hazardous-waste disposal can have on the ecosystem and on human life. Love Canal is an area just southeast of Niagara Falls in New York State. The area was named in honor of William T. Love, who in the 1800s had the idea of building a 7-mile-long canal from the upper Niagara River above Niagara Falls. This canal would terminate at a large hydroelectric generating plant, after which the water would be emptied back into the river below the falls. Love's dream was to build a large city that would attract industry by offering an inexpensive source of electrical energy. This idealistic plan never materialized, but a trench 50 feet wide, a mile long, and ranging in depth from about 9 to 100 feet remained as its monument.

During the 1800s, the population in this region grew. Spurred by inexpensive hydroelectric power, a large electrochemical industry developed. Vast quantities of many materials, including aluminum, chlorine, and chlorinated organic chemicals, were produced. In the 1940s, the Hooker Chemical Company used Love Canal as a site for the disposal of chemical waste, including byproducts from the manufacture of caustic soda, chlorinated organic pesticides, and plasticizers. During this decade, Hooker dumped over 20,000 metric tons of chemical waste into the canal.

In 1953 the Niagara Falls Board of Education was given the Love Canal dump site. The conditions of the gift stipulated that the soil could never be excavated, because the pit was capped with a relatively impervious clay soil. The Board of Education covered the dump with dirt and built a school, and the building of a school encouraged the development of neighborhood housing. Suburbia flourished atop the chemical waste dump, and the ban on excavation was relaxed. Later in the 1950s, children playing on top of the dump site often developed eye and respiratory problems.

In the 1970s, the Niagara Falls region experienced heavy rains. The Love Canal dump became saturated with water. Many of the liquid organic chemicals in the buried waste were lighter than water and floated to the surface. By the winter of 1978, several areas were oozing substantial masses of chemical sludge. Barrels originally buried in the site began to float to the top of the soil. In February of 1978, the EPA reported that 26 synthetic organic compounds could be found in the basements of homes on the site. A study later that year identified 82 compounds, 11 of which were suspected human carcinogens. At that point President Jimmy Carter declared part of the development a disaster area. The government instituted plans to evacuate 239 families. By the end of May of 1980, an additional 710 families were eligible for federal relocation assistance.

Love Canal was the disposal site of a large number of different types of haz-

Love Canal. (UPI/Bettman.)

ardous chemicals. Several appeared on the EPA's list of priority pollutants because of their toxicity to humans. In addition to toxic organic molecules, high levels of antimony, arsenic, cadmium, copper, chromium, lead, nickel, selenium, silver, and zinc were found.

By the end of 1981, over $80 million had been spent on the Love Canal problem. The EPA committed an additional $4 million of Superfund money to clean up waste chemical pollution in the general area. By 1987 the total spent on cleanup, relocation, and research exceeded $200 million. Later that year, the EPA announced plans to dredge the sewers and creeks in Love Canal to remove the contaminated sediments. 35,000 cubic meters of sediment were to be removed, burned, and reburied in a landfill. The EPA estimated that this cleanup effort would take 5 years and cost about $30 million.

"NIAGARA FALLS, N.Y., March 24, 1993 (AP) New York State closed its public information office at Love Canal today, declaring the neighborhood safe 15 years after it was evacuated after toxic wastes were found there." Love Canal is now repopulated. The subdivision, however, has been renamed Black Creek Village, after a brook that meanders through the area.

Love Canal has been the most widely publicized problem resulting from the disposal of hazardous industrial waste in the United States. However, it is not the biggest or the most hazardous. According to the EPA, there are about 14,000 abandoned hazardous-waste sites in the United States. As of 1989, 889 of these sites were slated for cleanup using the funds raised by taxes on the petroleum and chemical industry levied by the 1980 Superfund law. Unlike the time when Love Canal was used as a dumping site, strict laws now regulate the disposal of chemical waste. Presently, most large companies dispose of waste in a responsible manner.

Chemical Treatments of Cancer

One of the long-term goals of chemical treatments of cancer is to cure cancers that cannot be eradicated by surgery. Two general approaches are used. Radioactive isotopes are employed to expose cells to lethal amounts of gamma radiation. And chemical agents (chemotherapy) designed to interfere with cellular growth are administered. Depending on the type of cancer, one or both of these approaches are recommended.

Radiation

Two types of radiation employed in the treatment of cancer are X-rays and γ-rays. High-energy particles can be delivered by α-emitters, β-emitters, and electron accelerators. X-rays are produced by focusing a high-energy beam of electrons onto a metal target (tungsten). γ-rays result from the radioactive decay of atomic nuclei. Linear accelerators are used to produce electron beams. In contrast to chemotherapy, radiation can be localized to a particular part of the human body. Some cancer cells are very sensitive to radiation (leukemia), whereas others are resistant (melanoma). Linear accelerators and cobalt-60 units (a γ-ray source) are the most commonly employed radiation devices in the United States. Radiation is often applied in treating skin cancer, many head and neck cancers, localized breast cancer, and Hodgkin's disease. The malignant cells of Hodgkin's disease are very sensitive to radiation. Early stages of this disease can be treated effectively by external irradiation. Laryngeal carcinoma confined to the vocal cord is also highly curable with radiation.

In addition to external radiation sources, implants can be placed inside body cavities or directly in tumors, delivering large doses of high-energy particles to a specified area of the body. For example, colloidal suspensions of a radioisotope such as $^{32}_{15}P$ can be implanted in the lining of the lung (pleural cavity) or abdomen (peritoneal cavity). In this case, the cancerous growth to be irradiated must be less than 2 mm thick because of the limited range of penetration of the emitted β-particles. Radioisotopes are commonly brought in temporary contact with gynecological cancers using special applicators. The isotopes employed for this treatment include $^{226}_{88}Ra$ (an α-emitter), $^{60}_{27}Co$ (a β-emitter), and $^{137}_{55}Cs$ (a β-emitter). And radioactive materials may be planted temporarily or permanently in malignant tumors. Isotopes used in tumor implants include $^{192}_{77}Ir$ (a β-emitter), $^{131}_{53}I$ (β-emitter), $^{226}_{88}Ra$, and $^{137}_{55}Cs$. Radioactive seeds are employed for permanent implants; seeds, wires, and needles are used in temporary implants. Table 11.1 lists some of the radioisotopes applied in cancer therapy.

Research on new radiation therapies is an active field. Employing computerized tomography and magnetic resonance imaging, it is now possible to generate detailed three-dimensional images of the tumor and surrounding tissue. In a technique called dynamic radiation therapy, the radiation source is rotated around the patient, and the shape of the beam is adjusted to match that of the tumor. The efficacy of new radiation sources is also being explored. Neutron beam therapy has been found to be at least twice as effective as conventional sources for treating unresectable salivary gland cancer. It also appears that neutron beam therapy is superior to X-ray treatment of locally advanced prostate cancer. At present, the availability of neutron beam therapy is limited by the cost of such devices.

Table 11.1.   *Radioisotopes Used Internally in the Treatment of Cancer (Data from* The Merck Index, *11th ed., 1989)*

Element	Half-life	Form used	Therapeutic Uses
$^{241}_{95}$Am	432 years	Encapsulated	Intracavitary radiation source for malignancies
$^{137}_{55}$Cs	30 years	$CsCl$ or Cs_2SO_4, encapsulated in needles or applicator cells	External, intracavitary, or interstitial irradiation
$^{60}_{27}$Co	5.27 years	Metallic cobalt	External, intracavitary, or interstitial irradiation
$^{198}_{79}$Au	2.7 days	Colloidal gold or seeds	Treatment of widespread abdominal carcinomatosis with ascites; carcinomatosis of pleura; lymphomas; interstitially in metastatic tumors
$^{131}_{53}$I	8 days	NaI, administered orally or by i.v.	Cancer of the thyroid
$^{192}_{77}$Ir	73 days	Seed encased in nylon ribbon	Interstitial treatment of tumors
$^{32}_{15}$P	14 days	Chromic phosphate	Treatment for peritoneal or pleural effusions caused by metastatic disease
$^{226}_{88}$Ra	1600 years	$RaBr_2$	Treatment of malignancies such as cancer of the uterine cervix and fundus, and metastatic cancer of the lymph nodes

Oncologists often adopt an approach to the treatment of cancer that involves a combination of radiation and chemical therapies. Several chemical agents act as sensitizers, enhancing the effects of radiation when administered before, during, or after radiation treatment. Many chemotherapy drugs (fluorouracil, mitomycin, and cisplatin) possess this property. The ideal sensitizer would be nontoxic and would selectively enhance the effects of radiation at the tumor site. Unfortunately, such chemicals have not yet been developed.

A Man with a Mission

When the body's self-regulation of growth runs amok in a cell, that cell becomes cancerous. The cancer cells multiply and eventually take over the entire body. Because of the strong similarity between cancer cells and normal cells, it has been difficult to destroy cancer cells without damaging normal ones. Not even the body's sophisticated immune system can distinguish between the good and the bad cells. Therefore, it isn't surprising that progress in the fight against cancer has been slow. One of the most important chemotherapeutic advances was a result of serendipity and the perseverance of Barnett Rosenberg.

In 1964 Barney Rosenberg and his research group were studying the effects of an electric field on bacterial growth. They suspended *E. coli* in a growth medium between two inert platinum electrodes and applied AC power. The bacteria they saw under the microscope were like nothing they had seen before. Instead of single oblong *E. coli,* they saw strands of dead bacteria joined in filaments 300 long. Soon they showed that an extract from the bacterial broth in the electrolysis cell produced a similar result on fresh bacteria that had not been exposed to the electric field. Clearly, some chemical that was produced in solution must have caused the effect. Platinum electrodes were found to be a crucial ingredient; in a paper in *Nature* in 1965, Rosenberg and his colleagues proposed the hypothesis that a minute

Barney Rosenberg. (Courtesy of Barnett Rosenberg.)

amount of corrosion of the "inert" electrodes led to the change.

Although Rosenberg's basic research discovery was not directed specifically at the cancer problem, he had the good sense to follow it up and see whether it could be applied to cancer chemotherapy. One strategy for cancer control is to use drugs that interfere with cell replication, thereby selectively damaging the rapidly dividing cancer cells. Before Rosenberg could use his technique for blocking cell division, the exact chemical responsible had to be identified. This was not an easy task, because only small amounts were produced in the electric-field experiment. Given the presence of ammonia and chloride in their solutions, Rosenberg's group synthesized the known platinum compounds that contained ammonia and chlorine. A few showed the activity they sought, and *cis*-diamminedichloroplatinum(II), now called cisplatin, proved most effective. When its ability to block cell division was reported by Rosenberg in the *Journal of Biological Chemistry* (1967), cisplatin had been known for 122 years (it was originally prepared in 1845).

It is widely recognized that compounds containing heavy metals are toxic. Except for gold injections used in arthritis, and bismuth stomach medications, they have little medical use. But Rosenberg was persistent, and in 1969 his team at Michigan State University reported that cisplatin inhibits leukemia and solid cancerous tumors in mice. Several of the mice in his study appeared to be completely cured. Now Rosenberg had the ammunition to convince the National Cancer Institute to begin clinical trials of the drug. A successful Phase I trial was reported in 1974, and the drug passed Phase II and Phase III trials. Cisplatin (Platinol®) was distributed for general use in 1978. A derivative with reduced toxic side effects, called Paraplatin®, was released in 1985. Today they are among the most widely used anticancer drugs.

Chemotherapy

Chemotherapy is the treatment of cancer with chemicals. In many cases, chemotherapy is necessary because the cancerous tissue cannot be removed by surgery and the growth is resistant to radiation therapy. The purpose of treating cancer patients with chemotherapeutic agents is to prevent cancer cells from multiplying, invading, metastasizing, and ultimately killing the patient. Most chemicals used to treat cancer are designed to stop cell multiplication and tumor growth. However, because the human body is constantly producing new cells, chemotherapy has toxic effects on normal cells as well as on cancer cells. The cells most often affected by chemotherapeutic drugs are those that undergo fast growth. In addition to cancer cells, this generally includes healthy bone marrow, mucous membrane, and hair cells.

Chemotherapeutic agents interfere with cell multiplication and tumor growth in several ways. Many agents (bleomycin, cis-

368

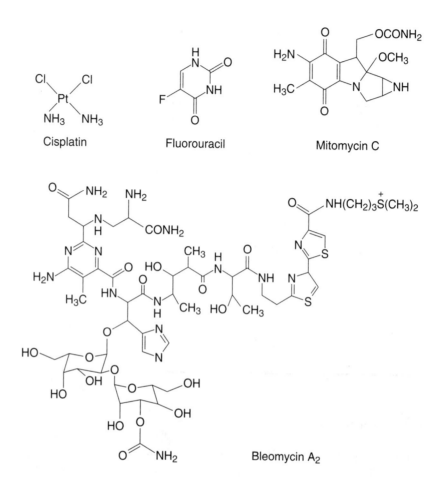

Figure 11.16.
Commonly used chemotherapy agents. Cisplatin and fluo-rouracil are synthetic agents. Mitomycin and bleomycin are natural products that are antitumor agents as well as antibiotics.

platin, carboplatin, mitomycin) react with the nucleic acids DNA and RNA. Others block metabolic pathways for the synthesis of needed proteins (asparaginase) or inhibit the ability of enzymes to carry out their functions within the cell (fluorouracil, methotrexate). Chemicals that stimulate the immune system (interferon-alpha, interleukin-2) are also used with the hope of initiating a response that inhibits tumor growth. For some drugs (androgens, progestins), the mechanism of action is not understood. The structures of several common chemotherapeutic agents are shown in Figure 11.16.

There are several chemical reactions that can occur between DNA and a chemotherapeutic drug. The binding of cisplatin to DNA is shown in Figure 11.17. Agents such as cisplatin bind to

DNA strands in a way that blocks DNA replication and prevents cell growth. Drugs such as bleomycin (an antibiotic) cleave strands of DNA, thereby destroying the genetic information stored in the DNA molecule.

Some chemotherapy drugs are designed to interfere with proteins and enzymes involved in the biochemical synthesis of DNA and RNA. These agents exert their effect by virtue of their structural or functional similarity to naturally occurring molecules involved in nucleic acid synthesis. Because the cells mistake these drugs for the naturally occurring molecules, they work either by inhibiting critical enzymes or by being incorporated into nucleic acids and thereby producing incorrect codes for protein synthesis. Both mechanisms result in the inhibition of DNA synthesis and cause cell death.

One enzyme that is commonly targeted by cancer drugs is dihydrofolate reductase (DHFR), an enzyme involved in the synthesis of thymidine nucleotide (one of the building blocks of DNA). To inhibit this enzyme from carrying out its function, a chemical is needed that can access and tie up the active site. To fool the enzyme and gain access to the active site, the drug must be structurally similar to the normal substrate, folic acid (vitamin B_9). Chemical analogs such as methotrexate are effective in inhibiting DHFR. The structures of folic acid and methotrexate are shown in Figure 11.18. Note the very subtle chemical differences between folic acid and methotrexate. One molecule is essential for sustaining healthy cells; the other is deadly.

Many chemicals are not active in their administered form but become powerful chemotherapeutic agents when metabolized. Fluorouracil (see Figure 11.16), is converted by enzymes into an active nucleotide. In that form, it inhibits the enzyme thymidylate synthetase, thereby blocking DNA synthesis.

Chemotherapy has many risks. In addition to the problem of toxicity, many chemotherapeutic drugs are themselves cancer-causing. High doses of these chemicals are often administered. In recent years, single-agent chemotherapy has been largely replaced by combination therapies. For example, physicians treating various gastrointestinal cancers administer a combination of cisplatin, bleomycin, and methotrexate on a three-week

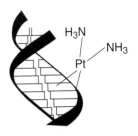

Figure 11.17. *Chemical reactions of cisplatin with DNA. The binding of the drug to two bases on a single strand prevents replication of the DNA, thereby killing the cell.*

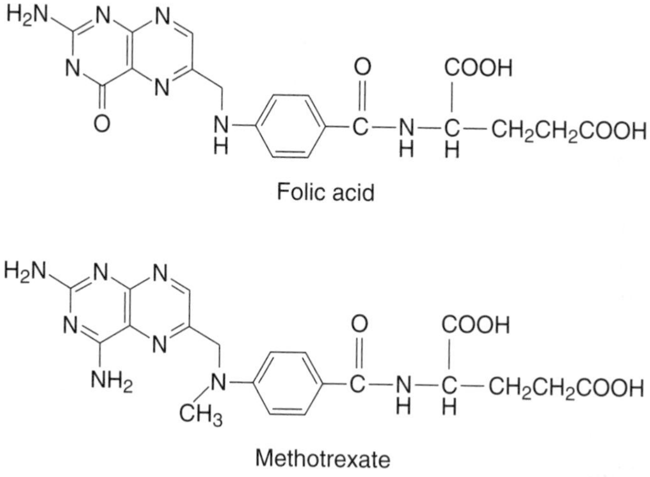

Figure 11.18. *The molecular structures of folic acid, the substrate for the enzyme DHFR, and methotrexate, a chemotherapy agent used to inhibit DHFR.*

cycle. Because of synergistic actions among these chemicals, lower doses of each compound can be given when they are used together instead of separately. This helps reduce harmful side effects, and also makes it difficult for the cancer cells to mutate and acquire resistance to the chemotherapy drug combination. Acquired resistance limits the effectiveness of many cancer drugs, including cispatin. Chemotherapy is an area where improved approaches are urgently needed.

Electronic Configurations of Atoms

Atomic No.	Element	1	2		3			4				5				6				7
							Principal Quantum Number of Electron Shell													
		s	s	p	s	p	d	s	p	d	f	s	p	d	f	s	p	d	f	s
1	H	1																		
2	He	2																		
3	Li	2	1																	
4	Be	2	2																	
5	B	2	2	1																
6	C	2	2	2																
7	N	2	2	3																
8	O	2	2	4																
9	F	2	2	5																
10	Ne	2	2	6																
11	Na	2	2	6	1															
12	Mg	2	2	6	2															
13	Al	2	2	6	2	1														
14	Si	2	2	6	2	2														
15	P	2	2	6	2	3														
16	S	2	2	6	2	4														
17	Cl	2	2	6	2	5														
18	Ar	2	2	6	2	6														
19	K	2	2	6	2	6*	..	1												
20	Ca	2	2	6	2	6	..	2												
21	Sc	2	2	6	2	6	1	2												

Atomic No.	Element	1	2		3			4				5				6				7
		s	s	p	s	p	d	s	p	d	f	s	p	d	f	s	p	d	f	s
22	Ti	2	2	6	2	6	2	2												
23	V	2	2	6	2	6	3	2												
24	Cr	2	2	6	2	6	5*	1												
25	Mn	2	2	6	2	6	5	2												
26	Fe	2	2	6	2	6	6	2												
27	Co	2	2	6	2	6	7	2												
28	Ni	2	2	6	2	6	8	2												
29	Cu	2	2	6	2	6	10*	1												
30	Zn	2	2	6	2	6	10	2												
31	Ga	2	2	6	2	6	10	2	1											
32	Ge	2	2	6	2	6	10	2	2											
33	As	2	2	6	2	6	10	2	3											
34	Se	2	2	6	2	6	10	2	4											
35	Br	2	2	6	2	6	10	2	5											
36	Kr	2	2	6	2	6	10	2	6											
37	Rb	2	2	6	2	6	10	2	6	..	..	1								
38	Sr	2	2	6	2	6	10	2	6	..	..	2								
39	Y	2	2	6	2	6	10	2	6	1*	..	2								
40	Zr	2	2	6	2	6	10	2	6	2	..	2								
41	Nb	2	2	6	2	6	10	2	6	4*	..	1								
42	Mo	2	2	6	2	6	10	2	6	5*	..	1								
43	Tc	2	2	6	2	6	10	2	6	6*	..	1								
44	Ru	2	2	6	2	6	10	2	6	7*	..	1								
45	Rh	2	2	6	2	6	10	2	6	8*	..	1								
46	Pd	2	2	6	2	6	10	2	6	10*	..	0								
47	Ag	2	2	6	2	6	10	2	6	10*	..	1								
48	Cd	2	2	6	2	6	10	2	6	10	..	2								
49	In	2	2	6	2	6	10	2	6	10	..	2	1							
50	Sn	2	2	6	2	6	10	2	6	10	..	2	2							
51	Sb	2	2	6	2	6	10	2	6	10	..	2	3							

		1	2		3			4				5				6				7
Atomic No.	**Element**	s	s	p	s	p	d	s	p	d	f	s	p	d	f	s	p	d	f	s
52	Te	2	2	6	2	6	10	2	6	10	..	2	4							
53	I	2	2	6	2	6	10	2	6	10	..	2	5							
54	Xe	2	2	6	2	6	10	2	6	10	..	2	6							
55	Cs	2	2	6	2	6	10	2	6	10	..	2	6	..	..	1				
56	Ba	2	2	6	2	6	10	2	6	10	..	2	6	..	..	2				
57	La	2	2	6	2	6	10	2	6	10	..	2	6	1*	..	2				
58	Ce	2	2	6	2	6	10	2	6	10	1	2	6	1*	..	2				
59	Pr	2	2	6	2	6	10	2	6	10	3	2	6	..	..	2				
60	Nd	2	2	6	2	6	10	2	6	10	4	2	6	..	..	2				
61	Pm	2	2	6	2	6	10	2	6	10	5	2	6	..	..	2				
62	Sm	2	2	6	2	6	10	2	6	10	6	2	6	..	..	2				
63	Eu	2	2	6	2	6	10	2	6	10	7	2	6	..	..	2				
64	Gd	2	2	6	2	6	10	2	6	10	7	2	6	1*	..	2				
65	Tb	2	2	6	2	6	10	2	6	10	9	2	6	..	..	2				
66	Dy	2	2	6	2	6	10	2	6	10	10	2	6	..	..	2				
67	Ho	2	2	6	2	6	10	2	6	10	11	2	6	..	..	2				
68	Er	2	2	6	2	6	10	2	6	10	12	2	6	..	..	2				
69	Tm	2	2	6	2	6	10	2	6	10	13	2	6	..	..	2				
70	Yb	2	2	6	2	6	10	2	6	10	14	2	6	..	..	2				
71	Lu	2	2	6	2	6	10	2	6	10	14	2	6	1	..	2				
72	Hf	2	2	6	2	6	10	2	6	10	14	2	6	2	..	2				
73	Ta	2	2	6	2	6	10	2	6	10	14	2	6	3	..	2				
74	W	2	2	6	2	6	10	2	6	10	14	2	6	4	..	2				
75	Re	2	2	6	2	6	10	2	6	10	14	2	6	5	..	2				
76	Os	2	2	6	2	6	10	2	6	10	14	2	6	6	..	2				
77	Ir	2	2	6	2	6	10	2	6	10	14	2	6	7	..	2				
78	Pt	2	2	6	2	6	10	2	6	10	14	2	6	9	..	1*				
79	Au	2	2	6	2	6	10	2	6	10	14	2	6	10	..	1*				
80	Hg	2	2	6	2	6	10	2	6	10	14	2	6	10	..	2				
81	Tl	2	2	6	2	6	10	2	6	10	14	2	6	10	..	2	1			

Principal Quantum Number of Electron Shell

Atomic No.	Element	Principal Quantum Number of Electron Shell																	
		1	2		3			4				5				6				7
		s	s	p	s	p	d	s	p	d	f	s	p	d	f	s	p	d	f	s
82	Pb	2	2	6	2	6	10	2	6	10	14	2	6	10	..	2	2			
83	Bi	2	2	6	2	6	10	2	6	10	14	2	6	10	..	2	3			
84	Po	2	2	6	2	6	10	2	6	10	14	2	6	10	..	2	4			
85	Pt	2	2	6	2	6	10	2	6	10	14	2	6	10	..	2	5			
86	Rn	2	2	6	2	6	10	2	6	10	14	2	6	10	..	2	6			
87	Fr	2	2	6	2	6	10	2	6	10	14	2	6	10	..	2	6	..	..	1
88	Ra	2	2	6	2	6	10	2	6	10	14	2	6	10	..	2	6	..	..	2
89	Ac	2	2	6	2	6	10	2	6	10	14	2	6	10	..	2	6	1	..	2
90	Th	2	2	6	2	6	10	2	6	10	14	2	6	10	..	2	6	2	..	2
91	Pa	2	2	6	2	6	10	2	6	10	14	2	6	10	2	2	6	1*	..	2
92	U	2	2	6	2	6	10	2	6	10	14	2	6	10	3	2	6	1*	..	2
93	Np	2	2	6	2	6	10	2	6	10	14	2	6	10	4	2	6	1*	..	2
94	Pu	2	2	6	2	6	10	2	6	10	14	2	6	10	6	2	6	..	..	2
95	Am	2	2	6	2	6	10	2	6	10	14	2	6	10	7	2	6	..	..	2
96	Cm	2	2	6	2	6	10	2	6	10	14	2	6	10	7	2	6	1*	..	2
97	Bk	2	2	6	2	6	10	2	6	10	14	2	6	10	9*	2	6	..	..	2
98	Cf	2	2	6	2	6	10	2	6	10	14	2	6	10	10	2	6	..	..	2
99	Es	2	2	6	2	6	10	2	6	10	14	2	6	10	11	2	6	..	..	2
100	Fm	2	2	6	2	6	10	2	6	10	14	2	6	10	12	2	6	..	..	2
101	Md	2	2	6	2	6	10	2	6	10	14	2	6	10	13	2	6	..	..	2
102	No	2	2	6	2	6	10	2	6	10	14	2	6	10	14	2	6	..	..	2
103	Lr	2	2	6	2	6	10	2	6	10	14	2	6	10	14	2	6	1	..	2
104	Rf	2	2	6	2	6	10	2	6	10	14	2	6	10	14	2	6	2	..	2

*Note irregularity in the filling pattern.

Lists of Proven and Suspected Human Carcinogens

A. California OSHA (Occupational Safety and Health Administration) Regulated Carcinogens

2-Acetylaminofluorene
4-Aminodiphenyl
Benzidine and its salts
3,3′-Dichlorobenzidine and its salts
4-Dimethylaminoazobenzene
alpha-Naphthylamine
beta-Naphthylamine
4-Nitrobiphenyl
N-Nitrosodimethylamine
beta-Propiolactone
Bis(chloromethyl) ether
Methyl chloromethyl ether
Ethyleneimine
Asbestos
Vinyl chloride
Coke oven emissions
Dibromochloropropane
Acrylonitrile
Inorganic arsenic
MBOCA (methylene bis(2-chloroaniline))
Formaldehyde
Benzene
Ethylene dibromide
Ethylene oxide

B. IARC (International Agency for Research on Cancer) Identified Carcinogens

1. Group 1. The Working Group concluded that the following agents are carcinogenic to humans.

Aflatoxins production
4-Aminobiphenyl
Analgesic mixtures containing phenacetin
Arsenic and arsenic compounds
Asbestos
Auramine, manufacture of
Azathioprine
Benzene
Benzidine
Betel quid with tobacco
N,N-bis(2-chloroethyl)-2-naphthylamine
    (Chlornaphazine)
Bis(chloromethyl)ether and chloromethyl methyl
    ether (technical grade)
Boot and shoe manufacture and repair
1,4-Butanediol dimethanesulfonate (Myleran)
Chlorambucil
l-(2-Chloroethyl)-3(4-methylcyclohexyl)-l-
    nitrosourea (Methyl-CCNU)
Chromium compounds, hexavalent
Coal gasification
Coal-tar pitches

375

Coal tars
Coke production
Cyclophosphamide
Diethylstibestrol
Erionite
Estrogen replacement therapy
Estrogens, nonsteroidal
Estrogens, steroidal
Furniture and cabinet making
Hematite mining, underground, with exposure to
   radon
Iron and steel founding
Isopropyl alcohol manufacture, strong-acid
   process
Magenta, manufacture of
Melphalan
8-Methoxypsoralen (Methoxsalen) plus ultraviolet
   radiation
Mineral oils untreated and mildly treated
MOPP (combined therapy with nitrogen mustard,
   vincristine, procarbazine,and prednisone) and
   other combined chemotherapy including alky-
   lating agents
Mustard gas (sulfur mustard)
2-Naphthylamine
Nickel and nickel compounds
Oral contraceptives, combined
Oral contraceptives, sequential
The rubber industry
Shale oils
Soots
Talc-containing asbestiform fibers
Tobacco products, smokeless
Tobacco smoke
Treosulphan vinyl chloride

2. Group 2A. *The Working Group concluded that the following agents are probably carcinogenic to humans.*

Acrylonitrile
Adriamycin
Androgenic (anabolic) steroids
Benz[a]anthracene
Benzidine-based dyes
Benzo[a]pyrene

Beryllium and beryllium compounds
Bis(chloroethyl) nitrosourea (BCNU)
Cadmium and cadmium compounds
1(2-Chloroethyl)-3-cyclohexyl-l-nitrosourea
   (CCNU)
Cisplatin
Creosotes
Dibenz[a,h]anthracene
Diethyl sulfate
Dimethylcarbamoyl chloride
Dimethyl sulfate
Epichlorohydrin
Ethylene dibromide
Ethylene oxide
N-Ethyl-N-nitrosourea
Formaldehyde
5-Methoxypsoralen
4,4'-Methylene bis(2-chloroaniline) (MBOCA)
N-Methyl-N'nitro-N-nitrosoguanidine (MNNG)
N-Methyl-N-nitrosourea
Nitrogen mustard
N-Nitrosodiethylamine
N-Nitrosodimethylamine
Phenacetin
Polychlorinated biphenyls
Procarbazine hydrochloride
Propylene oxide
Silica, crystalline
Styrene oxide
Tris(1-aziridinyl)phosphine sulphide (Thiotepa)
Tris(2,3-dibromopropyl) phosphate
Vinyl bromide

3. Group 2B. *The Working Group concluded that the following agents are possibly carcinogenic to humans.*

A-alpha, C (2-Amino-9H-pyrido[2,3-b]indole)
Acetaldehyde
Acetamide
Acrylamide AF-2[2{2-Furyl}-3-(5-nitro-2-
   furyl)acrylamide]
para-Aminoazobenzene
ortho-Aminoazotoluene
2-Amino-5-(5-nitro-2-furyl)-1,3,4-thiadiazole
Amitrole

ortho-Anisidine
Aramite
Auramine, technical grade
Azaserine
Benzo[b]fluoranthene
Benzo(j)fluoranthene
Benzo[k]fluoranthene
Benzyl violet 4B
Bitumens, extracts of stream-refined and air-
 refined
Bleomycins
Bracken fern
1,3-Butadiene
Butylated hydroxyanisole (BHA)
beta-Butyrolactone
Carbon-black extracts
Carbon tetrachloride
Carpentry and joinery
Carrageenan, degraded
Chloramphenicol
Chlordecone (Kepone)
alpha-Chlorinated toluenes
Chloroform
Chlorophenols
Chlorophenoxy herbicides
4-Chloro-*ortho*-phenylenediamine
para-Chloro-*ortho*-toluidine
Citrus Red No. 2
para-Cresidine
Cycasin
Dacarbazine
Daunomycin
DDT
N,N′-Diacetylbenzidine
2,4-Diaminoanisole
4,4′-Diaminodiphenyl ether
2,4-Diaminotoluene
Dibenz[a,h]acridine
Dibenz[a,j]acridine
7H-Dibenzo[c,g]carbazole
Dibenzo[a,e]pyrene
Dibenzo[a,h]pyrene
Dibenzo[a,l]pyrene
1,2-Dibromo-3-chloropropane
para-Dichlorobenzene

3,3′-Dichlorobenzidine
3,3′-Dichloro-4,4′-diaminodiphenyl ether
1,2-Dichloroethane
Dichloromethane
1,3-Dichloropropene (technical-grade)
Diepoxybutane
Di(2-ethylhexyl)phthalate
1,2-Diethylhydrazine
Diglycidyl resorcinol ether
Dihydrosafrole
3,3′-Dimethoxybenzidine (*ortho*-Dianisidine)
para-Dimethylaminoazobenzene
trans-2[(Dimethylamino)methylimino]-
 5[2(5-nitro-2-furyl)vinyl]-1,3,4-oxadiazole
3,3′-Dimethylbenzidine (*ortho*-Tolidine)
1,1-Dimethylhydrazine
1,2-Dimethylhydrazine
1,4-Dioxane
Ethyl acrylate
Ethylene thiourea
Ethyl methanesulfonate 2,2-formylhydrazino)-
 4,5-nitro-2-furyl)thiazole
Glu-P-1 (2-Amino-6-methyldipyrido[1,2-a:3′,2′-
 d]imidazole)
Glu-P-2 (2-Aminodipyrido[1,2-a:3′,2′-d]imidazole)
Glycidaldehyde
Griseofulvin
Hexachlorobenzene
Hexachlorocyclohexanes
Hexamethylphosphoramide
Hydrazine
Indeno[1,2,3-cd]pyrene
IQ (2-Amino-3-methylimidazo[4,5-f]quinoline)
Iron–dextran complex
Lasiocarpine
Lead and lead compounds, inorganic
MeA-*alpha*-C (2-amino-3-methyl-9H-pyrido[2,3-
 b]indole)
Medroxyprogesterone acetate
Merphalan
2-Methylaziridine
Methylazoxymethanol and its acetate
5-Methylchrysene
4,4′-Methylene bis(2-methylaniline)
4,4′-Methylenedianiline

Methyl methanesulfonate
2-Methyl-1-nitroanthraquinone (uncertain purity)
N-Methyl-N-nitrosourethane
Methylthiouracil
Metronidazole
Mirex
Mitomycin C
Monocrotaline
5-(Morpholinomethyl)-3-
  [(5-nitrofurfurylidene)amino]-2-oxazolidinone
Nafenopin
Niridazole
5-Nitroacenaphthene
Nitrofen (technical grade)
1-[(5-Nitrofurfurylidene)amino-2-imidazolidinone
N-N-[4(5-Nitro-2-furyl)-2-thiazoly]acetamide
Nitrogen mustard N-oxide
2-Nitropropane
N-Nitrosodi-*n*-butylamine
N-Nitrosodiethanolamine
N-Nitrosodi-*n*-propylamine
3-(N-Nitrosomethylamino)propionitrile
4-(N-Nitrosomethylamino)-1(3-pyridyl)-1-
  butanone (NNK)
N-Nitrosomethylethylamine
N-Nitrosomethylvinylamine
N-Nitrosomorpholine
N'-Nitrosonornicotine
N-Nitrosopiperdine
N-Nitrosopyrrolidine
N-Nitrososarcosine
Oil Orange SS
Panfuran S (containing dihydroxymethy-
  furatrizine)
Phenazopyridine hydrochloride
Phenobarbital
Phenoxybenzamine hydrochloride
Phenytoin
Polybrominated biphenyls
Ponceau MX
Ponceau 3R
Potassium bromate
Progestins
1,3-Propane sulfone
B-Propiolactone

Propylthiouracil
Saccharin
Safrole
Sodium *ortho*-phenylphenate
Sterigmatocystin
Streptozotocin
Styrene
Sulfallate herbicides
2,3,7,8-Tetrachlorodibenzo-para-dioxin (TCDD)
Tetrachloroethylene
Thioacetamide
4,4'-Thiodianiline
Thiourea
Toluene diisocyanates
ortho-Toluidine
Toxaphene (polychlorinated camphenes)
Trp-P-1 (3-Amino-1,4-dimethyl-5H-
  pyrido[4,3-b]indole)
Trp-P-2 (3-Amino-1-methyl-5H-pyrido[4,3-
  b]indole)
Trypan blue
Uracil mustard
Urethane

C. NTP (National Toxicological Program) Identified Carcinogens

1. *Substances Known to Be Carcinogens*

4-Aminobiphenyl
Analgestic mixtures containing phenacetin
Arsenic and certain arsenic compounds
Asbestos
Azathioprine
Benzene
Benzidine
Bis(chloromethyl)ether and technical grade
  chloromethyl methyl ether
1,4-Butanediol dimethylsulfonate (Myeran)
Chlorambucil
Chromium and certain chromium compounds
Conjugated estrogens
Cyclophosphamide
Diethylstibestrol
Melphalan
Methoxsalen with ultraviolet A therapy (PUVA)

Mustard gas
2-Naphthylamine
Thorium dioxide
Vinyl chloride

2. Substances Reasonably Anticipated to be Carcinogens

2-Acetylaminofluorene
Acrylonitrile
Adriamycin
Aflatoxins
2-Aminoanthraquinone
o-Aminoazotoluene
1-Amino-2-methylanthraquinone
Amitrole
o-Anisidine hydrochloride
Benzotrichloride
Beryllium and certain beryllium compounds
Bis(chloroethyl) nitrosourea
1,3,-Butadiene
Cadmium and certain cadmium compounds
Carbon tetrachloride
Chlorendic acid
Chlorinated paraffins (C12 60% chlorine)
1-(2-Chloroethyl)-3-cyclohexyl-1-nitrosourea
    (CCNU)
Chloroform
3-Chloro-2-methylpropene
4-Chloro-o-phenylenediamine
C.1. Basic Red 9 monohydrochloride
p-Cresidine
Cupferron
Dacarbazine
DDT
2,4-Diaminoanisole sulfate
2,4-Diaminotoluene
1,2-Dibromo-3-chloropropane
1,2-Dibromoethane (EDB)
1,4-Dichlorobenzene
3,3'-Dichlorobenzidine and 3,3'-dichlorobenzidine
    dihydrochloride
1,2-Dichloroethane
Dichloromethane (methylene chloride)
1,3-Dichloropropene (technical grade)
Diepoxybutane

Di(2-ethylhexyl)phthalate
Diethyl sulfate
Diglycidyl resorcinol ether
3,3'-Dimethoxybenzidine
4-Dimethylaminoazobenzene
3,3'-Dimethylbenzidine
Dimethylcarbamoyl chloride
1,1-Dimethylhydrazine
Dimethylvinyl chloride
1,4-Dioxane
Direct Black 38
Direct Blue 6
Epichlorohydrin
Estrogens (not conjugated): Estradiol-17(beta)
Estrogens (not conjugated): Estrone
Estrogens (not conjugated): Ethinylestradiol
Estrogens (not conjugated): Mestranol
Ethyl acrylate
Ethylene oxide
Ethylene thiourea
Formaldehyde (gas)
Hexachlorobenzene
Hexamethylphosphoramide
Hydrazine and hydrazine/ sulfate
Hydrazobenzene
Iron–dextran complex
Kepone (Chlordecone)
Lead acetate and lead phosphate
Lindane and other hexachlorocyclohexane
    isomers
2-Methylaziridine (Propyleneimine)
4,4'-Methylene bis(2-chloroaniline) (MBOCA)
4,4'-Methylene bis(N,N-dimethyl)benzenamine
4,4'-Methylene dianiline and its dihydrochloride
Metronidazole
Michler's ketone
Mirex
Nickel and certain nickel compounds
Nitrilotriacetic acid
5-Nitro-o-anisidine
Nitrofen
Nitrogen mustard hydrochloride
2-Nitropropane
N-Nitrosodi-n-butylamine
N-Nitrosodiethanolamine

N-Nitrosodiethylamine

N-Nitrosodimethylamine

p-Nitrosodiphenylamine

N-Nitrosodi-n-propylamine

N-Nitroso-N-ethylurea

N-Nitroso-N-methylurea

N-Nitrosomethylvinylamine

N-Nitrosomorpholine

N-Nitrosonornicotine

N-Nitrosopiperidine

N-Nitrosopyrrolidine

N-Nitrososarcosine

Norethisterone

4,4′-oxydianiline

Oxymetholone

Phenacetin

Phenazopyridine hydrochloride

Phenoxybenzamine hydrochloride

Phenytoin

Polybrominated biphenyls

Polychlorinated biphenyls

Polycyclic aromatic hydrocarbons

  Benz[a]anthracene

  Benzo[b]fluoranthene

  Benzo[j]fluoranthene

  Benzo[k]fluoranthene

  Benzo[a]pyrene

  Dibenz[a,h]acridine

  Dibenz[a,j]acridine

  Dibenz[a,h]anthracene

7H-Dibenzo[c,g]carbazole

Dibenzo[a,e]pyrene

Dibenzo[a,h]pyrene

Dibenzo[a,i]pyrene

Dibenzo[a,l]pyrene

Indeno[1,2,3-cd]pyrene 5-Methylchrysene

Procarbazine hydrochloride

Progesterone

1,3-Propane sulfone

(beta)-Propiolactone

Propylene oxide

Propylthiouracil

Reserpine

Saccharin

Safrole

Selenium sulfide

Streptozotocin

Sulfallate herbicides

2,3,7,8-Tetrachlorodibenzo-p-dioxin (TCDD)

Tetrachloroethylene (Perchloroethylene)

Thioacetamide

Thiourea

Toluene diisocyanate

o-Toluidine and o-toluidine hydrochloride

Toxaphene

2,4,6-Trichlorophenol

Tris(l-aziridinyl)phosphine sulfide

Tris(2,3-dibromopropyl)phosphate

Urethane

Glossary

A

acetaldehyde: CH_3CHO, industrial chemical, product of ethanol metabolism in humans.

acetaminophen: analgesic found in over-the-counter pain relievers such as Tylenol®.

acetic acid: CH_3COOH, weak acid, sour component in vinegar, an industrial solvent.

acetylcholine: an organic molecule that plays an important role in the transmission of nerve impulses.

acetylene: $HC{\equiv}CH$, a gas that forms highly explosive mixtures in air. Used as a component in the oxy-acetylene welding torch.

acetylsalicylic acid: aspirin, analgesic made from phenol.

acid: a molecule that releases protons (H^+) when dissolved in water.

acidic solution: a solution that contains an acid.

acidity: the behavior exhibited by an acid.

acid rain: precipitation that is acidic. The acidity results from reactions between water droplets and atmospheric sulfur dioxide (producing sulfuric acid) and nitrogen dioxide (producing nitric acid).

acrolein: $CH_2{=}CHCHO$, an irritating component of cigarette smoke, a poisonous volatile liquid.

acrylics: a class of synthetic polymers and plastics.

activated charcoal: a highly porous and adsorbent form of carbon, used in the purification of white cane sugar and as an absorbent in gas masks.

activation energy: the energy barrier that must be overcome for a chemical reaction to occur.

acyclovir: an organic drug used to treat herpes.

addition polymerization: the formation of a polymer by sequential addition of molecules.

adenine: one of the base molecules found in DNA and RNA. In helical DNA, it is paired with thymine. In helical RNA, it is paired with uracil.

adenosine: the adenine-ribose nucleoside.

adrenaline: hormone responsible for flight-or-fight response in mammals. Elevates blood pressure.

adsorption (molecular): the binding of a molecule to the surface of a solid.

aerosol: a fine mist of solid or liquid particles suspended in air.

aflatoxin B_1: a toxic molecule and suspected carcinogen produced by molds in many foods.

agent GA: a highly toxic nerve gas.

agent GB: a highly toxic nerve gas.

Agent Orange: a mixture of the herbicides 2,4-D and 2,4,5-T that was used as a defoliant during the Vietnam War. In 1985 the United States Environmental Protection Agency terminated all registered use of 2,4,5-T on rice fields, orchards, sugarcane, range land, and other non-crop sites as a result of the problems American veterans experienced from exposure to Agent Orange. However, it is still used in evergreen forests where hardwood trees are grown for commercial use.

alabaster: $CaSO_4{\cdot}2H_2O$, a fine-grained form of gypsum resembling marble.

alanine: one of the 20 amino acids used to construct proteins.

alar: herbicide that prevents ripe apples from falling off the tree. As a result of public outcries in 1990, it was phased out by the United States Food and Drug Administration. Political pressure forced a voluntary withdrawal before the official date.

alcohol dehydrogenase: an enzyme that breaks down alcohols.

alkali: a substance that behaves like a base.

alkali metals: the name given to metals in the column of the periodic table that includes lithium (Li), sodium (Na), potassium (K), rubidium (Rb), and cesium (Cs).

alkaline earth metals: the name given to the metals in the column of of the periodic table that includes beryllium (Be), magnesium (Mg), calcium (Ca), strontium (Sr), and barium (Ba).

alkaline solution: a solution containing a basic substance.

alkaloid: a nitrogen-containing organic compound found in plants that is basic, tastes bitter, and is often poisonous (examples include cocaine and quinine).

allergy: discomfort associated with the body's immune response.

alloy: a solid solution formed by melting a mixture of two or more metals and then cooling it.

alnico: a magnetic alloy of aluminum, nickel, and cobalt.

α-decay: radioactive decay process wherein the nucleus of an element ejects a helium nucleus.

α-helix: a coiled rod-like structure formed by segments of an amino acid chain in the folded three-dimensional structure of a protein.

α-particle: a helium nucleus (two protons and two neutrons) that is emitted from atomic nuclei undergoing α-decay.

alumina: Al_2O_3, high-melting oxide of aluminum. Used in ceramics and also as an adsorbent.

aluminosilicate: a mineral that contains aluminum, silicon, and oxygen.

aluminum: Al, the 13th element of the periodic table, common structural metal

aluminum oxide: product of the reaction between oxygen and aluminum metal; another name for alumina.

amino acid: $NH_2-CHR-COOH$, name of a general class of 20 organic molecules that serve as the building blocks of proteins.

ammonia: NH_3, a basic chemical and the main nitrogen source for fertilizers.

ammonium: NH_4^+, protonated form of ammonia.

ammonium nitrate: NH_4NO_3, an ionic salt that is a widely used fertilizer and a high explosive used in mining.

ammonium perchlorate: NH_4ClO_4, an ionic salt, a high explosive used, partially diluted, as the solid fuel in the space shuttle.

ammonium sulfate: $(NH_4)_2SO_4$, a common fertilizer.

amoxicillin: synthetic derivative of the antibiotic penicillin.

anabolic steroid: a synthetic hormone that aids in tissue building.

anerobic: refers to a process, usually in living organisms, that occurs in the absence of oxygen.

anethole: the molecule that gives licorice its fragrance.

angel dust: PCP, phenylcyclidine, an anesthetic that causes psychological disorders.

angstrom: 0.0000000001 meter, a unit of distance commonly used to describe matter on the atomic scale.

anion: a negatively charged atom or molecule.

annealing: the slow cooling of a heated material, used to reduce strain.

anode: the electrode where a chemical is oxidized (loses electrons to the electrode).

anthropogenic: arising from human activity.

antibiotic: a drug that kills bacteria.

antibody: a protein that attacks foreign substances that enter a living creature. An important molecule in the human immune response.

antifreeze: a mixture of ethylene glycol and water that has a low melting point and a high boiling point compared to pure water.

antihistamine: a drug that blocks the allergic inflammatory response.

antimony: Sb, the 51st element of the periodic table.

antioxidant: a molecule that blocks oxidation (an example is vitamin E).

aqua regia: a mixture of 3 parts hydrochloric acid and 1 part nitric acid. A highly corrosive liquid that will dissolve gold and platinum.

aquo (aq): abbreviation for water.

arginine: one of the 20 amino acids used to construct proteins.

argon: Ar, the 18th element of the periodic table, an inert gas and trace component of the atmosphere.

arsenic: As, the 33rd element of the periodic table, a poisonous heavy metal.

asparagine: one of the 20 amino acids used to construct proteins.

aspartic acid: one of the 20 amino acids used to construct proteins.

aspirin: acetylsalicylic acid, analgesic made from phenol.

atmosphere: the collection of gaseous chemicals that surround the Earth, consists primarily of water vapor, nitrogen, and oxygen.

atom: the fundamental chemical building block of an element.

atomic mass unit (amu): 1 amu is defined as the mass of the proton.

atomic number: the number of protons in the nucleus of an atom.

atomic weight: mass in grams of a mole of atoms of an element.

ATP: adenosine triphosphate, the energy currency of many biochemical reactions. Its conversion to adenosine diphosphate (ADP) releases energy, which is used to drive reactions in the cell.

azidodeoxythymidine (AZT): a drug that blocks DNA synthesis and is used to treat AIDS.

B

baking powder: a mixture of baking soda and a weak acid that causes rapid release of carbon dioxide when dissolved in water.

baking soda: $NaHCO_3$, sodium bicarbonate, a weak base used to neutralize stomach acid; releases carbon dioxide gas on protonation, which causes baked goods to rise.

barium: Ba, the 56th element of the periodic table, a poisonous heavy metal.

base: a molecule that accepts protons and neutralizes an acid. Bases generally have a bitter taste and are slippery to the touch.

basic solution: a solution that a contains a base.

battery: a device that converts chemical energy into electricity.

benzene: C_6H_6, the simplest aromatic compound.

benzene ring: a six-membered carbon ring in a molecular structure with alternating carbon–carbon double bonds.

benzocaine: topical anesthetic.

benzopyrene: potent human carcinogen found in smoke (such as cigarette smoke).

benzoyl peroxide: oxidizing or bleaching agent, used to control acne.

beryllium: Be, the 4th element of the periodic table, found in emeralds.

β-carotene: an antioxidant and precursor to vitamin A, found in carrots.

β-decay: radioactive decay process wherein the nucleus of an element ejects an electron.

β-particle: an electron ejected from an atomic nucleus during β-decay.

β-pleated sheet: a sheet-like structure formed by segments of an amino acid chain in the folded three-dimensional structure of a protein. Adjacent strands of amino acids run in opposite directions to create a flat region.

bicarbonate ion: HCO_3^-, ion found in baking powder and baking soda.

biochemicals: molecules that participate in the chemical reactions of living systems.

biochemistry: study of the chemical processes of living systems.

bioluminescence: biochemical reaction wherein chemical energy is converted into light. The light observed in the tail of a firefly is an example of bioluminescence.

bismuth: Bi, the 83rd element of the periodic table, component of Wood's metal and used in medications for the treatment of ulcers and indigestion.

blast furnace: furnace that uses a forced air flow to enhance combustion.

bleach: chemical that oxidizes and removes colored impurities. Household liquid bleach is a solution of sodium hypochlorite (NaOCl) in water.

bleaching powder: a solid bleaching agent, such as $Ca(OCl)_2$.

blood/brain barrier: a biochemical fatty membrane that protects the brain from entry of foreign chemicals but is permeable to some nonpolar compounds, such as anesthetics.

boiling point: the temperature at which the pressure of the vapor emitted by a liquid equals the pressure of the atmosphere.

bond: attractive force between the electrons and nuclei on two atoms, holding the two atoms together.

bond order: the number of electron pairs in a chemical bond that bind two atoms together.

bond strength: the energy needed to break a chemical bond.

borax: $Na_2B_4O_7 \cdot 10H_2O$, sodium tetraborate, used to make Pyrex glass.

boric acid: H_3BO_3, a weak acid used in water solution as an eye wash.

boron: B, the 5th element of the periodic table, a nonmetal found in borax mineral.

boron composite: a high-strength material made of epoxy and embedded fibers of boron.

brass: an alloy of copper and zinc.

breeder reactor: a nuclear reactor that uses plutonium and uranium as fuel and produces more plutonium than it consumes.

brimstone: biblical name for yellow sulfur.

British thermal unit (Btu): a measure of the energy needed to heat 1 pound of water 1°F. Equals 252 calories.

bromine: Br, the 35th element of the periodic table, exists naturally as Br_2 molecules, a brown liquid with noxious vapors.

bronze: an alloy of copper, tin, and zinc.

Brownian motion: the random path that a very small particle experiences, because of random collisions with individual molecules, when suspended in a liquid or gas.

buckminsterfullerene: a name given to the C_{60} molecule, whose structure is similar to the geodesic domes designed by R. Buckminster Fuller.

buckyball: a name given to the C_{60} molecule.

butadiene: C_4H_6, a molecule that contains two carbon–carbon double bonds and can be polymerized to make synthetic rubber.

butylated hydroxytoluene (BHT): an antioxidant used as a food preservative.

C

C_{60}: a molecule that consists of 60 carbon atoms joined in an approximately spherical (soccer ball) shape.

cadmium: Cd, the 48th element of the periodic table, the metal anode in the NiCad battery, and a toxic heavy metal found in polluted soils.

caffeine: a molecule found in colas, tea, and coffee that acts as a stimulant and diuretic.

calcite: $CaCO_3$, a crystalline form of calcium carbonate.

calcium: Ca, the 20th element of the periodic table, essential nutrient found in milk and vegetables.

calcium oxide: CaO, lime, used in the manufacture of cement and steel, as well as in water purification.

calcium phosphate: $Ca_3(PO_4)_2$, primary component in phosphate rock and primary mineral component in bones and teeth.

calcium propionate: an antimold food preservative.

calomel: slang for mercurous chloride, Hg_2Cl_2. Formerly used as a laxative and antisyphilitic. Leads to mercury poisoning.

calorie: amount of energy needed to heat 1 gram of water 1°C. A nutritional Calorie is 1000 of the chemist's calories.

carat alloy (gold): measure of the gold content by weight of an alloy, where 24 carat represents pure gold.

carbohydrate: an organic molecule (sugar, starch, and cellulose) that contains only carbon, hydrogen, and oxygen and has the general formula $(CH_2O)_n$.

carbon: C, the 6th element of the periodic table, found in all organic compounds. The element can exist as charcoal, graphite, C_{60}, and diamond.

carbon-14 dating: the technique of measuring the age of an object by determining the amount of radioactive carbon in it.

carbonate: CO_3^{2-}, a basic anion found in limestone.

carbonated water: water that contains dissolved carbon dioxide gas.

carbon black: finely divided soot used in inks and tires.

carbon cycle: a model for the movement of carbon in the environment.

carbon dioxide: CO_2, a greenhouse gas produced by combustion and respiration and used in photosynthesis.

carbon monoxide: CO, a poisonous gas found in cigarette smoke and automobile exhaust.

carborundum: slang for silicon carbide, SiC, an abrasive.

carboxypeptidase A: a digestive enzyme.

carcinogen: a substance that causes cancer.

cast iron: crude, brittle form of iron with a high carbon content.

catalysis: the speeding up of a chemical reaction by addition of a catalyst.

catalyst: a chemical that accelerates the rate of a reaction without being consumed.

catalytic converter: a device that is attached to cars to ensure complete oxidation of hydrocarbon fuels to carbon dioxide. These devices are partially effective at controlling nitric oxide emissions, thus reducing photochemical smog.

cathode: the electrode where a chemical is reduced (accepts electrons from the electrode).

cation: a positively charged atom or molecule.

caustic potash: slang for potassium hydroxide, KOH.

caustic soda: slang for sodium hydroxide, NaOH.

cell: the smallest structural unit of a living organism that is capable of functioning independently.

cellulose: a woody structural component of plants, a polymer of the sugar glucose.

Celsius: a temperature scale used in the metric system. Water freezes at 0°C and boils at 100°C.

centi-: a prefix meaning one hundredth.

centimeter: 0.01 meter, one hundredth of a meter.

ceramic: a brittle, inorganic solid with a high melting point.

CFC: a chlorofluorocarbon, a compound that contains carbon, chlorine, and fluorine. CFCs are used in refrigeration systems and as industrial solvents. These highly stable molecules migrate up to the stratosphere where they catalyze ozone destruction.

chalcogens: elements in the column of the periodic table that includes oxygen (O), sulfur (S), selenium (Se), tellurium (Te), and polonium (Po).

chalk: slang for calcium carbonate, $CaCO_3$.

charcoal: carbon residue that remains when plant or animal materials are heated in the absence of oxygen.

chemical bond: attractive force between the electrons and nuclei on two atoms, holding the two atoms together.

chemical energy: the energy that is stored in chemical bonds and can be released in reactions.

chemiluminescence: chemical reaction wherein chemical energy is converted into light. Retail light sticks are an example of chemiluminescence.

chemotherapy: the treatment of cancer with chemicals.

Chernobyl: location of a nuclear reactor in the former Soviet Union. In April 1986, an explosion at the site blew the roof off the building and released large amounts of radioactive materials into the atmosphere.

chloral hydrate: "knock-out drops," an anesthetic.

chloramine: a volatile poison formed when household bleach and ammonia cleansers are mixed.

chloride ion: Cl^-, a common form of chlorine found in ionic salts.

chlorinated drinking water: drinking water purified with chlorine gas as a disinfectant.

chlorine: Cl, the 17th element of the periodic table, exists naturally as Cl_2 molecules, a reactive green gas used to disinfect water and used as a chemical weapon in World War I.

2-chloroacetophenone: MACE or tear gas.

chlorofluorocarbon: *see* CFCs.

chloroform: $CHCl_3$ a toxic organic solvent with anesthetic properties.

chlorophyll: an organic molecule that is essential for plant photosynthesis.

cholesterol: a steroid that is manufactured in the liver and intestines.

chrome yellow: slang for lead chromate, $PbCrO_4$, a yellow paint pigment.

chromium: Cr, the 24th element of the periodic table, a metal added to iron to make stainless steel.

chymotrypsin: a digestive enzyme found in the small intestines.

cinnabar: HgS, mercuric sulfide mineral.

cinnamaldehyde: chemical responsible for the fragrance of cinnamon.

cisplatin: a widely used anticancer drug containing platinum.

citric acid: weak acid found in citrus fruits.

clay: a plastic material that contains aluminum, silicon, oxygen, and hydrogen.

Clean Air Act: legislation passed by the U.S. government in 1970 to regulate emissions of various pollutants into the atmosphere. This law has been modified several times during the past two decades to require further reductions in the emissions of specific molecules from automobile engines.

cobalt: Co, the 27th element of the periodic table. A metallic element and essential trace nutrient.

cocaine: an anesthetic and addictive drug isolated from the leaves of the coca plant.

codeine: an addictive narcotic drug used as a cough suppressant and pain killer.

coke: slang for cocaine, but to the chemist coke is a form of carbon made by heating coal in the absence of oxygen. Used industrially as a reducing agent.

collagen: a protein that makes up human skin. This protein is also used in preparing the gelatin for photographic film.

colloid: particles approaching molecular dimensions suspended in a liquid or a gas.

color: a property of light that is determined by its wavelength.

combustion: the heat-releasing reaction between oxygen in air and a substance. The combustion of fossil fuels and other organic compounds produces carbon dioxide and water.

composite: a material that consists of a plastic with embedded fiber particles.

computer-assisted tomography (CAT): computer-generation of a three-dimensional X-ray image from a series of two-dimensional X-ray photographs, commonly used in medical diagnosis.

concentration: a measure of the number of molecules of a substance dissolved in a fixed volume of solvent. Chemists usually specify concentration in moles of solute present in one liter of solution.

condensation polymerization: polymerization reaction that occurs when the addition of each monomer to the polymer chain also produces a small byproduct molecule such as water.

conduction band: an energy level in which electrons freely move throughout a solid. Generally the conduction band of a semiconductor is sparsely populated, so its conductivity is lower than that of a metal.

conductivity: quantitative measure of how well a material conducts electricity. Usually measured in units of mho (ohm^{-1}).

cone: an eye cell that provides humans with the ability to differentiate colors.

conservation of charge: the physical principle that charge cannot be created or destroyed.

conservation of mass: the physical principle that mass cannot be created or destroyed during a chemical reaction. Einstein demonstrated that this rule could be violated (in nuclear reactions a detectable loss in mass occurs, and a large amount of energy is produced).

copolymer: a polymer that is built from two different monomers.

copper: Cu, the 29th element of the periodic table, a soft, gold-colored metal of high electrical and thermal conductivity.

cortisol: a steroid hormone that influences the metabolism of carbohydrates and proteins in the body.

cortisone: a steroid used to reduce inflammation.

cosmic rays: high-energy particles and gamma rays that bombard the Earth from outer space.

coupler molecule: a chemical contained in photographic film that forms a bond with the developer to produce a colored molecule. Different coupler molecules create different colors, making possible the development of color negatives and the printing of color pictures.

covalent bond: a chemical bond wherein the bonding electrons are shared equally between the two atoms.

covalent solid: a solid wherein the atoms are linked together in three dimensions by covalent bonds; covalent solids usually are brittle, hard materials with poor electrical and thermal conductivity. Diamond, quartz, and silicon are covalent solids.

cracking: a process in which the components of oil are changed by heating in the presence of a catalyst. In this process, the large hydrocarbon molecules in oil are broken down into smaller hydrocarbon molecules that make up gasoline.

cream of tartar: KHC_4O_6, potassium acid tartrate, salt of tartaric acid used in cooking.

cross-linking: the process of connecting individual polymer chains to one another to add rigidity to a three-dimensional structure.

crystal: a solid formed by the regular stacking of atoms or molecules in a three-dimensional solid.

cubic centimeter: volume of a cube with edges 1 centimeter long, equals 1 milliliter.

cyanide: CN^-, a weak base that is highly toxic.

cyclonite: a high explosive used in plastique.

cysteine: one of the 20 amino acids used to construct proteins. It contains an SH group, so disulfide

(–S–S–) linkages can form between cysteine residues in protein chains.

cytidine: the cytosine-ribose nucleoside.

cytochrome P-450: an enzyme found in the liver that plays an important role in removing toxic molecules from the human body.

cytosine: one of the base molecules found in DNA and RNA. In a double helix, it is paired with guanine.

D

Dacron®: a polyester polymer used to make synthetic fibers.

dAMP: deoxyadenosine monophosphate, the adenosine (or A) nucleotide in DNA.

Darvon®: a narcotic pain reliever.

dCMP: deoxycytosine monophosphate, the cytosine (or C) nucleotide in DNA.

DDI: dideoxyinosine, an antiviral drug, similar to AZT, used in the treatment of AIDS.

DDT: a pesticide that caused great concern in the 1970s. It has been banned in the United States but is still used in many countries throughout the world. While it was in use, the worldwide incidence of malaria decreased greatly. Since its ban, the worldwide incidence of malaria has been on the increase.

deca-: a prefix meaning ten.

Demerol®: a narcotic pain reliever.

density: mass per unit volume of a substance. Chemists usually express density in grams per cubic centimeter, which is the same as grams per milliliter.

dental amalgam: alloy of silver, mercury, copper, and tin used to fill cavities in teeth.

deoxyribose: the sugar molecule used in the synthesis of nucleotides for DNA.

deprotonate: removal of a proton (H^+) from a molecule.

designer drug: a synthetic drug whose structure is very similar to that of an illegal drug.

desulfurization: a process by which sulfur compounds are removed from oil.

detergent: a bifunctional molecule that has a polar group attached to a long hydrocarbon tail. It forms micelles in water and is effective as a cleaning agent.

deuterium: ^2H or D, the isotope of hydrogen in which the nucleus contains one proton and one neutron.

developer: a chemical that creates the photographic image on photographic film or printing paper.

dextromethorphan hydrobromide: synthetic analog of codeine, a cough suppressant.

dGMP: deoxyguanosine monophosphate, the guanosine (or G) nucleotide in DNA.

di-: a prefix meaning two.

diamond: a covalent solid form of carbon.

diatomaceous earth: porous form of silica (SiO_2) that consists of the fossilized remains of microorganisms.

diazepam: valium, an addictive tranquilizer.

dideoxyinosine: DDI, an antiviral drug, similar to AZT that is used in the treatment of AIDS.

diethylenetriamine: an organic compound that contains nitrogen and can chelate metal ions, also used as a reductant in liquid rocket fuels.

diethyl ether: $(CH_3CH_2)_2O$, an extremely flammable solvent used in cocaine processing, one of the early medical anesthetics.

diethylzinc: $(CH_3CH_2)_2Zn$, a compound that spontaneously burns in air and has been used to neutralize acid paper in books.

dimethylnitrosoamine: a potent human carcinogen.

dinitrogen tetroxide: a liquid rocket fuel oxidant.

dipole (electric): the net separation of positive and negative charge in a molecule.

distillation: a process in which a liquid is vaporized by heating to its boiling point, then recondensed as a liquid by cooling the vapors over a different container.

disulfide bond: a bond between two sulfur atoms. An example is the bond formed between sulfur atoms of two cysteine molecules. These bonds are important in stabilizing the three-dimensional structures of proteins.

diuretic: a chemical that causes dehydration and frequent urination.

DNA: deoxyribonucleic acid, the molecule that stores all genetic information in cells. It is constructed from a set of four building blocks called nucleotides. Each nucleotide contains a base molecule (adenine, thymine, guanine, or cytosine), a sugar group (deoxyribose), and a phosphate group.

double bond: a strong bond between two atoms that consists of four shared electrons.

double helix: the name given to the helical structure of two interacting strands of DNA. It was deduced by James Watson and Francis Crick.

dry ice: solid form of carbon dioxide. Solid CO_2 does not melt but vaporizes directly when heated.

dTMP: deoxythymine monophosphate, the thymine (or T) nucleotide in DNA.

E

elastomer: a polymer that exhibits rubber-like elastic properties.

electrolysis: a nonspontaneous chemical reaction driven by the passage of electric current through an ionic solution.

electromagnetic spectrum: the entire collection of electromagnetic radiation. This includes radio waves, microwaves, infrared light, visible light, ultraviolet light, X-rays, and gamma rays.

electron: a small, negatively charged particle of low mass that makes up part of an atom.

electron-pair bond: covalent bond between atoms that consists of a pair of electrons.

electroplating: the process of depositing a thin film of metal on the cathode in an electrolysis cell.

emulsion: a fine suspension of liquid droplets in another liquid with which the droplets do not mix. An example is oil in shaken salad dressing. Photographic emulsion refers to the special case of silver halide crystals dispersed in a gelatin matrix.

endothermic reaction: a chemical reaction that proceeds with the absorption of heat.

enzyme: a protein that catalyzes a chemical reaction.

epinephrine: adrenalin, hormone evoking the flight-or-fight response in mammals. Elevates blood pressure.

epoxy: a rigid polymer that sets by a condensation polymerization of two components, the resin component of fiberglass.

epsom salt: slang for magnesium sulfate heptahydrate, $MgSO_4 \cdot 7H_2O$, a mineral bath component.

estrogen: a steroid hormone that is responsible for the development of sexual characteristics in females.

etching: the selective removal of material from a solid surface.

ethane: CH_3CH_3, a flammable hydrocarbon gas.

ethanethiol: CH_3CH_2SH, a chemical added to natural gas so that leaks in the gas line can be detected by their foul smells.

ethanol: CH_3CH_2OH, a gasoline additive and an intoxicating component in alcoholic beverages.

ether: slang for diethyl ether.

ethylene: CH_2CH_2, a flammable gas that contains a reactive carbon-carbon double bond and undergoes addition polymerization to form polyethylene film or plastic.

ethylene glycol: a viscous toxic di-alcohol used as a component in antifreeze and sometimes illegally added to wine to enhance its sweetness.

ethyl ether: slang for diethyl ether.

ethynylestradiol: a synthetic estrogen hormone.

evaporation: the process by which molecules on the surface of a liquid absorb heat and enter the gas phase.

excited state: a high-energy state of an atom or molecule that usually arises from the absorption of light. It is commonly denoted by raised star (*) immediately after the chemical formula.

exothermic reaction: a chemical reaction that proceeds with the liberation of heat.

F

Fahrenheit: a temperature scale used in the English system. Water freezes at 32°F and boils 212°F.

fats: naturally occurring organic molecules that consist primarily of hydrocarbon chains and contain about the same fuel content as gasoline. Common fats are butter, lard, margarine, olive oil, corn oil, and beef tallow.

fentanyl: a potent narcotic used to relieve postsurgical pain.

ferric oxide: Fe_2O_3 or hematite. Rust is hydrated ferric oxide.

ferritin: a protein that stores iron in the liver.

fertilizer: chemical applied to soil that provides the nitrogen, phosphorus, and potassium atoms that are essential for plant growth.

fiberglass: a high-strength material that consists of glass fibers suspended in an epoxy resin; commonly used in skis, tennis rackets, and boats.

filled shell: a complete set of outer electrons around an atom. Only the atoms of the noble gases [helium (He), neon (Ne), argon (Ar), krypton (Kr), xenon (Xe), and radon (Rn)] naturally have filled shells without forming chemical bonds with other atoms. Hence, they are the only elements that exist as individual atoms.

filter: a glass or plastic material that transmits selective colors of light.

fission: a nuclear reaction wherein an atomic nucleus heavier then iron splits into two smaller nuclei.

fixing solution: a chemical solution used to remove unreacted silver bromide crystals from photographic film.

fluorine: F, the 9th element of the periodic table. Fluorine gas (F_2) is a powerful oxidant.

fluorite: CaF_2, calcium fluoride mineral.

formalin: a water solution of formaldehyde, HCHO.

fossil fuels: coal, oil, and gas, all containing hydrocarbons derived from the decomposition of the remains of plants and animals.

fractional distillation: the purification of a mixture of liquids by slowly heating the mixture and condensing the vapor of each component at a different temperature. Used to separate the different boiling components in crude oil.

free radical: a reactive molecule that contains an unpaired electron.

freezing point: the temperature at which a liquid solidifies.

Freon: another name for a CFC.

fructose: a sugar found in fruits.

fusion: a nuclear reaction wherein two nuclei lighter than iron combine to form a heavier element.

G

gadolinium: the 64th element of the periodic table, the +3 ion Gd^{3+} is used as a contrast agent in MRI (magnetic resonance imaging).

galena: PbS, lead sulfide mineral.

γ-ray: a collection of photons at the high-energy end of the electromagnetic spectrum.

γ-ray spectrometer: a machine that measures the energies of γ-rays.

gas: one of the three states of matter, a low-density material that expands to fill its container.

gasoline: a mixture of hydrocarbons derived from oil that boils between 40°C and 180°C.

glacial acetic acid: pure acetic acid.

gland: an organ that extracts specific substances from the blood and concentrates or alters them for future secretion.

glass: a rigid material that has no well-defined structure. Common glass consists of silica (SiO_2), limestone ($CaCO_3$), and sodium carbonate (Na_2CO_3).

global warming: term used to describe the increases in the Earth's average temperature that result from increases in the concentrations of greenhouse gases.

glucose: a sugar found in blood.

glutamic acid: one of the 20 amino acids used to construct proteins.

glutamine: one of the 20 amino acids used to construct proteins.

glycerin: another name for glycerol.

glycerol: $(HO)H_2C-CH(OH)-CH_2(OH)$, a viscous liquid used in cosmetics; a component of animal fats and vegetable oils. Its three alcohol groups react with fatty acids to produce triglycerides.

glycine: the simplest of the 20 amino acids used to construct proteins.

gold: Au, the 79th element of the periodic table, used for treatment of rheumatoid arthritis.

Gore-Tex®: A Teflon®-coated fiber used to manufacture water-repellent, "breathable" garments.

gram: a measure of weight. One gram is equivalent to 0.0352 ounce.

graphite: a solid form of carbon in which the carbon atoms are linked in hexagonal planar arrays. Because these layered sheets are not covalently bonded to one another, this material is a good lubricant and constitutes the "lead" in pencils.

graphite composite: high-strength material made of epoxy and fibers of carbon.

greenhouse effect: the heating of the atmosphere by the trapping of infrared radiation by greenhouse gases.

greenhouse gases: chemicals in the atmosphere that absorb the infrared radiation (heat) emitted by the Earth. These include carbon dioxide, water, nitrous oxide, chlorofluorocarbons, and methane.

guanine: one the of the base molecules found in DNA and RNA. In helical structures, it is paired with cytosine.

guanosine: the guanine-ribose nucleoside.

guncotton: an explosive used in bullets that is produced by the action of nitric acid on cotton.

gunpowder: an explosive mixture of charcoal, sulfur, and sodium nitrate; the first explosive known to humans; invented by the Chinese.

gypsum: hydrated calcium sulfate ($CaSO_4 \cdot 7H_2O$), the primary component in wallboard.

H

Halcion®: a valium-like sleeping pill.

half-life: the time required for a reaction to proceed half-way to completion, commonly used as a measure of radioactive decay times.

halogens: the name given to the elements in the column of the periodic table that includes fluorine (F), chlorine (Cl), bromine (Br), iodine (I), and astatine (At).

halothane: an anesthetic gas.

heat of combustion: the heat released when a substance is burned in oxygen.

helium: He, the 2nd element of the periodic table, an inert gas that is lighter than air. It is produced in the Earth by α-decay.

hematite: Fe_2O_3, ferric oxide.

hemoglobin: an iron-containing protein that transports oxygen in the blood.

hepta-: a prefix meaning seven.

herbicide: a chemical that interferes with the growth of plants, used to control weeds.

heroin: an addictive narcotic made by adding acetic acid to morphine.

heterogeneous: in which more than one phase is involved.

heterogeneous catalysis: a reaction wherein the catalyst is in a different phase from the reactants or products.

hexa-: a prefix meaning six.

histidine: one of the 20 amino acids used to construct proteins. Often involved in binding metal ions.

homogeneous: in which only one phase is involved.

homogeneous catalysis: a reaction wherein the catalyst is in the same phase as the reactants or products.

hormone: a chemical that directs the activity of cells in the body.

hydrazine: N_2H_4, a liquid rocket fuel.

hydrocarbons: organic chemicals that contain only carbon and hydrogen.

hydrochloric acid: HCl, a strong industrial acid; also a stomach acid.

hydrofluoric acid: HF, a highly toxic acid used in oil refining and in etching glass and silicon.

hydrogen: H, the 1st element of the periodic table, exists natually as H_2, a highly flammable gas.

hydrogenation: addition of hydrogen to carbon–carbon double (or triple) bonds. This reaction converts unsaturated fats and oils to saturated ones.

hydrogen bomb: a bomb that uses the fusion reaction between deuterium nuclei to generate explosive energy.

hydrogen cyanide: HCN, a deadly gas.

hydrogen peroxide: H_2O_2, a strong oxidant and bleaching agent available in drugstores as a weak 3% solution in water.

hydrogen sulfide: H_2S, an odorous and highly toxic molecule.

hydrolysis: a chemical reaction wherein water is one of the reactants.

hydrosphere: the areas of the Earth that are composed of water. This includes oceans, rivers, streams, lakes, reservoirs, glaciers, and ground water.

hydroxide: OH^-, a strong base.

hygrometer: a device used to measure the density of liquids.

I

ibuprofen: an analgesic pain reliever found in Motrin® and Advil®.

inert: unreactive.

infrared radiation: the part of the electromagnetic spectrum that extends from just below the visible to the microwave. Hence it has less energy than visible light but more energy than microwave. Although it is invisible to the eye, humans can detect infrared radiation by the sensation of warmth.

inorganic chemistry: chemistry of all nonorganic compounds.

insulator: a substance that does not conduct electricity.

insulin: a hormone that is involved in many important biochemical processes in humans (such as storage of sugar and the synthesis of other proteins). A deficiency of insulin causes diabetes.

iodine: I, the 53rd element of the periodic table, an essential element for human nutrition.

ion: a charged atom or molecule.

ion exchange: a technique that is used to extract calcium and magnesium ions from hard water and replace them with sodium ions.

ionic bond: a bond between oppositely charged ions.

ionic solid: a solid that is held together by ionic bonds.

ionic solution: a solution that contains dissolved ions.

ionization: the dissociation of a neutral substance into positive and negative ions when it dissolves in water. For atoms or molecules in the gas phase, it refers to the creation of a positive ion by loss of an electron.

iron: Fe, the 26th element of the periodic table, an abundant metal essential for life.

isoleucine: one of the 20 amino acids used to construct proteins.

isooctane: a liquid hydrocarbon in gasoline that has an octane rating of 100.

isoprene: the fundamental repeating unit in natural rubber.

isopropanol: rubbing alcohol, flammable toxic alcohol.

isotopes: atoms of the same element that differ only in the number of neutrons in their nuclei (for example, tritium and deuterium are isotopes of hydrogen).

K

kaolin: a pure form of clay.

Kelvin: the same as the Celsius temperature scale, except that 0 K (−273°C) is equated with absolute zero. Water freezes at 273 K and boils at 373 K.

keratin: a tough, fibrous protein that is rich in the amino acid cysteine and forms the outer layer of hair and nails.

kerosene: a mixture of hydrocarbons derived from oil that boils between 175°C and 230°C.

Kevlar®: a high-strength condensation polymer.

kilocalorie: 1000 calories, equal one nutritional Calorie.

kilometer: a measure of distance. One kilometer is 1000 meters or 0.62 mile.

krypton: Kr, the 36th element of the periodic table, an inert gas.

L

L-dopa: a chiral drug used in the treatment of Parkinson's disease.

lactic acid: a weak acid that is found in milk and is a waste product of metabolism.

laminate: a material that is rolled or compressed into sheets.

laughing gas: nitrous oxide, N_2O, a greenhouse gas and an anesthetic.

lead: Pb, the 82nd element of the periodic table, the anode (reductant) in a car battery, a toxic heavy metal.

lead azide: $Pb(N_3)_2$, a solid explosive that detonates by shock, used in blasting caps and firearm rounds.

lead dioxide: PbO_2, the cathode (oxidant) in a car battery.

leaded gasoline: gasoline that contains the octane enhancer tetraethyl lead.

leucine: one of the 20 amino acids used to construct proteins.

lidocaine: a topical anesthetic.

lime: CaO, a weak base used in steel and cement manufacturing.

limestone: $CaCO_3$, may also contain $MgCO_3$; when heated, it makes lime and releases carbon dioxide.

lipid: a naturally occurring fat that is generally insoluble in water.

liquid: one of the three states of matter. A liquid adopts the shape of the container but does not expand to fill it. Liquids are nearly as dense as solids, but they differ in that the molecules are not arranged in a regular fashion.

liquid crystal: a special class of materials whose molecules align in an ordered fashion (as found in crystals) in their liquid state; used in liquid crystal displays (LCDs).

liquid rocket fuel: rocket fuel that consists of a separate liquid oxidant and reductant that are mixed together in the motor to cause a highly exothermic reaction.

liter: a measure of liquid volume. One liter corresponds to 1000 cubic centimeters or 1000 milliliters. One liter is the same as 1.06 quarts.

litharge: PbO, lead monoxide.

lithium: Li, the 3rd element of the periodic table. Lithium salts are used in the treatment of mental illness.

lithosphere: the solid regions of the Earth.

LSD: acronym for lysergic acid diethylamide.

Lucite®: trade name for Plexiglas® or polymethylmethacrylate, a clear plastic used for windows and lenses.

lye: slang for sodium hydroxide, a strong base.

lysergic acid diethylamide: LSD, a hallucinogenic drug that was popular in the 1960s; causes permanent psychological disorders.

lysine: one of the 20 amino acids used to construct proteins.

lysozyme: an enzyme that dissolves certain bacteria by cleaving the components of their cell walls.

M

Mace: 2-chloroacetophenone, tear gas.

magnesia: MgO, magnesium oxide.

magnesium: Mg, the 12th element of the periodic table, an essential trace element; forms a light alloy with aluminum and zinc that is used in mag wheels.

magnesium carbonate: $MgCO_3$, a component in dolomite limestone.

magnesium hydroxide: $Mg(OH)_2$, an active ingredient in Maalox®, a weak base.

magnetite: Fe_3O_4, a magnetic oxide of iron.

malachite: $Cu(OH)_2 \cdot CuCO_3$, basic copper carbonate mineral, the green film that forms on weathered bronze and copper statues.

manganese: Mn, the 25th element of the periodic table. This metal burns in air and is used in flares.

manganese dioxide: MnO_2, the oxidant in the dry-cell and alkaline-cell batteries.

Manhattan Project: the top-secret U.S. effort during World War II to build an atomic bomb.

marble: a hard, crystalline form of calcium carbonate.

mass: a quantity that depends on the number and type of atoms in a substance. Mass is a measure of an object's resistance to the imparting of motion to it. Although the term is often used interchangeably with weight, they are not the same quantity. Weight is a measure of the gravitational force on a mass, which differs from planet to planet.

mass number: the sum of the protons and neutrons in the nucleus of an atom.

melanins: polymeric materials responsible for the dark pigments in human skin. The tanning of skin involves the biosynthesis of black melanins in response to exposure to ultraviolet rays in sunlight.

melt-spinning: process by which a molten polymer is spun into a fiber.

melting point: the temperature at which a material changes from the solid to the liquid state.

meperidine: a narcotic pain reliever in Demerol®.

mercury: Hg, the 80th element of the periodic table, the only common metal that is a liquid at room temperature. Its vapor and compounds are highly toxic.

mescaline: a hallucinogenic drug isolated from the peyote plant.

metabolism: the process that living organisms use to generate energy. It generally involves the enzyme-catalyzed combustion of carbon-containing foods with oxygen from the air to form carbon dioxide and water as byproducts.

metal: a solid characterized by its metallic luster, high thermal and electrical conductivity, and flexibility. Its unique properties result from the electrons being delocalized throughout the entire piece of metal.

metal hydroxide: metal compound that contains OH^-.

metallic bonds: a delocalized bonding that occurs in metals.

metal oxide: metal compound that contains oxide (O_2^-), often formed when a metal reacts with oxygen in air.

methadone: a narcotic used in the treatment of drug addiction.

methamphetamine hydrochloride: chemical name for speed or crystal meth or meth; an addictive, central nervous system stimulant that dangerously elevates the blood pressure.

methane: CH_4, swamp gas, the simplest hydrocarbon and the primary component in natural gas.

methanol: CH_3OH, wood alcohol, the simplest alcohol; a proposed alternative to gasoline; causes blindness, convulsion, and death when ingested as an impurity in moonshine.

methionine: one of the 20 amino acids used to construct proteins. Contains sulfur.

methyl salicylate: the chemical responsible for the odor of wintergreen, an aspirin derivative.

methyl-t-butyl ether: MTBE, an octane enhancer in unleaded gasoline.

micelle: a spherical structure that detergent molecules form in water. It consists of a greasy interior and a charged surface. Oily grime dissolves in its interior and thus is washed away.

micro-: a prefix meaning one millionth.

microgram: 0.000001 gram or one millionth of a gram.

micrometer: 0.000001 meter or one millionth of a meter.

microwave radiation: the part of the electromagnetic spectrum between infrared light and radio waves, used to heat food by exciting water molecules; used in telecommunications.

milk of magnesia: a milky suspension of solid magnesium hydroxide in water, used to neutralize stomach acid.

milli-: a prefix meaning one thousandth.

milligram (mg): 0.001 gram or one thousandth of a gram.

milliliter (ml): 0.001 liter or one thousandth of a liter, equals 1 cubic centimeter in volume.

millimeter (mm): 0.001 meter or one thousandth of a meter.

mirror image: the reflected image of an object.

mole: defined as 6.02×10^{23} particles.

molecular crystal: a crystal wherein weak forces hold the molecules in the solid. Such crystals often undergo sublimation. Examples include dry ice and mothballs.

molecule: two or more atoms that are chemically bonded together.

molybdenum: Mo, the 42nd element of the periodic table, a metal added to steel alloys.

mono-: prefix meaning one.

monomethylhydrazine (MMH): a liquid rocket fuel.

monounsaturated fat: fat wherein the hydrocarbon chains contain one carbon–carbon double bond.

morphine: an addictive narcotic pain reliever.

MPPP: a designer drug, synthetic heroin.

MPTP: an impurity in synthetic heroin that causes permanent nerve damage.

MTBE: abbreviation for methyl-t-butyl ether.

MTD: maximally tolerated dose; the level at which chemicals are administered to test animals to determine toxicity, mutagenicity, and carcinogenicity.

muriatic acid: a 30% solution of HCl in water.

mustard gas: 2-chloroethyl sulfide, a chemotherapy agent and chemical weapon.

mutagen: a substance that causes permanent changes in DNA.

mutation: the change in DNA caused by a mutagen. Mutation may cause permanent changes in the function of the organism that may be beneficial, harmless, or deadly.

mylar: a plastic made from a polyester.

myoglobin: an iron-containing protein that stores and releases oxygen in muscle tissues.

N

n-type silicon: an electron-rich silicon containing small amounts of phosphorus or arsenic that is used in the manufacture of integrated circuits and solar cells.

nano-: a prefix meaning one billionth.

nanogram: 0.000000001 gram or one billionth of a gram.

nanometer (nm): 0.000000001 meter or one billionth of a meter.

natural gas: mixture of gaseous hydrocarbons found in underground deposits, consists primarily of methane.

neon: Ne, the 10th element of the periodic table, an inert gas used in lighting.

nerve gas: molecules that interfere with the transmission of nerve impulses. All "nerve gases" are high-boiling liquids, not gases.

network solid: a rigid solid that is linked in three dimensions by covalent bonds; usually a high-melting, brittle insulator (such as diamond or silicon).

neurotransmitter: chemical that relays nerve impulses in the body.

neutron: a component of the nucleus of an atom that has mass but no charge.

neutron activation analysis: a method for identifying the amount of each element present in a sample. It involves bombarding the sample with neutrons to create radioactive nuclei and analyzing the gamma rays produced.

nichrome: an alloy of nickel and chromium.

nickel: Ni, the 28th element of the periodic table, used in many common alloys.

nicotine: the addictive chemical in cigarettes, also a highly toxic chemical used as an insecticide.

niobium: Nb, the 41st element of the periodic table.

nitrate: NO_3^-, an anion with strong oxidizing properties.

nitric acid: HNO_3, a strong acid and strong oxidant used in the manufacture of fertilizers and explosives.

nitric oxide: NO, a component in automobile exhaust, a neurotransmitter in the brain, and a deadly gas when inhaled.

nitrilotriacetate (NTA): a chelating agent for metal ions.

nitrogen: N, the 7th element of the periodic table; naturally exists as the unreactive gas N_2; the primary constituent of the Earth's atmosphere.

nitrogen dioxide: NO_2, the brown gaseous component of smog.

nitroglycerine: a high explosive and heart medication.

nitrosoamines: a family of potent carcinogens.

nitrous oxide: N_2O, laughing gas, a propellent in canned whipped cream and a greenhouse gas.

noble gases: inert gases and rare gases are also names given to the rightmost column in the periodic table, which consists of the gases helium (He), neon (Ne), argon (Ar), krypton (Kr), xenon (Xe), and radon (Rn).

nona-: a prefix meaning nine.

norethindrone: synthetic analog of progesterone, used in birth control pills.

normal paraffins: hydrocarbons wherein all the carbon atoms are linked together in a straight chain.

novocaine: a local anesthetic.

nuclear chemistry: the study of nuclear reactions.

nuclear fission: a nuclear reaction that results in the splitting of the atomic nucleus, producing lighter elements.

nuclear fusion: a nuclear reaction in which two nuclei are joined to form a heavier element.

nuclear reactions: reactions that result in a change in the type of atom.

nuclear waste: radioactive byproducts from nuclear reactors.

nucleic acid: *see* RNA and DNA.

nucleoside: a nucleotide without its phosphate group.

nucleotides: the building blocks of RNA and DNA.

nucleus: tiny center of the atom that contains nearly all the atom's mass as protons and neutrons.

NutraSweet®: the dipeptide Asp-Phe (aspartic acid–phenylalanine), a sugar substitute.

nylon: a high-strength condensation polymer used as a fiber in apparel.

O

octa-: a prefix meaning eight.

octane: an arbitrary scale that rates fuels on the basis of their antiknocking properties. Isooctane, a good fuel, is assigned an octane rating of 100, and n-heptane, a fuel that knocks badly, an octane rating of 0.

oil of vitriol: slang for sulfuric acid, H_2SO_4.

opsin: the protein in the eye's rod and cone cells that binds retinal.

organic chemistry: the study of compounds that contain carbon, hydrogen, and other nonmetallic elements.

outermost shell: the highest-energy shell of an atom or molecule that is occupied by one or more electrons.

oxidant: any compound that removes electrons from another compound (the reductant) in a chemical reaction.

oxidation: a chemical process that removes electrons from an atom or molecule. Often one or more oxygen atoms are transferred from the oxidant to the reductant in such reactions.

oxide ion: O^{2-}, a doubly negatively charged ion present in many metal oxides. For example, aluminum oxide (Al_2O_3) contains two Al^{3+} ions and three O^{2-} ions.

oxygen: O, the 8th element of the periodic table, exists naturally as the molecules O_2 and O_3. Molecular oxygen (O_2) takes part in the combustion reactions in the human body as a strong oxidant.

ozone: O_3, a gas found in the stratosphere. Ozone protects the plants and animals on the surface of the Earth by filtering out the harmful ultraviolet light emitted by the Sun. Ozone is toxic to most forms of life. In the troposphere, it is formed in photochemical smog.

ozone hole: term for the depletion in the stratospheric ozone concentration above Antarctica and the Arctic. Photochemical reactions of CFC molecules lead to ozone depletion.

P

p-type silicon: an electron-poor silicon containing small amounts of boron that is used in the manufacture of integrated circuits and solar cells.

paraffin: saturated hydrocarbon molecules. All the carbon–carbon bonds in saturated hydrocarbons are single bonds.

PCP: "angel dust" or phenylcyclidine, an anesthetic street drug.

penicillin: an antibiotic that kills bacterial infections. The drug works by preventing the synthesis of bacterial cell walls.

penta-: a prefix meaning five.

periodic table: a chart that displays the elements in order of increasing atomic number. In the table, families of elements with similar properties are grouped in columns.

peroxide: abbreviation for hydrogen peroxide, HOOH, a strong oxidizing and bleaching agent. Also refers to compounds containing an O–O bond.

pesticide: chemicals that kill pests (such as rats, insects) that destroy crops.

petrochemicals: chemicals made from crude oil.

petroleum products: oil or products derived from oil.

pewter: an alloy of tin, antimony, and copper.

pH: a measure of the concentration of H^+ ions in a water solution.

phenobarbital: a common barbiturate sedative.

phenol: a benzene derivative that contains an acidic OH group. A common disinfectant and industrial chemical.

phenylalanine: one of the 20 amino acids used to construct proteins.

phenylcyclidine: *see* PCP.

phosgene: $COCl_2$, a highly toxic gas used in chemical warfare.

phosphate rock: $Ca_3(PO_4)_2$, a mineral with a composition similar to that of bones and teeth. Used to make a phosphate source for fertilizers.

phosphate: the phosphate ion, PO_4^{3-}; an essential nutrient for all life.

phosphoric acid: H_3PO_4, a moderately strong acid often used as an acidifying agent in foods and beverages.

phosphorus: P, the 15th element of the periodic table.

phosphorus pentoxide: P_2O_5, the product that forms when phosphorus burns. Dissolves in water to yield phosphoric acid.

photochemical smog: a brown haze observed in industrialized cities that results from the atmospheric chemistry of nitrogen dioxide. Nitrogen dioxide is generated from oxidation of the nitric oxide emissions from automobiles. Photochemical reactions of nitrogen dioxide produce ozone.

photochemistry: study of the chemical reactions that involve light.

photochromic glass: a type of glass that darkens when exposed to ultraviolet light. Once the source of ultraviolet light is blocked, the glass lightens.

photodissociation: a photochemical reaction in which light causes a chemical bond to break.

photoisomerization: a photochemical reaction in which light causes isomerization. The first step in vision is such a reaction.

photolithography: a photochemical technique that makes possible the creation of three-dimensional patterns on solid surfaces. It is extensively used to shape silicon substrates for use in manufacturing electronic devices.

photon: the smallest particle that makes up light, whose energy content is represented by the product $h\nu$, where h is Planck's constant and ν is the frequency of the light wave.

photoresist: a polymeric material that is used to protect a solid surface during the photolithography process.

photosynthesis: the biochemical process by which plants synthesize molecules. The energy for these reactions is provided by sunlight.

Photosystem I: a subassembly of proteins found in plants that is responsible for the generation of molecules that are important in the synthesis of carbohydrates. In addition, this unit generates molecules that work with products from Photosystem II to synthesize ATP.

Photosystem II: a subassembly of proteins found in plants that is responsible for splitting water, liberating O_2, and providing a source of hydrogen atoms for the plant. In addition, this unit generates molecules that work with products from Photosystem I to synthesize ATP.

physical chemistry: the study of the underlying physical principles of chemistry.

pico-: a prefix meaning one trillionth.

pig iron: crude, brittle iron from a blast furnace that has a high carbon content (it is 3.5–4.5% C).

Planck's constant: a proportionality constant (h) that relates the energy of a light photon to its frequency by the formula $E = h\nu$.

plasma: a high-temperature collection of gaseous ions and electrons.

plaster of Paris: $(CaSO_4)_2 \cdot H_2O$, a dehydrated form of gypsum used to make building plaster.

plastic: a solid, rigid polymer.

plasticizer: an organic compound added to a polymer to make it more flexible.

plastique explosive: RDX or cyclonite mixed with a polymer to make a flexible high explosive.

platinum: Pt, the 78th element of the periodic table.

Plexiglass®: Lucite®, polymethylmethacrylate, a clear plastic used in windows and lenses.

plutonium: Pu, the 94th element in the periodic table. A synthetic element made by neutron bombardment of uranium. A fissionable radioactive substance used in breeder reactors and nuclear weapons.

poison ivy: the plant *Toxicodendron radicans*. Oils from this plant cause an allergic reaction in the skin of many people. The active component in causing this response, urushiol, is a mixture of four structurally similar molecules.

polar bond: covalent bond with a lopsided charge distribution. One atom in the bond attracts electrons better and acquires a partial negative charge; the other atom is left with a partial positive charge.

polar vortex: A mass of very cold, stagnant air over the pole contained by the surrounding strong westerly winds. It is in the Antarctic vortex that the first symptoms of worldwide ozone depletion were observed.

polychlorinated biphenyls: a class of hazardous organic materials whose molecules contain two linked benzene rings with attached chlorine atoms. The manufacture of these materials has been banned in the United States since 1979. However, large amounts are still in use today in a variety of industrial applications.

polyester: a condensation polymer used to make synthetic fibers.

polyethylene: an inexpensive addition polymer with a low melting point used in food wrap and plastics.

polyisoprene: the addition polymer in natural rubber.

polymer: a long-chain molecule built up by chemically linking together many (typically 100–10,000) subunits called monomers.

polymerization: chemical reaction that converts monomers to a polymer.

polypeptide: a condensation polymer composed of amino acids.

polypropylene: an addition polymer of propylene that is stronger and higher melting than polyethylene.

polyprotic acids: acids that contain more than one ionizable proton.

polystyrene: an addition polymer of styrene. It is used in rigid plastics and can be blown into polystyrene foam.

polyurethane: a condensation polymer used in blown foam paddings and coatings for wood and metal.

polyvinyl chloride: an addition polymer of vinyl chloride ($H_2C=CHCl$) used to make rigid plastics.

porcelain: a glass made of silica (SiO_2), alumina (Al_2O_3), and lime (CaO).

porphyrin: an organic compound whose molecules contain four nitrogen atoms arranged in such a way that they can all simultaneously bind a metal ion (such as Fe^{2+}). This molecule occurs in heme proteins (such as myoglobin, hemoglobin, and cytochromes) and is responsible for their ability to transport and store O_2, as well as transfer electrons.

positron: a positively charged electron, an example of antimatter. A positron immediately reacts when it encounters a normal electron to annihilate both particles and create two gamma rays.

positron emission tomography (PET): medical imaging technique that measures gamma rays produced by positron-electron annihilation inside a patient's body to construct a three-dimensional image of organs.

potash: slang for potassium carbonate, K_2CO_3.

potassium: K, the 19th element of the periodic table, an element essential to life.

potassium chlorate: $KClO_3$, an oxidizing agent used in fireworks and matches.

pounds per square inch (psi): English unit of pressure that equals the force that a one-pound weight exerts when distributed over one square inch of area.

primary protein structure: the sequence of amino acids in a protein.

product: a substance resulting from a chemical reaction.

progesterone: a steroid hormone that prepares the uterine area for implantation of a fertilized egg.

proline: one of the 20 amino acids used to construct proteins; often appears at the termination of an α-helix in protein structures.

propylene: a hydrocarbon whose molecules contain three carbon atoms and one carbon–carbon double bond; monomer for polypropylene.

proteases: enzymes that digest proteins.

protein: a condensation polymer of amino acids.

proton: a hydrogen ion, H^+; the positively charged component of an atomic nucleus.

Prussian blue: $Fe_4(Fe(CN)_6)_3$, ferric ferrocyanide, a blue pigment used in printing inks, paints, and enamels.

prussic acid: slang for hydrogen cyanide, HCN.

pseudoephedrine hydrochloride: the active ingredient in Sudafed®.

purification: the isolation of a particular substance from a mixture.

PVC: slang for polyvinyl chloride.

Pyrex: a special glass characterized by a reduced tendency to expand when heated or contract when cooled; used in cookware.

Q

quartz: crystalline SiO_2, silicon dioxide.

quicksilver: slang for liquid mercury.

quinine: a bitter alkaloid used to flavor quinine water; an antimalarial drug.

R

radioactive waste: byproducts from a nuclear reactor.

radiocarbon dating: a technique used to determine the age of an object by measuring the content of the radioactive isotope of carbon, ^{14}C.

radium: Ra, the 88th element of the periodic table, the first radioactive element isolated by Marie Curie from pitchblende ore.

radon: Rn, the 86th element of the periodic table, a radioactive noble gas that is an environmental carcinogen.

random coil: a segment of a protein chain that has no ordered structure.

rare gases: inert gases and noble gases are also names given to the rightmost column in the periodic table, which consists of the gases helium (He), neon (Ne), argon (Ar), krypton (Kr), xenon (Xe), and radon (Rn).

rate of reaction: speed at which a chemical reaction occurs.

RDX: the high explosive in plastique.

reactant: a substance that undergoes a chemical reaction.

reaction coordinate: the path that atoms involved in a chemical reaction follow during the reaction.

red phosphorus: a red form of solid phosphorus that is stable in air.

redox reaction: oxidation–reduction reaction.

reductant: any substance that donates electrons to another substance (the oxidant) in a chemical reaction.

reduction: a chemical process that adds electrons to an atom or molecule. Often one or more oxygen atoms are removed from the substance being reduced.

reforming: treatment of a petroleum fraction over a catalyst to form aromatic hydrocarbons of high octane value.

refractory: a high-melting substance.

replication: the process by which a cell makes a copy of its DNA to pass on to its progeny.

repulsive force: a force that pushes two particles apart, as occurs between particles with the same electric charge.

retinal: molecule that absorbs light and triggers the sense of sight.

rhodium: Rh, the 45th element of the periodic table, an expensive metal used in industrial catalysts and automotive catalytic converters.

rhodopsin: the complex between retinal and the protein opsin that is found in rod and cone cells in the eye. Minor changes in the structure of opsin make color vision possible.

ribose: the sugar component of nucleotides for RNA.

RNA: ribonucleic acid, a genetic molecule similar to DNA. It is constructed from a set of four building blocks called ribonucleotides. Each ribonucleotide contains a base molecule (adenine, uracil, guanine, or cytosine), a sugar group (ribose), and a phosphate group. Many viruses use RNA instead of DNA for the storage of genetic information. In humans, RNA molecules play an important role in the synthesis of proteins according to the genetic information contained in DNA.

rod: eye cell that responds to low light levels but cannot distinguish colors.

RU 486: an abortion-causing drug.

rubber: an elastic polymer.

rubidium: Rb, the 37th element of the periodic table.

ruby: aluminum oxide (Al_2O_3) that contains a small amount of Cr^{3+} in place of Al^{3+} to give it a red color.

rust: a hydrated form of ferric oxide, $Fe_2O_3 \cdot nH_2O$.

rutile: TiO_2, a mineral form of the white pigment titanium dioxide.

S

salicylic acid: the active drug aspirin forms in the body; also a topical wart-removing agent.

saltpeter: slang for potassium nitrate, KNO_3. Sometimes confused with Chile saltpeter, which refers to natural deposits of sodium nitrate found in Chile.

sand: SiO_2, small grains of quartz.

sapphire: Al_2O_3, a crystalline gem form of aluminum oxide.

sarin: a nerve gas.

saturated fat: a fat whose molecules contain only single bonds between adjacent carbon atoms.

SBR rubber: acronym for styrene-butadiene rubber, a synthetic rubber invented during World War II and used in nonradial tires.

scanning tunneling microscope (STM): an instrument that cam image individual atoms in a material.

scrubber: a device that is attached to an industrial smokestack to remove particulates and various gases such as sulfur dioxide.

secondary protein structure: isolated segments of a protein chain that have characteristic three-dimensional structure: α-helical, β-pleated sheet, or random coil.

selenium: Se, the 34th element of the periodic table.

selenium sulfide: SeS, a compound found in anti-dandruff shampoos.

semiconductor: an electronic material that is used in the manufacture of integrated circuits and solar cells.

serine: one of the 20 amino acids used to construct proteins.

shell model: a crude theory for the arrangement of electrons around atoms in orbits of increasing distance from the nucleus.

sickle-cell anemia: a blood disorder that arises from a mutation in the protein hemoglobin.

silica: silicon dioxide, SiO_2.

silica gel: a porous form of SiO_2 that is highly adsorbent.

silicon: Si, the 14th element of the periodic table.

silicone: refers to a compound that contains a silicon–oxygen backbone with hydrocarbon side chains attached to silicon.

silicone gel: a mixture of silicone polymer and silicone oil used in breast implants.

silicone oil: a short polymer that contains a silicon–oxygen backbone with hydrocarbon side chains attached to silicon. Used as an oxidation-resistant oil.

silicone rubber: a long-chain silicone polymer that may be cross-linked; it contains SiO_2 as a solid additive. Used as a high-temperature-resistant rubber.

silver: Ag, the 47th element of the periodic table.

silver bromide: AgBr, an ionic solid widely used as the photosensitive component in photographic film and paper.

silver chloride: AgCl, an ionic solid also used in photographic emulsions.

simethicone: a short-chain silicone oil used to coat and soothe upset stomachs.

single bond: a bond between two atoms that consists of two shared electrons.

slag: a glass formed between lime (CaO) and silicate impurities in iron ore. It floats to the top of molten iron, so it can be separated easily.

slaked lime: $Ca(OH)_2$, calcium hydroxide. Formed by adding water to lime.

soap: alkali metal salts of fatty acids. These salts form micelles in water and behave as cleaning agents.

sodium: Na, the 11th element of the periodic table.

sodium azide: NaN_3, forms nitrogen gas on heating. Used to generate gas for rapid inflation of automotive airbags.

sodium benzoate: a food preservative.

sodium bicarbonate: $NaHCO_3$, a mild base used to neutralize stomach acid. Solutions in water form CO_2 gas on heating, causing certain baked goods to rise.

sodium hydroxide: NaOH or lye, a water-soluble ionic solid that is a strong base.

sodium pentothal: an anesthetic often called truth serum.

sodium silicate: Na_4SiO_4, called water glass. Forms when SiO_2 reacts with a concentrated solution of NaOH. Forms silica gel when acidified.

solanine: a natural insecticide found in potato skins.

solar cell: an electronic device that produces electricity from sunlight.

solder: a low-melting alloy of tin and lead.

solid: one of the three states of matter. A solid has a defined volume and space and is virtually incompressible. The atoms in a solid vibrate about fixed positions.

solid rocket fuel: a solid compound that consists of an intimate mixture of an oxidant and reductant. Once ignited, the reaction cannot be stopped until all the fuel is consumed.

solubility: a measure of how much solute can be dissolved in a solvent.

solute: a chemical substance dissolved in a liquid. In a water solution of table salt, the salt is the solute.

solvent: the liquid that dissolves a chemical substance. In a water solution of table salt, the water is the solvent.

soman: a nerve gas.

stainless steel: a corrosion-resistant alloy of iron and chromium.

steel: an alloy of iron, carbon, and other elements that is stronger and harder than pure iron.

steroid: a class of hormones whose molecules share a common structural backbone. Most are derivatives of testosterone, the male sex hormone.

stop solution: a chemical solution used to stop the development process of photographic negatives.

stratosphere: the region of the atmosphere that extends from the top of the troposphere to an altitude of about 50 km. This region contains the ozone layer.

strong acid: an acid that ionizes completely when dissolved in water.

strong force: the force between neutrons and protons that hold the nucleus together.

strontium: Sr, the 38th element of the periodic table. The isotope that contains 52 neutrons (strontium-90) is radioactive and is found in the wastes of nuclear power plants.

structural formula: a pictorial representation of the structure of a molecule that shows the bonds between the atoms.

styrene: a hydrocarbon whose molecular structure contains a carbon–carbon double bond linked to a benzene ring; readily polymerizes to polystyrene.

styrofoam: a foam form of polystyrene.

sublimation: conversion of a solid to a gas without going through a liquid state.

sucrose: common table sugar.

sugar: refers to a class of nonpolymeric carbohydrates of the general formula $(CH_2O)_n$.

sulfate: SO_4^{2-}, an ion that occurs in many ionic solids and minerals.

sulfide ion: S^{2-}, an ion that occurs in minerals.

sulfur: S, the 16th element of the periodic table.

sulfur dioxide: SO_2, a gas released in the burning of sulfur-containing coal and oil; reacts in the atmosphere to form acid rain.

sulfuric acid: H_2SO_4, a strong acid and the largest-volume industrial chemical in the world; the chief component in acid rain.

sunblocks: commercial chemical products that completely absorb or reflect ultraviolet light. They shield the skin from harmful solar radiation.

sunscreens: commercial chemical products that absorb ultraviolet light. The degree of protection depends on the concentration of the ultraviolet-absorbing molecule.

superconductor: a solid material in which an electron can move without experiencing any resistance.

superphosphate: $(NH_4)_2HPO_4$, a solid fertilizer that supplies nitrogen and phosphorus to the soil.

surfactant: a chemical that stabilizes the suspension of nonpolar molecules (such as grease and oil) in water.

synthetic element: a heavy element that is not found naturally on the Earth. This term usually refers to elements of the periodic table that have atomic numbers greater than 92.

T

tabun: a nerve gas.

taxol: an anticancer drug that is isolated from the bark of the yew tree.

tear gas: MACE or 2-chloroacetophenone; behaves as a strong eye irritant.

technetium: Tc, the 43rd element of the periodic table. The only element below atomic number of 92 that is not found naturally on Earth in appreciable amounts. Radioactive ^{99m}Tc is used extensively in medical imaging agents.

Teflon®: addition polymer of tetrafluoroethylene, $F_2C=CF_2$, used as a nonstick coating and lubricant.

tellurium: Te, the 52nd element of the periodic table.

tertiary protein structure: the overall three-dimensional structure of a protein.

testosterone: the male sex hormone, a steroid.

tetra-: a prefix meaning four.

tetrachloroethylene: $Cl_2C=CCl_2$, a dry-cleaning fluid and industrial solvent.

tetraethyl lead: $Pb(CH_2CH_3)_4$, an additive that increases the octane rating of gasoline.

tetrahydrocannabinol: THC, the active ingredient in marijuana.

thalidomide: a sedative that causes birth defects in pregnant women.

thermal conductivity: the ability of a material to conduct heat.

thermal energy: heat energy.

thermonuclear reaction: a nuclear fusion process that requires tremendous amounts of thermal energy to initiate.

Three Mile Island: location of a nuclear reactor outside Harrisburg, Pennsylvania. In March of 1979, there was a partial loss of coolant water from the reactor core. Radioactive gas was released into the atmosphere, and 2.4 million liters of radioactive water spilled into the reactor's containment building.

threonine: one of the 20 amino acids used to construct proteins.

thrombin: an enzyme that plays a role in the formation of blood clots.

thymidine: the thymine-ribose nucleoside.

thymine: one of the base molecules found in DNA. In helical structures, it is paired with adenine.

tin oxide: SnO_2, a transparent, electrically conductive oxide used in electronic devices.

tin: Sn, the 50th element of the periodic table.

titanium dioxide: TiO_2, a white pigment.

titanium: Ti, the 22nd element of the periodic table.

TNT: trinitrotoluene, an explosive.

toluene: an aromatic hydrocarbon solvent and an ingredient in unleaded gasoline.

tomography: imaging technique that creates a three-dimensional picture of an organ in the body.

transferrin: a protein that transports iron in the blood.

translation: the name given to the process by which the base sequence in DNA is translated into the amino acid sequence in a protein.

tri-: a prefix meaning three.

trichloroethylene: $Cl_2C=CHCl$, a dry-cleaning fluid and industrial solvent.

triglyceride: chemical name for a fat molecule.

trinitrotoluene: TNT, an explosive.

triple bond: a strong bond between two atoms that consists of six shared electrons.

trisodium polyphosphate: TSP, an industrial-strength cleaning agent and builder in detergents.

tritium: ^3H, or T, the radioactive isotope of hydrogen. A tritium nucleus contains one proton and two neutrons.

troposphere: the layer of the atmosphere that is in contact with the Earth's surface. It extends to an altitude of 10 to 15 km above sea level.

tryptophan: one of the 20 amino acids used to construct proteins.

TSP: trisodium polyphosphate, an industrial-strength cleaning agent and builder in detergents.

tungsten: W, the 74th element of the periodic table. Commonly used as the filament in electric light bulbs.

tyrosine: one of the 20 amino acids used to construct proteins.

U

ultraviolet radiation: electromagnetic radiation with energy higher than that of visible light but less than that of X-rays.

unleaded gasoline: gasoline that does not contain the additive tetraethyl lead.

unsaturated fat: a fat whose molecules have carbon–carbon double bonds in their hydrocarbon chains.

unsymmetrical dimethylhydrazine (UDMH): a liquid rocket fuel.

uracil: one of the base molecules found in RNA. In helical structures, it is paired with adenine.

uranium: U, the 92nd element of the periodic table. The isotope uranium-235 is used as a nuclear fuel and explosive.

urea: an inorganic molecule found in urine; a nitrogen fertilizer.

uric acid: an acid, unrelated to uracil and urea, found in the urine of birds. Excess levels in human blood cause gout.

uridine: the uracil-ribose nucleoside.

V

valence shell: the outermost shell of electrons in an atom.

valence: the number of bonds an atom "prefers" to form.

valine: one of the 20 amino acids used to construct proteins.

Valium®: diazepam, an addictive tranquilizer.

vanadium: V, the 23rd element of the periodic table.

vasodilator: a chemical that causes blood vessels to dilate and lowers the blood pressure.

vegetable oil: fat derived from a plant.

vinegar: 5% solution of acetic acid in water.

vision: the photochemical process by which humans and animals see.

vitamin A: retinol. Deficiency causes night blindness. Vitamin A is the precursor to retinal, the light-absorbing molecule in rhodopsin.

vitamin B: a complex set of molecules. Deficiency causes a range of nervous disorders and anemia.

vitamin C: ascorbic acid; deficiency causes scurvy.

vitamin D: calciferol; deficiency causes rickets.

vitamin E: tocopherol; deficiency causes a lack of hemoglobin in the blood.

vulcanization: the addition of sulfur to natural rubber, which cross-links and produces a usable elastomer.

W

washing soda: Na_2CO_3, sodium carbonate, used as a water softener.

water cycle: a model for the movement of water in the environment.

water softener: a chemical that is added to water supplies to remove calcium and magnesium cations.

water sterilization: the process of killing pathogens that may be in water supplies.

wavelength: the distance between successive crests of a wave.

wax: a high-melting hydrocarbon.

weak acid: an acid that partially ionizes in water.

weight: a measure of the gravitational force on a mass.

white lead: $(PbCO_3)_2 \cdot Pb(OH)_2$, basic lead carbonate, a toxic pigment used in some outdoor paints. Formerly used in indoor paints.

white phosphorus: a low-melting form of elemental phosphorus that spontaneously ignites and burns on contact with air.

wood alcohol: another name for methanol, a volatile liquid given off when wood is heated in the absence of air.

Wood's metal: an alloy of bismuth (Bi), lead (Pb), tin (Sn), and cadmium (Cd) that melts below the temperature of boiling water. It finds use in electrical fuses and as the plug in fire sprinkler systems.

wrought iron: pure iron, softer and weaker than steel.

X

X-ray diffraction: a technique that measures the deflection of X-rays when they pass through a solid substance. This information can be analyzed to deduce the positions of atoms in solids.

X-rays: electromagnetic radiation with energy higher than that of ultraviolet light but less than that of γ-rays. The wavelengths of X-rays are about the same as the distances between atoms.

xenon: Xe, the 54th element in the periodic table.

xylene: an aromatic hydrocarbon solvent found in crude oil.

Y

yield (chemical): the efficiency of conversion, from 0 to 100%, of reactants to products in a chemical reaction.

Z

zeolites: a family of aluminosilicate minerals with pores of molecular dimensions. Important as catalysts and molecular sieves.

zinc: Zn, the 30th element of the periodic table.

zinc blende: ZnS, zinc sulfide mineral.

zinc oxide: ZnO, a mild base and a white pigment used in sunblocks.

zirconium: Zr, the 40th element of the periodic table.

Suggested Readings

Chapter 1

Aldersey-Williams, H. 1994. The third coming of carbon. *Technology Review* 97 (January): 54. [buckminsterfullerene]

Asimov, I. 1965. *A short history of chemistry.* Garden City, NY: Anchor Books.

Kotz, J. C., Joeston, M. D., Wood, J. L., and Moore, J. W. 1994. *The chemical world.* New York: Saunders College Publishing.

Kuroda, P. K. 1982. *The origin of the chemical elements and the Oklo phenomenon.* New York: Springer-Verlag.

Levi, P. 1984. *The periodic table.* Translated by Raymond Rosenthal. New York: Schocken Books.

Mason, S. F. 1991. *Chemical evolution: Origin of the elements, molecules, and living systems.* New York: Oxford University Press.

Naeye, R. 1994. An island of stability. *Discover* 15 (August): 22. [Chemists add elements to the periodic table.]

Partington, J. R. 1989. *A short history of chemistry.* 3rd ed. New York: Dover Publications.

Puddephatt, R. J., and Monaghan, P. K. 1986. *The periodic table of the elements.* 2nd ed. New York: Oxford University Press.

Chapter 2

Armbruster, P., and Münzenberg, G. 1989. Creating superheavy elements. *Scientific American* (May): 66.

Conn, R. W., Chuyanov, V. A., Inoue, N., and Sweetman, D. R. 1992. The international thermonuclear experimental reactor. *Scientific American* (April): 102.

Greiner, W., and Sandulescu, A. 1990. New radioactivities. *Scientific American* (March): 58.

Häfele, W. 1990. Energy from nuclear power. *Scientific American* (September): 136.

Hileman, B. 1994. U.S. and Russia face urgent decision on weapons plutonium. *Chemical & Engineering News* (June) 13: 12.

Macklis, R. M. 1993. The great radium scandal. *Scientific American* (August): 94.

Marples, D. 1993. Chernobyl's lengthening shadow. *Bulletin of Atomic Scientists* 49: 38.

Rhodes, R. 1986. *The making of the atomic bomb.* New York: Simon and Schuster.

Segre, E. G. 1989. The discovery of nuclear fission. *Physics Today* (July): 38.

Shulman, S. 1989. Legacy of Three Mile Island. *Nature* 338:190.

Chapter 3

Gillespie, R. J., Eaton, D. R., Humphreys, D. A., and Robinson, E. A. 1994. *Atoms, molecules, and reactions.* Englewood Cliffs, NJ: Prentice-Hall.

Gray, H. B. 1994. *Chemical bonds.* Mill Valley, CA: University Science Books.

Pauling, L. 1967. *The chemical bond: A brief introduction to modern structural chemistry.* Ithaca, NY: Cornell University Press.

Chapter 4

Beebe, D. K., and Walley, E. 1991. Substance abuse: The designer drugs. *American Family Physician* (January): 1689.

Bindra, J. S., and Lednicer, D. 1993. *Chronicles of drug discovery.* New York: Wiley.

Bird, C., and Sattaur, O. 1991. Medicines from the rainforest. *New Scientist* 131 (August 17): 34.

Burger, A. 1986. *Drugs and people: Medications, their history and origins, and the way they act.* Charlottesville: University Press of Virginia.

Douglass, J. D., and Livingstone, N. C. 1987. *America the vulnerable: The threat of chemical and biological warfare.* Lexington, MA: Lexington Books.

Goodpasture, H. C. 1991. Antiviral drug therapy. *American Family Physician* 43 (January); 197.

Haber, L. F. 1986. *The poisonous cloud: Chemical warfare in the First World War.* New York: Oxford University Press.

Resistance to antibiotics. 1994. *Science* 264 (April 15): 359.

Taking vitamins: Can they prevent disease? 1994. *Consumer Reports* 59 (September): 561.

Chapter 5

Antacids: Which beat heartburn best? 1994. *Consumer Reports* 59 (July): 443.

Gardner, R. 1982. *Kitchen chemistry: Science experiments to do at home.* New York: J. Messner.

Illman, D. L. 1994. Automakers move toward new generation of greener vehicles. *Chemical & Engineering News* 72 (August 1): 8.

Kurti, N., and Thisbenckhard, H. 1994. Chemistry and physics in the kitchen. *Scientific American* 270 (April): 66.

Richardson, S. 1994. The war on radicals. *Discover* 15 (July): 27.

Woodruff, D. 1994. Electric cars: Will they work; and who will buy them? *Business Week* 3374 (May 30): 104.

Chapter 6

Howe-Grant, M., ed. 1991. *Encyclopedia of chemical technology.* 4th ed. Executive editor J. I. Kroschwitz. New York: Wiley.

Reichhardt, T. 1995. A New Formula for fighting urban ozone. *Environmental Science and Technology* 29 (January): 36A.

Ross, L. R. 1994. Science in American life. *Chemical & Engineering News* 72 (March 7): 30.

Thomas, J. M. 1992. Solid acid catalysts. *Scientific American* 266 (April): 112.

Werth, B. 1994. The billion-dollar molecule: One company's quest for the perfect drug. New York: Simon & Schuster.

Wiseman, P. 1979. *An introduction to industrial organic chemistry.* 2nd ed. London: Applied Science.

Chapter 7

Adams, R. 1939. *Biographical memoir of Wallace Hume Carothers, 1986–1937.* Washington, DC: National Academy of Sciences.

Boretos, J. W. 1987. Bioceramics. *Chemtech* (April): 224.

Campbell, I. M. 1994. *Introduction to synthetic polymers.* New York: Oxford University Press.

Herbert, V., and Bisio, A. 1985. *Synthetic rubber: A project that had to succeed.* Westwood, CT: Greenwood Press.

Rochow, E. G. 1987. *Silicone and silicones: About stone-age tools, antique pottery, modern ceramics, computers, space materials, and how they all got that way.* New York: Springer-Verlag.

Seymour, R. B., and Kauffman, G. B. 1992. Polyurethanes: A class of modern versatile materials. *Journal of Chemical Education* 69 (November): 909.

Chapter 8

Berg, P. and Singer, M. 1994. *Dealing with genes.* Mill Valley, CA: University Science Books.

Bud, R. 1993. *The uses of life: A history of biotechnology.* New York: Cambridge University Press.

Dressler, D., and Parker, H. 1991. *Discovering enzymes.* New York: Freeman.

Fruton, J. F. 1972. *Molecules and life.* New York: Wiley.

Pääbo, S. 1993. Ancient DNA. *Scientific American* (November): 86.

Radman, M., and Wagner, R. 1988. The high fidelity of DNA duplication. *Scientific American* (August): 40.

Rusting, R. L. 1992. Why do we age? *Scientific American* (December): 130.

Stryer, L. 1988. *Biochemistry.* New York: Freeman.

Watson, J. D. 1968. *The double helix.* New York: Atheneum.

Chapter 9

Anderson, I. 1994. Sunny days for solar power. *New Scientist* 143 (July 2): 21.

Borman, S. 1992. New light shed on mechanism of human color vision. *Chemical & Engineering News* (April 6): 27.

Deisenhofer, J., Epp, O., Miki, K., Huber, R., and Michel, H. 1984. X-ray structure analysis of a membrane protein complex. *Journal of Molecular Biology* 180: 385.

Fox, M. A., Jones, W. E., and Watkins, D. M. 1993. Light-harvesting polymer systems. *Chemical & Engineering News* (March 15): 38.

Govindjee, and Coleman, W. J. 1990. How plants make oxygen. *Scientific American* (February): 50.

Madronich, S., and de Gruijl, F. R. 1993. Skin cancer and UV radiation. *Nature* 366: 23.

Simon, M. S. 1994. New developments in instant photography. *Journal of Chemical Education* 71: 132.

Trotter, Jr., D. M. 1991. Photochromic and photosensitive glass. *Scientific American* (April): 124.

Weinberg, C. J., and Williams, R. H. 1990. Energy from the sun. *Scientific American* (September): 146.

Chapter 10

Charlson, R. J., and Wigley, T. M. L. 1994. Sulfate aerosol and climate change. *Scientific American* (February): 48.

Hileman, B. 1992. Web of interactions makes if difficult to untangle global warming data. *Chemical & Engineering News* (April 27): 7.

Jones, P. D., and Wigley, T. M. L. 1990. Global warming trends. *Scientific American* (August): 84.

Lents, J. M., and Kelley, W. J. 1993. Clearing the air in Los Angeles. *Scientific American* (October): 32.

Mohner, V. A. 1988. The challenge of acid rain. *Scientific American* (August): 30.

Roan, S. L. 1989. *Ozone crisis*. New York: Wiley.

Rowland, F. S., and Molina, M. J. 1994. Ozone depletion: 20 years after the alarm. *Chemical & Engineering News* (August 15): 8.

Toon, O. B., and Turco, R. P. 1991. Polar stratospheric clouds and ozone depletion. *Scientific American* (June): 68.

Wayne, R. P. 1991. *Chemistry of atmospheres*. New York: Oxford University Press.

White, R. M. 1990. The great climate debate. *Scientific American* (July): 36.

Zurer, P. S. 1993. Ozone depletion's recurring surprises challenge atmospheric scientists. *Chemical & Engineering News* (May 24): 8.

Chapter 11

Beardsley, T. 1994. A war not won. *Scientific American* (January): 130.

Boon, T. 1993. Teaching the immune system to fight cancer. *Scientific American* (March): 82.

EPA's dioxin reassessment (highlights). 1995. *Environmental Science and Technology* 29: 26A.

Forman, C. 1993. Getting practical about pesticides. *Time* 141 (February 15): 52.

Garnick, M. B. 1994. The dilemmas of prostate cancer. *Scientific American* (April): 72.

Hileman, B. 1993. Concerns broaden over chlorine and chlorinated hydrocarbons. *Chemical & Engineering News* (April 19): 11.

Hileman, B. 1994. Environmental estrogens linked to reproductive abnormalities, cancer. *Chemical & Engineering News* (January 31): 19.

Link of radon to lung cancer looks loopy. 1994. *Science News* 145 (March 19): 188.

Liotta, L. A. 1992. Cancer cell invasion and metastasis. *Scientific American* (February): 54.

Marshall, E. 1991. Breast cancer: Stalemate in the war on cancer. *Science* 254: 1719.

National Research Council (U.S.), committee report. 1993. *Pesticides in the diets of infants and children*. Washington, DC: National Academy Press.

Ravage, B. 1994. A battlefield called cancer. *Current Health* 20: 6.

Steen, R. G. 1993. *A conspiracy of cells: The basic science of cancer*. New York: Plenum Press.

Index